Communications and Control Engineering Series

Editors: A. Fettweis · J. L. Massey · M. Thoma

Peter Young

Recursive Estimation and Time-Series Analysis

An Introduction

With 54 Figures

Springer-Verlag
Berlin Heidelberg New York Tokyo 1984

PETER YOUNG
Professor and Head of Department of Environmental Sciences
University of Lancaster
Lancaster, LA1 4YQ, Lancs. England

ISBN 3-540-13677-0 Springer-Verlag Berlin Heidelberg New York Tokyo
ISBN 0-387-13677-0 Springer-Verlag New York Heidelberg Berlin Tokyo

Library of Congress Cataloging in Publication Data
Young, Peter C.
Recursive estimation and time-series analysis.
(Communications and control engineering series)
Based on lecture notes originally prepared for a NATO summer school on
"The theory and practice of systems modelling and identification" held July 17–28, 1972,
at the Ecole nationale supérieure de l'áeronautique et de l'espace.
Bibliography: p.
Includes indexes.
1. Estimation theory. 2. Time-series analysis.
I. Title. II. Series.
QA276.8.Y68 1984 519.5'44 84-14078

Printed in Germany

Offsetprinting: Mercedes-Druck, Berlin
Bookbinding: Lüderitz & Bauer, Berlin
2061/3020 5 4 3 2 1 0

To
Wendy, Timothy, Melanie and Jeremy

Preface

This book has grown out of a set of lecture notes prepared originally for a NATO Summer School on *"The Theory and Practice of Systems Modelling and Identification"* held between the 17th and 28th July, 1972 at the Ecole Nationale Superieure de L'Aeronautique et de L'Espace. Since this time I have given similar lecture courses in the Control Division of the Engineering Department, University of Cambridge; Department of Mechanical Engineering, University of Western Australia; the University of Ghent, Belgium (during the time I held the IBM Visiting Chair in Simulation for the month of January, 1980), the Australian National University, and the Agricultural University, Wageningen, the Netherlands. As a result, I am grateful to all the recipients of these lecture courses for their help in refining the book to its present form; it is still far from perfect but I hope that it will help the student to become acquainted with the interesting and practically useful concept of recursive estimation. Furthermore, I hope it will stimulate the reader to further study the theoretical aspects of the subject, which are not dealt with in detail in the present text. The book is primarily intended to provide an introductory set of lecture notes on the subject of recursive estimation to undergraduate/Masters students. However, the book can also be considered as a "theoretical background" handbook for use with the CAPTAIN Computer Package. This 'Computer Aided Program for Time Series Analysis and the Identification of Noisy Systems' was originally conceived by the author in the mid nineteen sixties and was developed initially in the Control Division of the Engineering Department at the University of Cambridge in 1971 on the basis of recursive algorithms developed during the previous six years (Young *et al.*, 1971; Shellswell, 1971). A *command mode* version of the package was developed as an alternative to the original interactive, *conversational mode* version, by the Institute of Hydrology in England (Venn and Day, 1977) and in this form, it has been acquired by the Commonwealth Scientific and Industrial Research Organisation (CSIRO) of Australia for use in its CSIRONET nationwide computing system (Freeman, 1981). More recently, advanced versions of CAPTAIN, based on some of the more sophisticated procedures discussed in the present book, have been developed by the author and his colleagues at the Centre for Resource and Environmental Sciences, (CRES), ANU, for use in the UNIVAC 1100 series computer. A more comprehensive, "user friendly" version of the package is currently under development by the author and Mr. John Hampton in the Department of Environmental Sciences, University of Lancaster, Lancaster, LA1 4YQ, England. This version will be available initially for use on the VAX 11/780 computer. Also a microcomputer version MICROCAPTAIN has recently been developed by the author (see Epilogue) for use on the APPLE II and EPSON HX-20 microcomputers. Any enquiries about any of these CAPTAIN programs should be addressed to me at the above address.

Acknowledgements

The final appearance of this book has been delayed for various reasons. Foremost amongst these are the various changes in location of my family and myself during the preparation of the text. It was begun in 1979, while I was spending a period of study leave at my old College, Clare Hall, in Cambridge; it was continued, very much as a spare time activity between numerous environmental projects, at the Centre for Resource and Environmental Studies, Australian National University, Canberra, between 1980 and 1981; and it has been completed, whenever time has allowed, at the Department of Environmental Science, University of Lancaster, between 1981 and 1983.

The overlong gestation period has meant that, at various times, a number of people have read draft chapters of the book and suggested improvements. In particular, I would like to thank Dr. Paul Steele, Dr. Tony Jakeman, and Miss Christina Sirakoff for their help and encouragement. Mrs. June Harries (Canberra) and Mrs. Glenys Booth (Lancaster) have typed most of the manuscript, often from poorly handwritten drafts, and I am extremely grateful to both of them for their patience and the quality of their work. A number of the diagrams were drawn by Elizabeth Barta.

Of course all the errors, omissions and other deficiencies of the book are my sole responsibility. Like most authors, if I was writing the book again, I would change many aspects of the presentation. Having gone through the not inconsiderable labours of preparing a "camera-ready" manuscript, however, I find such a possibility daunting in the extreme. I hope, therefore, the reader will not be too disappointed in the final outcome.

Finally, in dedicating this book to my wife and family, I am acknowledging their unfailing support, without which the book would never have appeared.

Peter Young
Lancaster, England.
January 1984.

Contents

Algorithms

1 Introduction

Nous traiterous particulièrement le problème suivant, tant à cause de son utilite pratique que de la simplicité de la solution

K.F. Gauss
Méthode des Moindres Carrés
Translation by J. Bertrand, 1855.

The concept of least squares estimation is inextricably linked with the name of Karl Friedrick Gauss. Although Legendre was responsible for the first published account of the theory in 1805 and, indeed, first coined the term "least squares", it was Gauss who developed the method into a statistical tool, embedding it within a statistical framework involving a probabilistic treatment of observational errors.

Gauss's first published exposition on least squares appeared in his famous *Theoria Motus Corporum Coelestum* which appeared in 1809 when he was 31 years of age. But, as D.A. Sprott (1978) has pointed out recently in his excellent review of Gauss's contributions to statistics, the basic ideas were most probably formulated while he was still in his twenties.

In the *Theoria Motus*, the discussion on least squares appears in relation to an important practical problem; namely the estimation of the six constant coefficients or "parameters" that determine the elliptical orbit of a planetary body, on the basis of $n > 6$ observations. His second exposition on the subject was presented in a series of papers (1821, 1823, 1826) which were collected together under the title *Theoria Combinationis Erroribus Minimum Obnoxiae*. Here he abandoned the previous "inferential" treatment delineated in the *Theoria Motus* and concentrated on a "decision theoretic" approach, in which he restricted attention to estimates that are a linear function of the observations. And it was here that he presented perhaps his most famous statistical theorem that, *Among all linear error-consistent estimates, the least squares estimate has minimum mean square error*.

But our current interest lies elsewhere in the *Theoria Combinationis:* in a quite short and apparently little known section of some five pages (Bertrand, 1855, p. 53-58; Trotter, 1957, p. 67-72) Gauss shows how it is possible *to find the changes which the most likely values of the unknowns undergo when a new equation (observation) is adjoined and to determine the weights of these new determinations*. In other words, and to use more contemporary terminology, he developed an algorithm for sequentially or *recursively* updating the least squares parameter estimates on receipt of additional data.

Bertrand's French translation of Gauss's work appeared in 1855 under the appropriate title "Méthode des Moindres Carrés" and was authorised by Gauss himself.

This translation, together with a commentary provided by the present author, appears in Appendix 2. In addition to its importance in historical terms, Gauss's analysis is interesting because it demonstrates the elegance of his approach and the power of his mind: without the advantages of matrix algebra which, as we shall see, considerably simplify the derivation, Gauss was able to obtain the recursive least squares algorithm with consummate ease.

Gauss's analysis represents the birth of recursive least squares theory; a theory so much ahead of its time that it would lie dormant for almost a century and a half before it was rediscovered on two separate occasions: first by the statistician R.L. Plackett in 1950; and then later and in a more sophisticated form, as the core of the linear filtering and prediction theory evolved by the control and systems theorist R.E. Kalman (1960).

Not surprisingly, perhaps, Plackett's paper went almost unnoticed in the pre-computer age of the early nineteen fifties. Harking back to Gauss, he re-worked the original results in more elegant vector-matrix terms and developed an algorithm for the general case in which additional observations occur in sets $S > 1$. In the present book, like Gauss, we restrict the analysis largely to $S = 1$, although the extension to $S > 1$ is straightforward and is discussed in certain special cases.

Kalman's results, almost certainly obtained without knowledge of either Gauss's or Plackett's prior contributions, were developed within the context of state variable estimation and filter theory, using an argument based on orthogonal projection. Not only were Kalman's results mathematically elegant in providing a computationally straightforward solution to the optimal filtering problem, which had a number of advantages over the earlier Wiener solution (Wiener, 1949), but they also had good potential for practical application. Not surprisingly, therefore, they caused quite a revolution in the automatic control and systems field providing, during the next ten years, a rich source of research material for control and systems analysts. Subsequently, the term *Kalman Filter* has become widely used, not only in the academic and industrial world of automatic control but also in other disciplines such as statistics and economics.

It is now well known that the Kalman filter estimation algorithm can be derived in various ways; via orthogonal projection, as in Kalman's exposition; as well as from the standpoint of maximum likelihood or Bayesian estimation. It can also be developed in various different forms, for application to both discrete (Kalman, 1960) and continuous (Kalman and Bucy, 1961) time-series. But,in all its forms, it has had a profound effect on data processing during the last two decades, being used in applications ranging from trajectory and orbit estimation to the forecasting of economic time-series.

Sprott (1978) has questioned whether the Kalman filter is really a significant 'development' of the Gauss-Plackett recursive least squares algorithm. While it is true that the Gauss-Plackett recursion formulae are an essential aspect of the

Kalman filter equations, it is also clear, as we shall see in this book, that Kalman considerably extended the theory both to allow for the estimation of *time-variable* parameters or *states*, and to handle the analysis of statistically *non-stationary* time-series. Nevertheless the Gauss-Plackett recursion is undoubtedly the central component of the Kalman filter and the basis of most other recursive least squares algorithms. Thus a good understanding of its function in a data processing sense is an essential pre-requisite for the practical application of the algorithm. It is the provision of such understanding, therefore, which is one of the primary aims of this book.

As Gauss pointed out so succinctly in the quotation at the beginning of this chapter, recursive least squares theory is both simple and useful. Here we will exploit this simplicity and take the reader gently through the mysteries of the subject, avoiding wherever possible undue rigour and complexity. In the spirit of Gauss, we will concentrate on mathematical analysis which, while it is often algebraic in form, also has sufficient statistical content to ensure that the reader is fully aware of the important statistical aspects of the results. We will, however, allow ourselves one luxury not available to Gauss and simplify the analysis still further by resort to matrix algebra, assuming that the reader is already acquainted with such analysis; has access to a good text on the subject; or finds that the background notes in Appendix 1 of the book provide sufficient revision.

Finally, to emphasize the practical utility of the various recursive least squares algorithms which will emerge during the analysis, we will provide a number of simulation and practical examples, with applications which range from the man-made world of engineering to the more natural world of ecology and the environment. Many other applications in diverse areas, from economics to hydrodynamics, are discussed in a variety of technical papers produced by the author and his colleagues over the past few years and these are either referred to in the text or listed in the bibliography. Some more recent references are discussed in a short Epilogue at the end of the book.

The text is divided into two major parts: the first is primarily concerned with the estimation of constant or time-variable parameters in general models which are *linear in the-parameters;* the second shows how the procedures developed in the first part can be modified to handle the analysis of stochastic time-series and so provide algorithms for the recursive estimation of parameters and states in stochastic dynamic systems. In sympathy with the introductory nature of this book, there has been a conscious attempt to simplify the mathematical analysis as much as possible, particularly in the early chapters, so as to enhance the readability of the book and avoid an overly esoteric presentation. For the reader unfamiliar with some of the mathematics used in the book, Appendix 1 provides background notes, not only on matrix algebra but also on probability and statistics, as well as some very simple concepts in dynamic systems. In all cases, the results in Appendix 1 are chosen because of their relevance to the material in the book and are quoted without proof. They should, however, be sufficient for the reader either as an aid to his memory, or as a guide to his reading of standard texts in the various areas (e.g. Johnston, 1963; Brownlee, 1966; Kendall and Stuart, 1961; Bryson and Ho, 1969; Takahashi *et al.*, 1972).

The analytical treatment in the main text starts at a very basic level requiring only a knowledge of simple algebra and it then increases in difficulty with each new chapter, hopefully in a manner which matches the reader's increasing understanding of the subject. However, the chapters also become increasingly skeletal: detailed tutorial material is not included, again in sympathy with the notion that the book should provide a guide to introductory lectures rather than a comprehensive presentation of the subject. Finally, Chapters 8 and 9 are included mainly for the more advanced reader and can be omitted if necessary. Essentially the book is an introductory tutorial text for undergraduate/masters students, which can also be considered as a primer for a more advanced book on the subject which is at present in preparation (Young, 1984).

Chapter 2 provides an introduction to recursive analysis by way of the simplest example, namely the recursive estimation of the constant mean value of a stationary random variable, and goes on to consider both the physical nature of the recursive algorithm and its relationship to the linear least squares algorithm for a single unknown parameter (or regression coefficient).

Chapter 3 develops the deterministic, recursive least squares (RLS) algorithm for multiple unknown parameters and shows how it can be viewed within the context of the procedure known to mathematicians as *stochastic approximation* (SA).

In Chapter 4 we see how, with statistical assumptions about the nature of the stochastic disturbances, it is possible to modify the deterministic algorithm so that it constitutes the recursive version of linear least squares regression analysis for the *general linear regression model*. It also includes an introduction to two important problems associated with statistical regression analysis - namely *multicollinearity* and *errors-in-variables* - and discusses briefly one of the simplest solutions to the latter problem which is based on the concept of *instrumental variables*.

Chapter 5, which completes the first part of the book, shows how the recursive algorithms are in an ideal form for extension to the estimation of *time variable parameters*, by the introduction of either a *fading memory* capability or, in the statistical versions of the algorithms, modifications based on stochastic modelling of the parameter variations. This latter approach leads naturally to the development of the Kalman filter and the estimation of *time-variable states* in stochastic dynamic systems.

The Kalman Filter provides a natural approach to the analysis of *time-series data* that are assumed to be generated from stochastic *state-space* equations. His analysis is concerned particularly with the estimation of the time-variable, but often unmeasurable, *state variables* of the system; whilst assuming perfect prior knowledge of the parameters which characterise the chosen state-space mathematical model of the system. In Chapter 6, we consider the alternative *observation space* or *polynomial matrix description* (PMD) models, where the dynamic relationships are usually characterized in *transfer function* terms. It is on the estimation of *parameters* in such models that much of the most recent research on recursive analysis has been concentrated and the chapter goes on to discuss why the simple recursive least squares algorithms developed in the previous chapters are not particularly appropriate to the solution of these more complex time-series estimation problems.

Chapters 7 and 8 describe in some detail a particular approach to time-series model estimation that has been developed by the author over the past fifteen years and which is based on the method of instrumental variables introduced in Chapter 3. This technique is chosen as a vehicle for introducing time-series estimation to the reader simply because it only requires a fairly straightforward modification of the recursive least squares algorithm and is relatively insensitive to assumptions about the statistical characteristics of the noise on the measured time-series data. Other methods, such as those discussed in Chapter 9, tend to be more complicated and more reliant for their success on assumptions about the statistical nature of the noise contamination. The *ordinary instrumental variable (IV) algorithm* is discussed in Chapter 7 while the more sophisticated, but correspondingly more complex, *refined IV algorithm* is developed in Chapter 8. Also discussed in Chapter 8 is the concept of *Optimal Generalized Equation Error* (OGEE) minimization. This is a general approach to algorithmic design for time-series estimation which is suggested by the refined IV method.

The OGEE approach to the recursive solution of the time-series estimation problems is compared and contrasted with other alternative recursive approaches in Chapter 9. These include the closely related *Prediction Error* (PE) minimization method for PMD models, the *Extended Kalman Filter* (EKF) approach for state-space (SS) models, and the very sophisticated but extremely complex *maximum likelihood* (ML) method, also designed for use with SS models.

Chapter 10 outlines how recursive estimation procedures of the types described in this book can prove useful as general tools in data analysis and stochastic model building: this includes *brief* discussion of the recursive approach to time-series modelling, including data pre-processing, model structure identification, model parameter estimation, and self adaptive (or self tuning) estimation, forecasting and control. In effect, this final chapter provides an introduction to the various stages of time-series analysis and indicates how such analysis can benefit from the use of recursive methods. As such, the Chapter serves two purposes: it emphasizes the potentially wide practical utility of recursive estimation; and it sets the scene for a more complete treatment of recursive time-series analysis which will describe, *in detail*, how the recursive algorithms described in the present book are employed in all aspects of signal processing and stochastic modelling (Young, 1984)

Also, in Chapter 10, it is revealed that the 'optimum' approaches to recursive time-series analysis, such as the refined IV and Prediction Error algorithms, have an added advantage. *Implicit* in the estimation of the model parameters is the estimation of the state variables associated with a chosen state-space model of the system. And such state variable estimates can be recovered simply as a linear function of the outputs of the adaptive prefilters which differentiate the optimum methods from their simpler 'sub-optimum' relatives. This would seem, at least partially, to satisfy Kalman's requirement (1960, page 40) that "the two problems (parameter and state estimation) should be optimized jointly if possible".

Finally, in an Epilogue to the book, reference is made to a number of recent publications which should help the reader both to further evaluate the recursive methods discussed in this book and put them in a wider context.

It is not possible to consider a subject such as recursive estimation without resort to mathematical analysis and certain of this analysis may be unfamiliar to some readers. But we urge such readers to persevere, for the mathematics is inherently quite simple, requiring only a knowledge of straight-forward matrix algebra and basic statistical concepts of the kind dealt with thoroughly in most good statistical texts (e.g., Brownlee , 1966; Johnston, 1963; Kendall and Stuart, 1961 and Bryson and Ho, 1969). And such effort will be amply rewarded, for the reader will be introduced to a much expanded world of parameter estimation and time-series analysis.

The recursive procedures described in this book are ideally suited to the digital computer: they are computationally elegant and yet obtained at little additional computational cost in relation to the more conventional *en-bloc* methods which have dominated least squares and time-series analysis heretofore. And the recursive estimates provide a new dimension to the analysis, providing not only the possibility of *on-line* data processing and forecasting, but also procedures for adaptive control, model structure (order) identification and adaptive state variable estimation, as discussed briefly in Chapter 10.

Of course, the importance of recursive least squares analysis will depend, to some extent, on the nature of the problem under consideration. But there is no doubt of its importance in general time-series analysis. Indeed we would speculate that recursive estimation will become commonplace in the next few years, taking its position as an essential component in most degree courses concerned with either control and systems theory or statistical estimation. Already computer program packages such as CAPTAIN and MICROCAPTAIN (see preface), which exploit recursive estimation in all areas of time-series analysis, are available for general use. And there can be little doubt that such packages will increase in popularity as their advantages over similar packages based on more conventional *en-bloc* computational methods become more widely known. Such popularity will arise, in part, because of the undoubted elegance and flexibility of the recursive solution to estimation problems in mathematical terms. But it will arise primarily because of its demonstrable *practical* utility, as exemplified by the practical examples discussed in this book and related publications. In this sense, we have attempted to follow the example of Gauss who says, in the preface to the *Theoria Motus:*

> *The more important problems are, for the most part, illustrated by appropriate examples, taken, wherever it was possible, from actual observations. In this way not only is the efficacy of the methods more fully established and their use more clearly shown, but also, care, I hope, has been taken so that inexperienced computers [analysts] should not be deterred from the study of these subjects, which undoubtedly constitute the richest and most attractive part of [the subject].*†

† words in parentheses added by the present author.

Part 1
Recursive Parameter Estimation

2 Recursive Estimation: A Tutorial Introduction

In Figure 2.1 we see a plot of the annual rainfall y_i, $i = 1,2,\ldots.,95$, for the town of Walgett in New South Wales, Australia, over the period 1880 to 1975. It is a typical example of time-series data and the human eye (which is an extremely good and underrated filter of data) can discern various characteristics

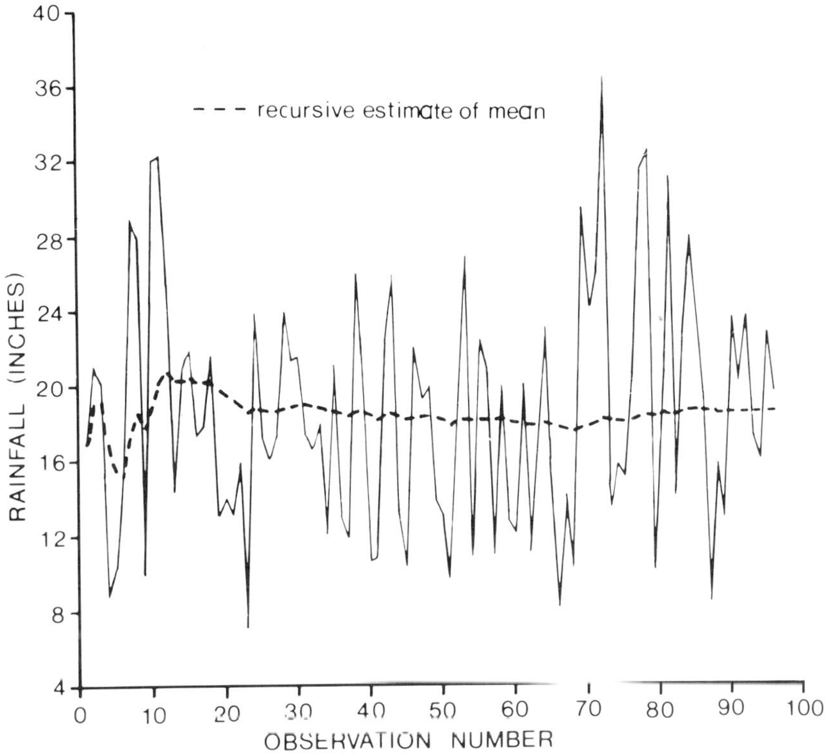

Figure 2.1 Rainfall at Walgett, N.S.W., 1880 to 1975.

and patterns which could be described verbally: for example, the rainfall is extremely variable; it averages about 18 to 20 inches a year; and there seems to be indications of some increase in the mean and variance over the last 20 to 30 years.

But such a description is largely qualitative; if we are to be more quantitative and precise in our evaluation, and particularly if we wish to compare the rainfall at Walgett with that at other stations in N.S.W. or elsewhere, then we must compress the data in some manner to yield a reduced and hopefully small number of 'statistics' which can be computed easily from the original data and which collectively provide a good description of that data.

The most obvious statistics in the case of the Walgett data are the first two statistical moments[†], in the form of the sample mean or average rainfall $\bar{a}_T$ and the sample variance $\bar{\sigma}^2_T$ about this mean value, where $\bar{a}_T$ and $\bar{\sigma}^2_T$ are defined as follows

$$\left.\begin{aligned} \bar{a}_T &= \frac{1}{T}\sum_{i=1}^{T} y_i &\quad \text{(i)} \\ \bar{\sigma}^2_T &= \frac{1}{T}\sum_{i=1}^{T} (y_i - \bar{a}_T)^2 &\quad \text{(ii)} \end{aligned}\right\} \qquad (2.1)$$

and T is the total number of samples available, in this case 95.

There is an implicit assumption in the formulae (2.1) that the true mean and variance, say a and σ^2 are constant. If we trust the evidence of our eyes, we might suspect that this is not the case and that there are possibly changes in both the mean and variance over the observation interval. We shall see later that there is good evidence to support this supposition. For the moment, however, let us take on the role of the naïve analyst who either did not bother to look at the graphical plot of the data or, as is quite common these days, merely consulted that most sterile of visual presentations, the computer data listing.

The computations involved in the calculation of $\bar{a}_T$ and $\bar{\sigma}^2_T$ from (2.1) are extremely simple, so that the evaluation of these statistics for any subset of the T samples is no problem. If we had computed the statistics over some k-1 samples within the data set, however, and then wished to obtain them for k samples, we might ask ourselves whether there was a simple *recursive* relationship between $\bar{a}_k$ and $\bar{a}_{k-1}$ or $\bar{\sigma}^2_K$ and $\bar{\sigma}^2_{k-1}$. Similarly, if we had the estimates for T samples and received an additional sample, as we might in the present example when the total rainfall measurement for 1976 became available, then we might suspect that some combination of the statistic already computed for T samples and the new *sample* at T+1 would yield the new value of the *statistic* at T+1.

2.1 Recursive Estimation of the Mean Value of a Random Variable

Let us examine the possibility of obtaining a recursive relationship for

[†]For the present, we will not consider the sample covariance (Appendix 1).

the mean estimate $\bar{a}_k$ in relation to the *prior* calculated mean estimate $\bar{a}_{k-1}$. There are basically two ways in which we can obtain such a relationship: first by simple manipulation of the *en bloc* solution (2.1); and the second by considering the problem within the wider context of recursive least squares estimation.

We can treat y_i as an observation of the true mean rainfall, a, at the ith sampling instant i.e.,

$$y_i = a + e_{yi} \tag{2.2}$$

where e_{yi}, which is the difference between a and y_i, can be considered as the error or 'noise' on the observation. The sample estimate of the mean at the (k-1)th instant is then given by definition as $\bar{a}_{k-1}$, where

$$(k-1)\bar{a}_{k-1} = \sum_{i=1}^{k-1} y_i$$

while at the kth instant

$$k\,\bar{a}_k = \sum_{i=1}^{k} y_i$$

It is clear, therefore, that $\bar{a}_k$ is related to $\bar{a}_{k-1}$ by

$$\bar{a}_k = \frac{1}{k}\,[(k-1)\,\bar{a}_{k-1} + y_k]$$

or

$$\bar{a}_k = \frac{k-1}{k}\,\bar{a}_{k-1} + \frac{1}{k}\,y_k.$$

Multiplying by k/k-1 we obtain

$$\frac{k}{k-1}\,\bar{a}_k = \bar{a}_{k-1} + \frac{1}{k-1}\,y_k$$

or

$$\frac{k}{k-1}\,\bar{a}_k = \bar{a}_{k-1} + \frac{1}{k-1}\,(y_k - \bar{a}_{k-1}) + \frac{1}{k-1}\,\bar{a}_{k-1}.$$

Finally, collecting terms, we see that

$$\bar{a}_k = \bar{a}_{k-1} + \frac{1}{k}(y_k - \bar{a}_{k-1})$$

or

$$\bar{a}_k = \bar{a}_{k-1} - \frac{1}{k}(\bar{a}_{k-1} - y_k) \tag{2.3}$$

Equation (2.3) is the recursive algorithm for estimating the sample mean value of the observations up to the kth instant from the sample mean computed at the previous (k-1)th instant and the new observation y_k received at the kth instant. A similar recursive relationship could be obtained for the sample variance. We can see from the diagrammatic representation of the data analysis in Figure 2.2 that the recursive estimates are obtained by working serially through the data, one sample at a time, using the recursion (2.3). This can be compared with the more conventional *en bloc* method of analysis, where a single estimate is obtained by operating on the *whole* set of data in one operation according to equation (i) of (2.1). And it can be contrasted with *iterative* data processing, which we will utilize here to mean the sequential processing of a *complete* set of data, as shown in Figure 2.2. Here, at each iterative step, the data base remains the same at T samples and only some estimated variable is modified. Using this terminology, the *en bloc* solution

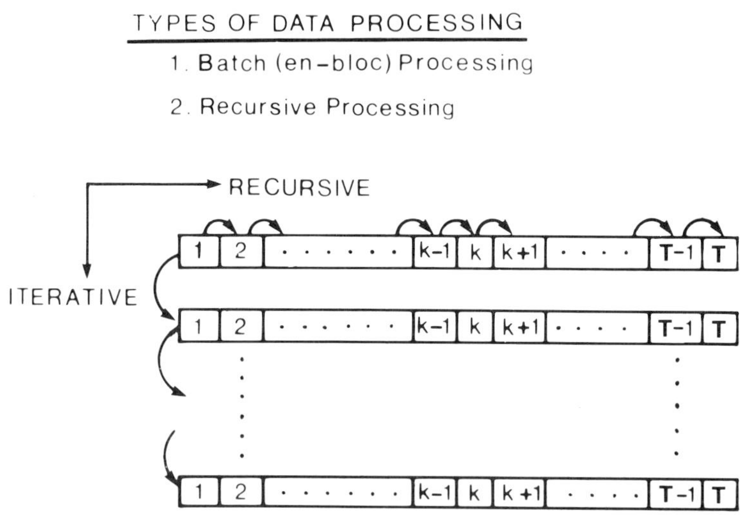

Figure 2.2 Recursive and Iterative Data Processing

(2.1) can be considered as a single iteration on the data. Iterative processing involving multiple iterations is not relevant to the simple problem under discussion here but it will assume some significance later when we consider more complex parameter estimation problems.

While the *en bloc* (non-recursive or single iteration) method of determining the mean value is well known, the recursive algorithm is comparatively little known. And yet the algorithm is significant in a number of ways: not only is it elegant and computationally attractive, but it also exposes, in a most vivid manner, the physical nature of the estimate for increasing sample size and so provides insight into a mechanism which, as we shall see, is useful in many more general problems.

Referring to equation (2.3) we see that the estimate of the mean after k samples is equal to the previous estimate $\bar{a}_{k-1}$ obtained after k-1 samples plus a correction term which is the product of 1/k and the difference between the new sample observation y_k and $\bar{a}_{k-1}$. In effect, therefore, the previous estimate $\bar{a}_{k-1}$ is modified in proportion to the *error* between the observation of the random variable and the latest estimate of its mean value.

Another way of looking at the problem of estimating the mean value of a random variable is to consider it as a problem of least squares estimation (Young, 1974). Here the least squares estimate $\hat{a}$ of a is simply that value of $\hat{a}$ which minimizes a cost function J defined as the sum of the squares of the differences between y_i and $\hat{a}$ over the observation period, i.e.

$$J \triangleq \sum_{i=1}^{k} [y_i - \hat{a}]^2 \tag{2.4}$$

The extremum (minimum in this case) of J is obtained in the usual manner by differentiating J with respect to $\hat{a}$ and equating the result to zero, i.e.

$$\nabla J_{\hat{a}} = \frac{\partial J}{\partial \hat{a}} = 2(k\hat{a}_k - \sum_{i=1}^{k} y_i) = 0 \tag{2.5}$$

where $\hat{a}_k$ denotes the least squares estimate based on an observation size of k samples.

From equation (2.5) we see that

$$\hat{a}_k = \frac{1}{k} \sum_{i=1}^{k} y_i \tag{2.6}$$

which is, of course, the arithmetic mean utilized previously. But continuing within the context of least squares estimation, we can write equation (2.6) in the form

$$\hat{a}_k = p_k b_k \tag{2.7}$$

where

$$p_k = \frac{1}{k} \text{ and } b_k = \sum_{i=1}^{k} y_i . \tag{2.8}$$

It is easily seen that p_k and b_k are related to their previous values p_{k-1} and b_{k-1} by the equations

$$\frac{1}{p_k} = \frac{1}{p_{k-1}} + 1 \tag{2.9}$$

and

$$b_k = b_{k-1} + y_k . \tag{2.10}$$

Multiplying throughout equation (2.9) by $p_k p_{k-1}$ we obtain

$$p_{k-1} = p_k + p_k p_{k-1} \tag{2.11}$$

so that

$$p_k = \frac{p_{k-1}}{1+p_{k-1}} \tag{2.12}$$

and finally substituting (2.12) into (2.11)

$$p_k = p_{k-1} - \frac{p_{k-1}^2}{1+p_{k-1}} . \tag{2.13}$$

This equation and equation (2.10) can now be substituted into equation (2.7) to yield

$$\hat{a}_k = [p_{k-1} - \frac{p_{k-1}^2}{1+p_{k-1}}] \; [b_{k-1} + y_k]$$

which, on reduction and using the fact that $\hat{a}_{k-1} = p_{k-1} b_{k-1}$, can be written as

$$\hat{a}_k = \hat{a}_{k-1} - \frac{p_{k-1}}{1+p_{k-1}} [\hat{a}_{k-1} - y_k] . \tag{2.14}$$

Equations (2.13) and (2.14) taken together constitute the recursive least squares algorithm for the estimation of the mean of the random variable y_k. Not surprisingly, they are exactly equivalent to equation (2.3), as can be seen if we note that $p_{k-1} = 1/k-1$ and then substitute into equation (2.14) to yield

$$\hat{a}_k = \hat{a}_{k-1} - \frac{1}{k}[\hat{a}_{k-1} - y_k] = \hat{a}_{k-1} - p_k[\hat{a}_{k-1} - y_k].$$

or

$$\hat{a}_k = \hat{a}_{k-1} + p_k\,[y_k - \hat{a}_{k-1}] \qquad (2.15a)$$

Finally, note that similar analysis to the above could be carried out for the problem of estimating the variance of y_k, in which case the recursive algorithm would be of the form

$$\hat{\sigma}_k^2 = \hat{\sigma}_{k-1}^2 - p_k\,[\hat{\sigma}_{k-1}^2 - e_k^2] \qquad (2.15b)$$

where $e_k = y_k - \bar{a}_T$ is the deviation of y_k from the sample mean $\bar{a}_T$ and again $p_k = 1/k$.

The exercise of obtaining (2.15) in what appears to be a rather roundabout fashion is not wasted for now we see how the algorithm fits within the context of least squares optimization theory. Moreover, it is now clear that the error term $[y_k - \hat{a}_{k-1}]$ can be interpreted as being proportional to the negative gradient of the *instantaneous* cost $[y_k - \hat{a}_{k-1}]^2$, i.e.,

$$\frac{\partial}{\partial \hat{a}_{k-1}}\{(y_k - \hat{a}_{k-1})^2\} = -2(y_k - \hat{a}_{k-1})$$

In other words, equation (2.15) can be viewed as a "gradient algorithm" (Wilde, (1964) in which the estimate $\hat{a}_{k-1}$ is updated in a direction defined by the negative gradient of the instantaneous cost and with a magnitude or step size dictated by the weighting factor p_k. As a result, the algorithm can be interpreted as a 'descent' procedure aimed at locating the estimates which define the minimum of the cost function (e.g. Young, 1965a).

But p_k is not constant; it is, in fact, in inverse proportion to the number of observations k. Thus as the algorithm proceeds and confidence in the estimate increases, so less and less notice is taken of the gradient measure, since it is more likely to arise from the noise e_{yk} than from error in the estimate of the mean value. And it is the harmonic sequence 1/k, i.e., 1,1/2,1/3,...., which provides the mechanism for the attenuation of the corrective action; indeed it can be seen that, as the data base becomes very large, new observations have little effect since, under the basic assumption that the mean value is constant, they provide essentially little 'new' information on which to base statistical inference.[†] In effect, the variable weighting factor p_k acts to smooth or 'filter' out the inaccuracy injected by the observation noise e_{yk}.

This interpretation of equation (2.15) as a discrete-time filtering algorithm becomes even more transparent if a block diagram of the algorithm is constructed as shown in Figure 2.3(a) where z^{-1} is the "backward shift"[††] operator, i.e.,

$$z^{-1}\hat{a}_k = \hat{a}_{k-1}.$$

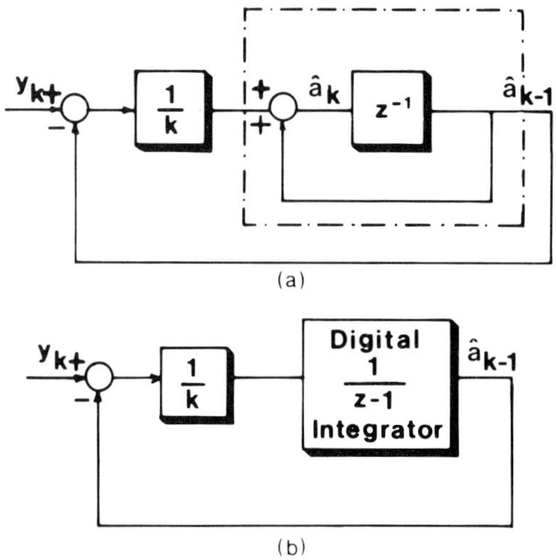

Figure 2.3 The Recursive Algorithm as a First Order, Digital Low-Pass Filter
a) Basic Block Diagram
b) Reduced Block Diagram Showing Presence of Digital Integrator

---

[†]Needless to say, if the assumption of a stationary mean value is not justified then such a procedure would be dangerous and could lead to poor estimation performance with, for example, heavily biased estimates. But we shall discuss this further in subsequent sections.

[††]The z^{-1} notation is used here in preference to alternatives such as B (Box and Jenkins, 1970) to emphasize the close relationship between the forward shift z and the z-transform operator (e.g., Ogata, 1970; Takahaski *et al.*, 1970).

The reader can easily verify that Figure 2.3(a) is a representation of equation (2.15) and, following the rules of block diagram manipulation (e.g., Ogata, 1970) it is also straightforward to manipulate the diagram into the form of Figure 2.3(b), where z is the "forward shift" operator, i.e.,

$$z\,\hat{a}_k = \hat{a}_{k+1}$$

We now see that the algorithm can be considered as a discrete feedback system or filter characterized by a unity feedback loop together with a forward path which includes a *time variable* gain $p_k = 1/k$ and a discrete integrator or summation device $1/z-1$. The feedback system of Figure 2.3 is, in fact, a variable gain "low pass" filter mechanism of a type often used in communication and control systems design (Tsypkin, 1971; and Ogata, 1970). Here the term "low pass" is applied because the filter in its fixed gain form (i.e. p_k = constant) "passes" low frequency variations in the input signal y_k but "filters off" any high frequency changes. When p_k is made a strictly decreasing function of time by defining it as $1/k$, the effect is simply to reduce sequentially the "pass-band" of the filter until, with $k = \infty$, it passes no change in signal at all and the output of the integrator remains fixed at the final asymptotic estimate of the mean, $\hat{a}_\infty$.

With this interpretation in mind, it also seems fairly clear, by physical intuition, that the averaging effect of the gain p_k will only work successfully if the 'noise' e_{yk} is, in fact, a zero mean random variable with finite variance. And, as we shall see later, it is necessary to put even stricter requirements on the statistical properties of e_{yk} if the estimate is to be relatively efficient in a statistical sense, i.e., if it is to possess low variance when compared with the theoretical minimum variance estimate.

The performance of the algorithm given by equation (2.15) is illustrated by the dotted line in Figure 2.1, which shows the recursive estimate of the mean for the Walgett rainfall data starting with an initial condition of $\hat{a}_0 = 0$. We remind the reader that, at any sampling instant k, this recursive estimate is precisely the same as the *en bloc* mean estimate obtained from equation (2.6) using *all the data up to that point*. The increasing smoothing effect of the variable gain p_k is quite clear from this diagram: for low k the recursive estimate has quite high variability but this is quickly attenuated so that by the end of the data the estimate is changing very little.

2.2 Recursive Least Squares Estimation for a Single Unknown Parameter in a Regression Relationship

When viewed within the context of least squares estimation, the recursive algorithm (2.15) for estimating the mean value of a random variable can be considered as a special example of regression analysis (e.g., Kendall and Stuart, 1961) for a single unknown parameter (the mean value) in which the regression variable or "regressor" is unity. Let us consider a somewhat more general situation where the regressor is an *exactly known* but variable quantity. For example, suppose we observe the distance from some known datum of a body moving in a straight line with constant velocity v. From basic physical principles, we should expect the distance s at time t to be related to the velocity v by the equation

$$s = t.v \tag{2.16}$$

In practice, however, we are not able to observe the distance exactly; normally there will be some unavoidable error in the observation, so that we actually measure y where

$$y = s + e_y = t.v + e_y \tag{2.17}$$

and e_y is the random error or noise associated with the measurement. A set of observations based on some data in Graybill (1961) is given in Table 1. Here, it is assumed, to begin with, that the datum distance s_0 at t = 0, is known to be 5.71 feet (see Graybill, p. 119).

TABLE 1

distance, y(ft)	9	15	19	20	45	55	78
time, t (sec)	1	2	3	4	10	12	18

By plotting the measured distance against time, as in Figure 2.4, we would expect that the observed data should lie, very approximately, on a straight line with slope, v. No straight line, however, will pass exactly through all the data points because of the measurement errors. Consequently, we are forced to ask which of the many straight lines that could be drawn through the points should be chosen as being the best representation of the data?

To obtain the least squares solution to this problem, we denote the observations in Table 1 by y_i, i = 1,2,....,7 and represent the equation (2.17) by

$$y_i = x_i a + e_{yi}; \quad i = 1,2,....,7 \tag{2.18}$$

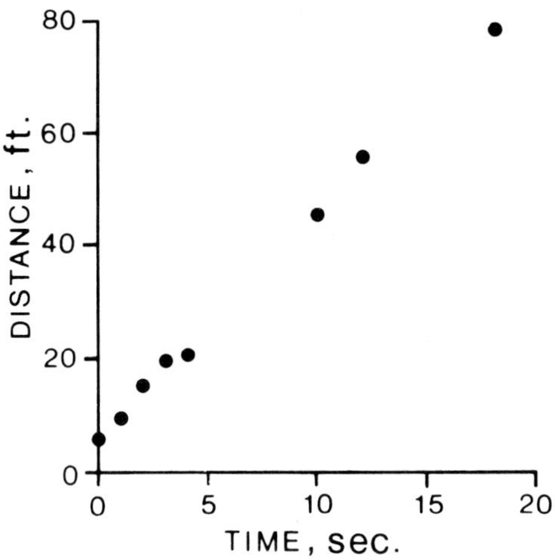

Figure 2.4 Data for Moving Body Example

where x_i denotes the time appropriate to the observation y_i.

Following the same procedure as in the previous section, the least squares cost function J is then formulated in the form

$$J = \sum_{i=1}^{k} [y_i - x_i\hat{a}]^2 = \sum e_i^2 \tag{2.19}$$

where, in this case, k=7. In other words, we are simply going to try and find the estimate of the unknown velocity, denoted by $\hat{a}$, which minimises the sum of the errors e_i^2, i = 1,2,....,k between the measurements y_1 and the 'best' (in a least squares sense) estimate of the measurement $\hat{y}_i = x_i\hat{a}$. The condition for minimum of J with respect to $\hat{a}$ is then

$$\nabla J_{\hat{a}} = \frac{\partial J}{\partial \hat{a}} = -2 \sum_{i=1}^{k} [y_i - x_i\hat{a}]\, x_i = 0$$

or

$$\{\sum_{i=1}^{k} x_i^2\}\hat{a} = \sum_{i=1}^{k} x_i y_i \tag{2.20}$$

so that the least squares estimate $\hat{a}_k$ for k samples is given by

$$\hat{a}_k = [c_k]^{-1} b_k = p_k b_k \tag{2.21}$$

where

$$c_k = \sum_{i=1}^{k} x_i^2 \quad \text{and} \quad b_k = \sum_{i=1}^{k} x_i y_i$$

Once again, we see the great simplicity of this formulation and solution; its simplicity appealed to Gauss and it has appealed to almost everyone concerned with the analysis of observations ever since. Other criterion functions such as the least magnitude (see e.g. Young, 1965a).

$$J_1 = \sum_{i=1}^{k} |y_i - x_i\hat{a}|,$$

could also be utilized but are less convenient in theoretical terms.

But suppose now that we wish to develop a recursive form of (2.21) in which, as before, the estimate at the kth instant is a linear sum of the estimate at the previous (k-1)th instant plus a corrective term based on the data at the kth instant Proceeding in a similar manner to that used in the case of the recursive estimation of the mean, we note that

$$p_k^{-1} = \sum_{i=1}^{k} x_i^2 = p_{k-1}^{-1} + x_k^2 \tag{2.22}$$

$$b_k = \sum_{i=1}^{k} x_i y_i = b_{k-1} + x_k y_k \tag{2.23}$$

Then rearranging (2.22) we obtain

$$p_{k-1} = p_k + p_k x_k^2 \, p_{k-1} \tag{2.24}$$

$$\therefore \quad p_{k-1} x_k = p_k x_k + p_k x_k^3 \, p_{k-1}$$

$$= p_k x_k [1 + p_{k-1} x_k^2]$$

so that

$$p_{k-1} x_k [1 + p_{k-1} x_k^2]^{-1} = p_k x_k$$

Now, multiplying by $p_{k-1}x_k$ and using (2.24) we find that

$$p_{k-1}^2 x_k^2 [1 + p_{k-1}x_k^2]^{-1} = p_k x_k^2 p_{k-1} = p_{k-1} - p_k$$

Consequently,

$$p_k = p_{k-1} - p_{k-1}^2 x_k^2 [1 + p_{k-1}x_k^2]^{-1} \qquad I(1)$$

By substituting this result in (2.21) and using (2.23)

$$\hat{a}_k = \{p_{k-1} - p_{k-1}^2 x_k^2 [1 + p_{k-1}x_k^2]^{-1}\} \{b_{k-1} + x_k y_k\}$$

Noting that $\hat{a}_{k-1} = p_{k-1}b_{k-1}$, we can expand this expression to yield

$$\hat{a}_k = \hat{a}_{k-1} - k_k \{x_k \hat{a}_{k-1} - y_k\} \qquad I(2)$$

where

$$k_k = p_{k-1}x_k [1 + p_{k-1}x_k^2]^{-1} \qquad I(3)$$

I(3) can be written

$$k_k = [p_k p_k^{-1}] p_{k-1} x_k [1 + p_{k-1}x_k^2]^{-1}$$

so that by substituting for p_k^{-1} from (2.22) we obtain the following alternative definition of k_k

$$k_k = p_k[p_{k-1}^{-1} + x_k^2]\, p_{k-1}x_k[1 + p_{k-1}x_k^2]^{-1}$$

$$= p_k[x_k + x_k^2 p_{k-1}x_k][1 + p_{k-1}x_k^2]^{-1} \qquad I(4)$$

$$= p_k x_k$$

The estimation algorithm given by equations I(1) through I(3) or I(4) is the required recursive version of equation (2.21). The similarity with the recursive algorithm for estimating mean value is obvious on inspection: the reader can easily verify that I(2) is a discrete-step gradient algorithm, as before, with the weighting factor or gain p_k a strictly decreasing function of k. The initiation of the gradient algorithm is again rather obvious: a_1 and p_1 can be computed easily from the non-recursive relationships in (2.21) and then the recursive algorithm can be used thereafter.

There is, however, an alternative initiation procedure which, while providing no real advantage in the present simple context, will assume major significance when we consider later the estimation of multiple unknown parameters. Here the algorithm is initiated with $\hat{a}_0$ set to some arbitrary finite value, e.g. zero, and p_0 set to some large number, say 10^2 in the present example. While this is less obvious from a numerical standpoint, it yields equivalent results provided p_0 is chosen large enough. And, as we shall see later, it can be justified on statistical grounds when the algorithm is considered from a statistical standpoint.

The performance of the algorithm I is illustrated in Figure 2.5 which shows the results obtained in the moving body example.

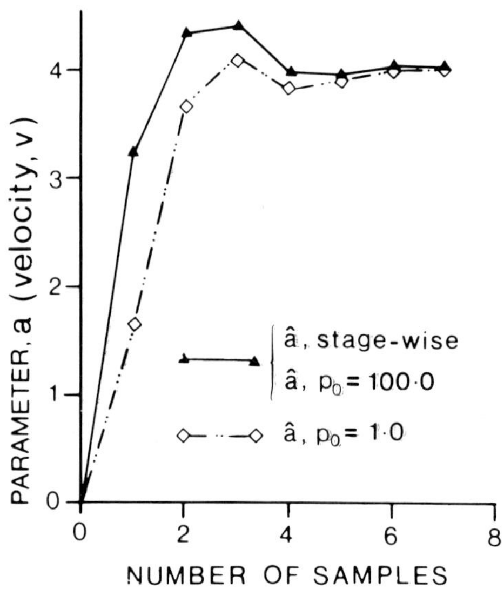

Figure 2.5 Recursive Estimates of Velocity in Moving Body Example

The recursive estimates are plotted for initial conditions $\hat{a}_0 = 0$, $p_0 = 100$ and $\hat{a}_0 = 0$, $p_0 = 1.0$ to show the effect of choosing different p_0 values. Also shown are the results obtained from the 'stage-wise' solution of the problem, i.e. the repeated solution of equation (2.21) for increasing k. It is clear that, whereas the results for $p_0 = 100$ are identical with the stage-wise results, those for p_0 = 1.0 show some discrepancy: this emphasises the need to select large values of p_0 to ensure equivalence with the *en bloc results*.

2.3 Summary

In this chapter we have considered at some length the simple problem of recursively estimating a single unknown parameter. At this point the reader may well see little advantage in the recursive procedure, since he will argue rightly that the stage-wise solution of the equivalent *en bloc* problem yields the same results and is computationally no more demanding. But our reason for treating the problem in detail will become clearer as we proceed to consider more complicated examples of recursive estimation. We will see that the analytical framework provided in this chapter provides a firm basis for solving these more complex problems and that the physical insight into the nature of the simple recursive algorithms is valuable in understanding the functioning of their more complicated and sophisticated relatives. In particular, the interpretation of the recursive algorithms as simple gradient procedures with time varying gain factors should help the reader to better appreciate the functioning of the recursive algorithms and counteract, to some extent, the superficial complexity of the recursive algorithms in the multiple parameter situation discussed in the next and subsequent chapters.

3 Recursive Estimation and Stochastic Approximation

So far, we have considered the least squares solution to a particularly simple estimation problem in a single unknown parameter. A more general problem is the estimation of the set of n unknown parameters a_j; $j = 1 \rightarrow n$, which appear in a linear "regression" relationship of the form (see e.g. Kendall and Stuart, 1961)

$$x_o = a_1x_1 + a_2x_2 + \ldots\ldots + a_nx_n \tag{3.1}$$

where, once again, the observation y of x_o is contaminated by noise e_y; while the x_j; $j = 1 \rightarrow n$, are *exactly known, linearly independent* variables which are also statistically independent of the measurement noise on y. In this case, the minimization of the least squares criterion function for k samples, i.e.,

$$J_2 \triangleq \sum_{i=1}^{k} [\sum_{j=1}^{n} x_{ji}\hat{a}_j - y_i]^2$$

requires that all the partial derivatives of J_2 with respect to each of the parameter estimates $\hat{a}_j$ should be set simultaneously to zero. Such a procedure yields a set of n linear, simultaneous algebraic equations that are sometimes termed the 'normal equations' and which can be solved for the parameter estimates $\hat{a}_{jk}$, $j=1,2,\ldots,n$, all at the kth instant.

3.1 The Recursive Least Squares Algorithm

A simple and concise statement of the least squares results in the multi-parameter case can be obtained by using a vector-matrix formulation: thus by writing (3.1) in the alternative vector form

$$x_o = \underline{x}^T\underline{a} \tag{3.2}$$

where $\underline{x}^T = [x_1x_2\ldots\ldots x_n]$; $\underline{a}^T = [a_1a_2\ldots\ldots a_n]$ with the superscript T denoting the vector/matrix transpose, and $\underline{x}^T\underline{a}$ is the vector inner product, we are able to define J_2 as

$$J_2 \triangleq \sum_{i=1}^{k} [\underline{x}_i^T\hat{\underline{a}} - y_i]^2 = \sum_{i=1}^{k} e_i^2 \tag{3.3}$$

where

$$y_i = \underline{x}_i^T \underline{a} + e_{yi} \qquad (3.4)$$

Now, referring to Appendix 1 (page 253)

$$\frac{\partial}{\partial \hat{\underline{a}}} \{(\underline{x}_i^T \hat{\underline{a}} - y_i)^2\} = 2\underline{x}_i(\underline{x}_i^T \hat{\underline{a}} - y_i)$$

so that the normal equations become

$$\frac{1}{2} \nabla_{\hat{\underline{a}}}(J_2) = [\sum_{i=1}^{k} \underline{x}_i \underline{x}_i^T] \hat{\underline{a}}_k - \sum_{i=1}^{k} \underline{x}_i y_i = 0 \qquad (3.5)$$

where $\nabla_{\hat{\underline{a}}}(J_2)$ denotes the gradient of J_2 with respect to all the elements of $\hat{\underline{a}}$. The reader well acquainted with matrix algebra can verify easily that equation (3.5) is correct by expanding the various terms. Those readers less familiar with matrix algebra will, at this point, need to consult a standard text on the subject (Johnston, 1963) or Appendix 1 before proceeding, but they will see both in equations (3.4) and (3.5) and the subsequent analysis, a great similarity with the equivalent scalar analysis pursued in the previous Chapter.

Now, provided that the matrix $\Sigma \underline{x}_i \underline{x}_i^T$ is non-singular† then the solution to the equations (3.5) takes the form

$$\hat{\underline{a}}_k = P_k \underline{b}_k \qquad (3.6)$$

where

$$P_k = [\sum_{i=1}^{k} \underline{x}_i \underline{x}_i^T]^{-1}; \quad \underline{b}_k = \sum_{i=1}^{k} \underline{x}_i y_i$$

The scalar relationships (2.22) and (2.23) of the single parameter example are now replaced by the following vector-matrix expressions

$$P_k^{-1} = P_{k-1}^{-1} + \underline{x}_k \underline{x}_k^T \qquad (3.7)$$

and

$$\underline{b}_k = \underline{b}_{k-1} + \underline{x}_k y_k \qquad (3.8)$$

†The need for linear independence between the x_j now becomes clear since linear dependence would result in singularity of the data covariance matrix $C_k = [\sum_{i=1}^{k} \underline{x}_i \underline{x}_i^T]$ with det.$[C_k] = 0$ and the inverse "blowing up" (see Appendix 1).

In order to develop a recursive version of equation (3.6), we now pre-multiply equation (3.7) by P_k and then post-multiply by P_{k-1} to give

$$P_{k-1} = P_k + P_k\underline{x}_k\underline{x}_k^T P_{k-1} \qquad (3.9)$$

Post multiplying by $\underline{x}_k$ then yields,

$$P_{k-1}\underline{x}_k = P_k\underline{x}_k + P_k\underline{x}_k\underline{x}_k^T P_{k-1}\underline{x}_k$$

$$= P_k\underline{x}_k[1 + \underline{x}_k^T P_{k-1}\underline{x}_k]$$

Then, post-multiplying by $[1 + \underline{x}_k^T P_{k-1}\underline{x}_k]^{-1}\ \underline{x}_k^T P_{k-1}$, we obtain

$$P_{k-1}\underline{x}_k[1 + \underline{x}_k^T P_{k-1}\underline{x}_k]^{-1}\ \underline{x}_k^T P_{k-1} = P_k\underline{x}_k\underline{x}_k^T P_{k-1}$$

Substituting from (3.9) we obtain finally

$$P_k = P_{k-1} - P_{k-1}\underline{x}_k[1 + \underline{x}_k^T P_{k-1}\underline{x}_k]^{-1}\ \underline{x}_k^T P_{k-1} \qquad \text{II}(1)$$

which is often termed the 'matrix inversion lemma' (Bodewig, 1956; Ho, 1962).

By substituting in equation (3.6) from II(1) and (3.8) and then continuing in a manner analogous to that used in the single parameter case (but taking care to obey the rules of matrix algebra), it is a simple matter to obtain the equivalent recursive equation for $\hat{\underline{a}}_k$: i.e.

$$\hat{\underline{a}}_k = \{P_{k-1} - P_{k-1}\underline{x}_k\ [1 + \underline{x}_k^T P_{k-1}\underline{x}_k]^{-1}\ \underline{x}_k^T P_{k-1}\}\{\underline{b}_{k-1} + \underline{x}_k y_k\}$$

$$= \hat{\underline{a}}_{k-1} - P_{k-1}\underline{x}_k[1 + \underline{x}_k^T P_{k-1}\underline{x}_k]^{-1}\ \underline{x}_k^T\hat{\underline{a}}_{k-1} + P_{k-1}\underline{x}_k y_k$$

$$- P_{k-1}\underline{x}_k[1 + \underline{x}_k^T P_{k-1}\underline{x}_k]^{-1}\underline{x}_k^T P_{k-1}\underline{x}_k y_k$$

$$= \hat{\underline{a}}_{k-1} - P_{k-1}\underline{x}_k\ [1 + \underline{x}_k^T P_{k-1}\underline{x}_k]^{-1}\{\underline{x}_k^T\hat{\underline{a}}_{k-1} + \underline{x}_k^T P_{k-1}\underline{x}_k y_k$$

$$- \ [1 + \underline{x}_k^T P_{k-1}\underline{x}_k]y_k\}$$

So that finally,

$$\hat{\underline{a}}_k = \hat{\underline{a}}_{k-1} - K_k\ \{\underline{x}_k^T\hat{\underline{a}}_{k-1} - y_k\} \qquad \text{II}(2)$$

where

$$K_k = P_{k-1}\underline{x}_k[1 + \underline{x}_k^T P_{k-1}\underline{x}_k]^{-1}$$

An alternative expression for K_k can be obtained straightforwardly by manipulation, i.e.

$$K_k = [P_k P_k^{-1}]P_{k-1}\underline{x}_k[1 + \underline{x}_k^T P_{k-1}\underline{x}_k]^{-1}$$

$$= P_k[P_{k-1}^{-1} + \underline{x}_k\underline{x}_k^T]P_{k-1}\underline{x}_k[1 + \underline{x}_k^T P_{k-1}\underline{x}_k]^{-1}$$

$$= P_k[\underline{x}_k + \underline{x}_k\underline{x}_k^T P_{k-1}\underline{x}_k][1 + \underline{x}_k^T P_{k-1}\underline{x}_k]^{-1}$$

$$= P_k\underline{x}_k[1 + \underline{x}_k^T P_{k-1}\underline{x}_k][1 + \underline{x}_k^T P_{k-1}\underline{x}_k]^{-1}$$

$$= P_k\underline{x}_k \tag{3.10}$$

Thus an alternative form of the recursion II(2) is the following

$$\hat{\underline{a}}_k = \hat{\underline{a}}_{k-1} - P_k\{\underline{x}_k\underline{x}_k^T\hat{\underline{a}}_{k-1} - \underline{x}_k y_k\} \qquad \text{II(3)}$$

This second form exposes nicely the nature of the algorithm but II(2) is usually preferred in computational terms.

The equation set II constitute the recursive form of the least squares solution (3.6). It is interesting to note the simplicity of the above derivation in vector-matrix terms which constrasts with the complexity of Gauss's original derivation using ordinary algebra. It provides an excellent example of the advantages of matrix algebra and will hopefully stimulate the reader not well acquainted with matrix analysis to examine its mysteries further.

Since the algorithm II is recursive, it is necessary once again to specify starting values $\hat{\underline{a}}_0$ and P_0 for the vector $\hat{\underline{a}}$ and the matrix P respectively. As in the scalar case, however, this presents no real problem; it can be shown that the criterion function-parameter hypersurface is unimodal (e.g. Young, 1965a) and that an arbitrary finite $\hat{\underline{a}}_0$ (say $\hat{\underline{a}}_0 = 0$) coupled with a P_0 having large diagonal elements (say 10^6 in general) will yield convergence and performance commensurate with the stage-wise solution of the same problem (Lee, 1964).

Another similarity with the scalar case is that the algorithm II can clearly be interpreted as a special form of gradient procedure. In this multi-parameter case, the gradient of the instantaneous part of the cost function, i.e.

$$\nabla_{\hat{\underline{a}}}(\hat{e}_k^2) = 2[\underline{x}_k\underline{x}_k^T\hat{\underline{a}}_{k-1} - \underline{x}_k y_k]$$

is an n vector so that the scalar weighting factor is replaced by the time variable weighting matrix, P_k.† However, the strictly decreasing nature and consequent smoothing effect of this matrix is directly analogous to that of p_k in the single parameter case, as pointed out by Ho (1962).

In contrast to the scalar situation, the algorithm II does provide some considerable advantage over the stage-wise solution of (3.6). In addition to the now convenient recursive form, which provides for a minimum of computer storage, note that the term

$$[1 + \underline{x}_k^T P_{k-1} \underline{x}_k]$$

is simply a *scalar* quantity. As a result, there is no requirement for direct matrix inversion even though the repeated solution of the equivalent classical solution (3.6) entails inverting an n x n matrix for each solution.

In order to demonstrate the use of the recursive least squares algorithm II, consider a slightly expanded form of the moving body problem discussed in the previous Chapter. Suppose that the datum distance s_o is also unknown, so that equation (2.16) has to be written

$$s = s_o + t.v \qquad (3.11)$$

This equation can be written in the vector form (3.2), where

$$\underline{x}^T = [1 \; t]; \quad \underline{a}^T = [s_o \; v]$$

Consequently, the simultaneous estimation of s_o and v by reference to the measurements y_i represents a two parameter estimation problem which can be solved in the above recursive least squares manner. Figure 3.1(a) shows the estimation results obtained from algorithm II with $\underline{\hat{a}}_o = [0]$ and

$$P_o = \begin{bmatrix} 10^4 & 0 \\ 0 & 10^4 \end{bmatrix}.$$

Figure 3.1(b) provides the time history of the P matrix elements; their strictly decreasing nature is quite apparent from this diagram. While it is a little

---

†We utilise $\hat{e}_k$ to differentiate the recursive errors (or residuals) $\hat{e}_i = \underline{x}_i^T \underline{\hat{a}}_{i-1} - y_i$ $i = 1,2,\ldots,k,\ldots.T$, from the least squares residuals e_i obtained in the same manner but with $\hat{a}_{i-1}$ replaced by $\hat{a}_T$; $\hat{e}_k$ can be interpreted as the *a priori* estimate of e_k at k given the estimate $\underline{\hat{a}}_{k-1}$ and could be written $\hat{e}_{k/k-1}$ (see later).

laborious, we would encourage the reader new to recursive analysis to try to reproduce at least the first five recursions in this example since it will provide an excellent feel for the nature of the computations and, in particular, the behaviour of the P_k gain matrix.

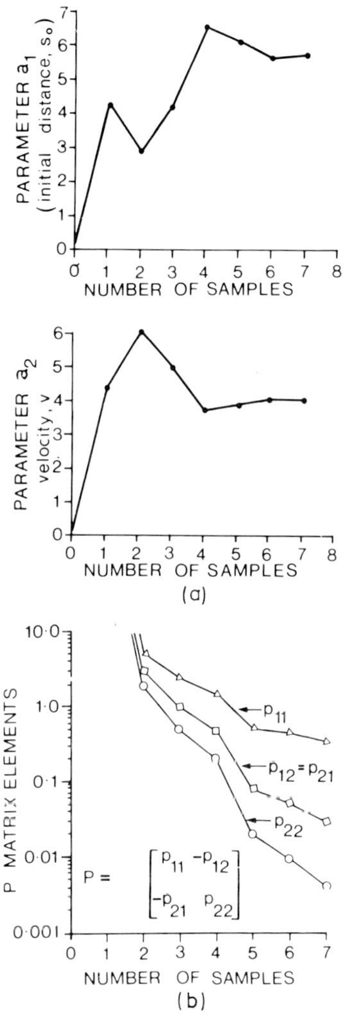

Figure 3.1 Estimation Results for Moving Body Example with Two Unknown Parameters
a) Parameter Estimates
b) P Matrix Elements

3.2 Connection with Stochastic Approximation

The scalar recursive algorithms discussed in the last Chapter and the multi-dimensional recursive algorithm II in the previous section of this Chapter are examples of a technique known to mathematicians as a *stochastic approximation* (SA) or *stochastic gradient* procedure. The procedure is approximate in the sense that the noisy data are used recursively to generate better and better 'approximations' or estimates of the unknown parameters; it is stochastic because it is able to handle the random character of the additive noise effects; finally, it is sometimes referred to as a stochastic gradient procedure because, as we have seen, it represents the stochastic equivalent of the well known deterministic hill climbing procedures in which the search direction at any point is chosen by reference to the measured gradient of the cost function at that point. This latter remark is significant because such deterministic gradient techniques have been used both for the optimization of *functions* in a *parameter* space (Wilde, 1964) and the optimisation of *functionals* in a *function* space (Lasdon *et al.*, 1967): they are, in other words, of wide practical significance.

In this latter sense, SA is a technique for general optimization in the presence of noise and is not restricted to the kind of recursive data processing discussed in the previous section. It is specifically this kind of application with which we are primarily concerned in this book, however, and we will continue to concentrate on the interpretation of stochastic approximation in these terms.

Much has been written of stochastic approximation in the technical literature and it would serve little useful purpose to review the topic in detail since several useful essays on the subject are available elsewhere for the interested reader (see e.g., Wilde, 1964; Sakrison, 1966; Mendel and Fu, 1970; Tsypkin, 1971; and Young, 1976b). It will suffice here merely to point out the major characteristics of stochastic approximation which are relevant to the present discussion in sufficient detail for the reader to appreciate its significance as a general tool for recursive analysis.

Search techniques which are able to reach a pre-assigned goal despite the effects of noise or uncertainty were first termed stochastic approximation procedures in a seminal paper by Robbins and Munro (1951). But they were primarily concerned with the problem of finding the root of a noisy function, and it was left to Kiefer and Wolfowitz (1952) to point out that a similar approach could be applied to general problems of estimating the maximum of a unimodal function obscured by noise. Blum (1954) extended the approach to the multidimensional case and Dvoretzky (1956) considered in detail the conditions for the convergence of a stochastic approximation scheme; conditions which, as Wilde (1964) has pointed out, are of special significance in helping the practitioner to design his own SA procedure.

In order to generalize our treatment of stochastic approximation let us use the approach of Tsypkin (1971) who considers the problem of finding the value of a vector $\underline{\hat{a}} = [\hat{a}_1,\hat{a}_2,\ldots,\hat{a}_n]^T$ which is optimal in the sense that it defines the extremum (maximum or minimum) of a performance index $J(\underline{\hat{a}})$[†] which is given in the form of an expectation (Appendix 1).

$$J(\underline{\hat{a}}) = \int_X Q(\underline{x},\underline{\hat{a}})p(\underline{x})dx \tag{3.12}$$

or, in shorthand form

$$J(\underline{\hat{a}}) = E_X\{Q(\underline{x},\underline{\hat{a}})\}$$

where $E_X\{\cdot\}$ is the expectation operation and the realization $Q(\underline{x},\underline{\hat{a}})$ is a function of the vector $\underline{\hat{a}}$ which also depends upon a vector of random sequences or processes $\underline{x} = [x_1,x_2,\ldots x_n]^T$ with (usually unknown) probability density function $p(x)$.

A simple form of such a performance index or criterion of optimality is one based on the averaging of the realization $Q(\underline{x},\underline{\hat{a}})$ with respect to time: if $\underline{x}$ is a random sequence $\{\underline{x}_i;\ i = 1,2,\ldots,\infty\}$ such a criterion takes the form

$$J(\underline{\hat{a}}) = \lim_{k\to\infty} \frac{1}{k} \sum_{i=1}^{k} Q(\underline{x}_i,\underline{\hat{a}}) \tag{3.13}$$

while if $\underline{x}$ is a random process $\{\underline{x}(t);\ 0 \leq t < \infty\}$ it becomes

$$J(\underline{\hat{a}}) = \lim_{T\to\infty} \frac{1}{T} \int_0^T Q(\underline{x}(t),\ \underline{\hat{a}})dt. \tag{3.14}$$

A typical example of the discrete-time criterion function (3.13) is the least squares cost function J_2 of equation (3.3) where $Q(\underline{x}_i,\underline{\hat{a}})$ is simply a quadratic function in $\underline{\hat{a}}$, i.e.,

$$e_i^2 = [\underline{x}_i^T\underline{\hat{a}} - y_i]^2,$$

and the infinite summation is replaced by the finite summation over k samples; in other words, the true probability density function is replaced or approximated by the empirical one based on the finite data set $\underline{x}_i$, y_i, $i = 1,2,\ldots,k$.

†Tsypkin refers to these general procedures of stochastic approximation as "probabilistic iterative methods".

Posed in the above terms the solution to the problem of optimality is basically one of finding, in some manner, a vector $\hat{\underline{a}} = \underline{a}^*$ which we shall call the optimal vector, that defines the extremum (maximum or minimum) of the function

$$J(\hat{\underline{a}}) = E_x\{Q(\underline{x},\hat{\underline{a}})\} \tag{3.15}$$

The condition of optimality for the function $J(\hat{\underline{a}})$ in equation (3.15) can be written as

$$\nabla J(\hat{\underline{a}}) = E_x\{\nabla_{\hat{\underline{a}}} Q(\underline{x},\hat{\underline{a}})\} = 0 \tag{3.16}$$

where J represents the gradient of the cost function and where

$$\nabla_{\hat{\underline{a}}} Q(\underline{x},\hat{\underline{a}}) = \left[\frac{\partial Q(\underline{x},\hat{\underline{a}})}{\partial \hat{a}_1}, \frac{\partial Q(\underline{x},\hat{\underline{a}})}{\partial \hat{a}_2}, \ldots, \frac{\partial Q(\underline{x},\hat{\underline{a}})}{\partial \hat{a}_n}\right]^T \tag{3.17}$$

is the gradient of $Q(\underline{x},\hat{\underline{a}})$ with respect to $\hat{\underline{a}}$.

The basic idea of solving equation (3.16) using stochastic approximation is to define a sequence of vectors $\hat{\underline{a}}_k$ by the recursive algorithm

$$\hat{\underline{a}}_k = \hat{\underline{a}}_{k-1} - \gamma_k\{\nabla_{\hat{\underline{a}}} Q(\underline{x}_k,\hat{\underline{a}}_{k-1})\} \tag{3.18}$$

which can be compared directly with the simple recursive algorithm for estimating the mean value of a random variable given in equation (2.15) of the previous chapter and the recursive least squares algorithm II(3) in the previous section of this chapter. It will also be noted that the algorithm is similar in form to the equivalent gradient algorithms used in deterministic optimization (Wilde, 1964). Unlike the deterministic algorithm, however, we know that even for $\hat{\underline{a}} = \underline{a}^*$

$$\nabla_{\hat{\underline{a}}} Q(\underline{x}_k,\hat{\underline{a}}) \not\equiv 0$$

because of the random effects due to the presence of $\underline{x}_k$.

We see that it is the presence of $\underline{x}_k$ which introduces 'noise' effects into the sample measure of the gradient

$$\nabla_{\hat{\underline{a}}} Q(\underline{x}_k,\hat{\underline{a}}_{k-1})$$

at each recursive step and which makes the vector sequence $\hat{\underline{a}}_k$, $k = 0,1,2,\ldots,$ itself a random sequence. As a result, it is no longer possible to consider the convergence of equation (3.18) in the ordinary (deterministic) sense; rather it is necessary to introduce new concepts of *stochastic* convergence.

This can be accomplished by introducing the concept of *almost sure* convergence; in the sense that the sequence of random vectors $\hat{\underline{a}}_k$ converges to $\underline{a}^*$ as $k \to \infty$ *almost certainly* or *with probability one*, if the probability of ordinary convergence of $\hat{\underline{a}}_k$ to $\underline{a}^*$ is equal to unity, i.e. if

$$P\{\lim_{k\to\infty} ||\hat{\underline{a}}_k - \underline{a}^*|| = 0\} = 1 \tag{3.19}$$

where $||.||$ denotes the L_2 norm of the error vector $\tilde{\underline{a}}_k^* = \hat{\underline{a}}_k - \underline{a}^*$ (see Appendix 1; Fröberg (1970)). Normally, of course, the cost function $J(\hat{\underline{a}})$ is chosen so that asymptotically as $k \to \infty$ so $\underline{a}^*$ tends to the 'true' value of the vector, say $\underline{a}$. In this case, the estimation error vector will be denoted by $\tilde{\underline{a}}_k = \hat{\underline{a}}_k - \underline{a}$. Clearly the *rate* of convergence depends on the particular sample sequence generated with specific $\underline{x}_k$ sequences and, therefore, it also has a random character.

As Tsypkin (1971) points out, to design SA algorithms of the form (3.18) we must conform with the necessary and sufficient conditions for convergence in the above sense. If we consider (3.18) in the following form

$$\Delta\hat{\underline{a}}_k = -\gamma_k \{\nabla_{\hat{\underline{a}}} Q(\underline{x}_k, \hat{\underline{a}}_{k-1})\} \tag{3.20}$$

where Δ is the difference operator, $\Delta\hat{a}_k = \hat{\underline{a}}_k - \hat{\underline{a}}_{k-1}$, then it is clearly necessary that, for $\hat{\underline{a}}_k$ to converge almost surely to $\underline{a}$, then the right hand side of (3.20) should converge to zero as $k \to \infty$ for practically any $\underline{x}_k$. And since the gradient is generally different from zero because of the presence of $\underline{x}_k$, it is necessary that *the SA gain sequence* γ_k *should tend towards zero as* k *increases*.

This is consistent with our experience of the recursive least squares estimation algorithms considered previously and is evidenced by the strictly decreasing nature of the gain factors which occur naturally in these algorithms. Indeed on the basis of this experience we might guess that the gain sequence should be a strictly decreasing function of k which reduces in some specific manner, perhaps as a harmonic sequence 1/k.

In fact, such intuition is confirmed by theoretical analysis: as Tsypkin shows, the algorithm (3.18) will converge *almost surely* if the following conditions (which are basically derived from the conditions of Dvoretzky (1956)) are satisfied (the introductory reader can ignore conditions (b) and (c))

(a) (i) $\gamma_k > 0$; (ii) $\sum_{k=1}^{\infty} \gamma_k = \infty$; (iii) $\sum_{k=1}^{\infty} \gamma_k^2 < \infty$

(b) $$\inf_{\varepsilon<||\hat{\underline{a}}-\underline{a}^*||<1/\varepsilon} E_X\{[\hat{\underline{a}}-\underline{a}^*]^T \nabla_{\hat{\underline{a}}} Q(\underline{x},\underline{a})\} > 0 \quad (\varepsilon>0) \tag{3.21}$$

(c) $$E_X\{\nabla_{\hat{\underline{a}}}{}^T Q(\underline{x},\hat{\underline{a}}) \cdot \nabla_{\hat{a}} Q(\underline{x},\hat{\underline{a}})\} \leqslant \alpha(1 + \hat{\underline{a}}\hat{\underline{a}}^T) \quad (\alpha>0)$$

In physical terms, condition (a) requires that γ_k tends towards zero and the rate of decrease in γ_k is such that the variance of the estimate of $J(\hat{\underline{a}})$ is also reduced to zero. In particular, although we want the corrective effect of γ_k to be small when $\hat{\underline{a}}_k$ is close to $\underline{a}$, we do not want it to vanish for $k < \infty$, for otherwise the procedure could stop before it had reached its goal. Thus we require condition (a)(ii), i.e., that the sum of the γ_k should be infinite so that if $\hat{\underline{a}}_k$ tried to converge to other than the optimum, an infinite amount of corrective effort would be available. Of course this is not to say that convergence might not take an inordinately long time.

For a good heuristic discussion on the need for condition (a)(iii) namely that the infinite sum of the variances is finite, see Wilde (1964; Ch.6). Basically, however, this property is required because, together with the requirement that the noise variance is finite, it ensures that the residual fluctuations caused by noise will die out in the long term, with the individual random errors caused by the noise at each step in the recursive algorithm, tending to cancel each other out.

Condition (b) is required to define the behaviour of the surface $E_X\{\nabla_{\hat{\underline{a}}} Q(\underline{x},\underline{a})\}$ near zero, and thus the sign of the increments in $\hat{\underline{a}}_k$; while condition (c) ensures that the rate of increase of the expectation of the quadratic form (Johnston, 1963)

$$\nabla_{\hat{\underline{a}}}{}^T Q(\underline{x},\hat{\underline{a}}) \nabla_{\hat{\underline{a}}} Q(\underline{x},\underline{a})$$

with $\hat{\underline{a}}$ is smaller than in a parabola of second degree. For further consideration of these conditions, however, see Tsypkin (1971) or Mendel and Fu(1970).

From the design standpoint, perhaps the most important of the convergence conditions discussed above is (a) which defines the characteristics of the SA gain γ_k. The harmonic sequence 1/k:1,1/2,1/3,.... etc., is, as we conjectured, the best known example of a sequence which satisfies these properties. Other examples exist, of course, but it is necessary to be careful in choosing sequences to ensure that they do satisfy conditions (a). Some sequences, e.g., $1/k^2$:1,1/4, 1/9.... are not divergent in the sense of conditions (a) (ii); indeed for all $p > 1$

$$\sum_{k=1}^{\infty} \frac{1}{k^p} < \infty$$

and an SA procedure based on such a gain sequence might well not converge under all circumstances since the total correction effort is limited. On the other hand, the sequence $1/k^{\frac{1}{2}}$:1,1/√2, 1/√3, etc., would converge too slowly to dissipate the cumulative error, although a faster sequence with the power of $k > 1/2$ would be suitable.

One final comment on the basic principles of SA concerns the nature of the realization $Q(\underline{x},\hat{\underline{a}})$. Although Tsypkin, for example, gives only brief information on how the realization should be chosen, it is clearly extremely important in relation to the performance of the SA algorithm. In most of the algorithms discussed in this book $Q(\underline{x},\hat{\underline{a}})$ is the form of a squared error function, as in the least squares regression example examined in the previous section. In this situation, we have a general cost function of the form

$$J(\hat{\underline{a}}) = \frac{1}{k}\sum_{i=1}^{k}[\hat{e}(\underline{x}_i,\hat{\underline{a}})]^2 \tag{3.22}$$

and the SA algorithm (3.18) takes the specific form

$$\hat{\underline{a}}_k = \hat{\underline{a}}_{k-1} - \gamma_k\{\nabla_{\hat{\underline{a}}}\hat{e}(\underline{x}_k,\hat{\underline{a}}_{k-1})[\hat{e}(\underline{x}_k,\hat{\underline{a}}_{k-1})]\} \tag{3.23}$$

where the contents of the $\{\cdot\}$ parentheses represent the gradient of $Q(\underline{x}_k,\hat{\underline{a}}_{k-1})$ with $\hat{e}$ denoting the *a priori* error at the kth instant $(\hat{e}_{k/k-1})$, i.e.

$$\nabla_{\hat{\underline{a}}}Q(\underline{x}_k,\hat{\underline{a}}_{k-1}) = 2\nabla_{\hat{\underline{a}}}\hat{e}(\underline{x}_k,\hat{\underline{a}}_{k-1})[\hat{e}(\underline{x}_k,\hat{\underline{a}}_{k-1})]$$

with the constant multiplier 2 simply a scale factor which is absorbed into the definition of γ_k. We see therefore, that the gradient term in the SA algorithm is simply the gradient of the error function multiplied by the error function itself, with both evaluated by substitution of the latest value of $\underline{x}_k$ and estimate $\hat{\underline{a}}_{k-1}$ obtained at the previous, (k-1)th recursive update. A special scalar example of this is the recursive mean estimator (2.15) where

$$\nabla_{\hat{a}}\hat{e}(\underline{x}_k,\hat{\underline{a}}_{k-1}) = 1.0, \quad \hat{e}(\underline{x}_k,\hat{\underline{a}}_{k-1}) = [\hat{\underline{a}}_{k-1} - y_k]$$

and γ_k defined as the harmonic sequence $1/k$.

We shall refer to $\hat{e}_{k/k-1} = \hat{e}(\underline{x}_k,\hat{\underline{a}}_{k-1})$ as the *recursive residual* (or *a priori* prediction error), since it represents the latest residual error computed in the recursive algorithm: its appearance in the algorithm makes obvious sense since the update of parameter estimates should clearly be a function of this latest residual error, which will be due to a combination of estimation error and noise on the data. The function of the SA gain is then to 'filter' the random noise progressively from the recursive residual so that estimation error tends to zero as $k \to \infty$.

From our previous discussion it seems reasonable that the recursive residual should possess certain desirable statistical properties: for example, it should be zero mean and have finite variance so that the smoothing effect of the SA gain can be fully effective. In addition, and by consideration of other methods of statistical inference, we might also assume that the recursive residuals should form a serially uncorrelated sequence of random variables since the problem becomes much simpler in statistical terms. Also, as we shall see, it is much easier to make the solution statistically efficient, in the sense that the estimation error variance is minimized.

With these intuitive notions in mind, it is not surprising to find that all of these properties are desirable in theory and that one of the primary functions of the SA algorithm designer is to ensure that the realization $Q(\underline{x}_k,\hat{\underline{a}})$ is such that, if possible, the recursive residuals constitute a zero mean, serially uncorrelated sequence of random variables; in other words, they represent a discrete "white noise" sequence.

In the control and systems literature (Kailath and Frost, 1968) this function of a recursive estimator in yielding a white noise sequence of recursive residuals is termed the "innovations property" and the sequence itself is termed the "innovations sequence". The term "innovations" has arisen because at each recursive step, the term represents the new information on which statistical inference is to be based. The usual innovations approach is, therefore, first to operate on the observed process by means of a linear transformation (although it is possible to conceive of analogous operations which might be nonlinear) such that the resultant transformed or "filtered" noise sequence has discrete white noise properties. This "pre-whitened" noise sequence then provides the basis for the definition of e in the cost function and determines the nature of the innovations sequence $\hat{e}_k$ in the SA algorithm.

In fact, theoretical analysis shows (Kumar and Moore, 1979) that convergence can be assured if the noise process satisfies rather more general requirements than those of zero mean, white noise: in particular, for standard convergence theory, the sequence $\hat{e}_k$ is taken to be a sample sequence from a noise process with

$$E[\hat{e}_k|A_{k-1}] = 0 \quad \text{and} \quad E[\hat{e}_k^2|A_{k-1}] \leq \sigma^2$$

for all k, where A_{k-1} represents past information and is in the form of a σ-algebra (Goodwin and Payne, 1977; Loeve, 1963) generated by $\hat{e}_1,\hat{e}_2,\ldots,\hat{e}_{k-1}$. The less mathematical reader need not be confused by this, however, since zero mean, white noise satisfies these conditions and serves to exemplify the requirements in more digestible and well known terms.

The problem of statistical inference with white noise innovations is much easier to solve and, if it is required, the solution to the simpler problem can then be expressed in terms of the original observations by means of the inverse of the original "whitening" filter. Similar arguments apply in the continuous-time case considered briefly in a subsequent section.

As a specific example of this innovations approach, consider again the mean value estimator: here the innovations sequence is simply $\hat{a}_{k-1} - y_k$ and it is clear that asymptotically as $\hat{a}_k \to a$ the sequence $\hat{e}_k$ is, by definition, serially uncorrelated if e_y is itself serially uncorrelated. What we are saying here is that, in a good estimator, if e_y is serially uncorrelated then $\hat{a}_{k-1} - y_k$ will also be serially uncorrelated for all k, although it will not necessarily have constant variance because of the effects of convergence and the varying gain $\gamma_k = p_k = 1/k$ (Kailath and Frost, 1968; Brown *et al.*, 1975).

It is interesting to note that, in recent years, a considerable amount of research effort has been expended on the development of "adaptive" procedures (e.g., Carew and Belanger, 1973; and Neethling and Young, 1974) in which the recursive algorithm (usually a state variable estimation algorithm of the type discussed later in Chapter 5) is adjusted until the sample statistical properties of the innovations sequence indicate that it is indeed zero mean and serially uncorrelated and that the algorithm is, therefore, performing optimally in this sense.

3.3 Some Extensions to Stochastic Approximation

So far we have discussed the basic elements of "scalar gain" SA applied to optimization problems and tried to provide a physical interpretation of the SA procedure. There are, however, a number of generalizations and extensions to the basic procedure which make it extremely flexible in application.

Matrix gain SA and optimum algorithms

The first point to note is that the algorithm (3.18) although multidimensional in the sense that it updates the n vector $\hat{\underline{a}}_k$, is characterized by a scalar SA gain γ_k. By reference to the recursive least squares algorithm II, it seems reasonable that we might, in general, replace this by a matrix gain Γ_k, i.e.,

$$\hat{\underline{a}}_k = \hat{\underline{a}}_{k-1} - \Gamma_k \nabla_{\hat{\underline{a}}} Q(\underline{x}_k, \hat{\underline{a}}_{k-1}) \qquad (3.24)$$

The problem is, of course, how to determine the form of Γ_k, since it could have as many as n^2 elements. Heuristically, we might simplify the problem by constraining Γ_k to the form,

$$\Gamma_k = \begin{bmatrix} \gamma_{1k} & & & & \mathbf{0} \\ & \gamma_{2k} & & & \\ & & \cdot & & \\ & & & \cdot & \\ \mathbf{0} & & & & \gamma_{nk} \end{bmatrix}$$

and then make each of the elements γ_{ik}, $i = 1,2,\ldots,n$ satisfy the conditions (a). But even so, we do not have any rules for choosing the differences between the elements and it might be a hazardous trial and error business.

Tsypkin (1971) attempts to solve this problem in theoretical terms and define an SA gain matrix sequence which is optimal in a statistical sense. We will refer to this work in later Chapters when we consider statistical interpretations of recursive least squares analysis. For the moment we will restrict our attention to deterministic algorithms[†] in which *a priori* assumptions and information about the noise are not used to infer anything about the statistical nature of the estimates other than that $\hat{\underline{a}}_k$ converges almost surely to $\underline{a}^*$ as $k \to \infty$.

There has been a great deal of research on the design of deterministic matrix gain SA algorithms of this general kind (e.g. Saridis, 1974) but the paper by Kumar and Moore (1979) probably represents the latest thinking on the subject. They consider a specific class of problems which are related to the RLS algorithm II. With insight gained both from the nature of the RLS solution and a useful geometric interpretation of the recursive procedure (see also Young, 1965a), they suggest SA algorithms in which the scalar SA gain is retained with its usual characteristics, but is modulated by a time variable matrix so that the product can be considered as a crude approximation to P_k, and is computed by reference to the $\underline{x}_k$ vector.

We will refer to Kumar and Moore's work again later in Chapter 10. Their approach serves to emphasize, however, that the RLS algorithm can be considered as a naturally occurring SA algorithm with optimum properties. There is, perhaps, a moral in this observation: if we are dealing with a stochastic

[†]For example, in deterministic algorithms such as (2.3) no estimates of the second statistical moment (or variance) of the estimation errors are generated. And more generally, in multi-parameter algorithms such as the recursive least squares (RLS) procedures (II), no estimates of second moment or covariance matrix of the estimation errors are obtained.

optimization problem in which a non-recursive analytic solution already exists, we may do better to look for a recursive version of this solution, in which the stochastic approximation properties are implicit, rather than concocting an SA algorithm and then attempting to make it optimal in some manner. This is particularly important in the matrix gain situation. We will exploit such an approach in subsequent Chapters of this book and synthesize recursive algorithms on the basis of analogies with RLS and similar procedures. This does not, of course, negate the utility of SA both as a conceptual basis for recursive analysis and a useful starting point for stochastic convergence analysis (e.g. Solo, 1979, 1980).

Continuous-time algorithms

Although in this age of the digital computer we are often concerned with sampled data and discrete-time formulations, it may sometimes be advantageous to use a continuous-time approach. To obtain a continuous-time version of the SA algorithm (3.24) we first note that it can be written in the form of equation (3.20). It is then easy to see by a limiting argument that a continuous version of the algorithm will be of the form

$$\frac{d\hat{\underline{a}}(t)}{dt} = -\Gamma(t)\nabla_{\hat{\underline{a}}}Q(\underline{x}(t),\hat{\underline{a}}(t)) \tag{3.25}$$

or, in the scalar gain case,

$$\frac{d\hat{\underline{a}}(t)}{dt} = -\gamma(t)\nabla_{\hat{\underline{a}}}Q(\underline{x}(t),\hat{\underline{a}}(t)) \tag{3.26}$$

Algorithms such as (3.25) and (3.26) are analogous to the constant gain continuous-time steepest (or steep) descent algorithms that have been in use in the control and computer field for some time (Young, 1965a). For example, such equations with constant gains are often used on analog computers to solve sets of algebraic equations. The importance of the link between these well known deterministic algorithms and the lesser known continuous-time SA algorithms is that it may often be advisable to incorporate such a variable gain device to counter the inevitable noise that affects any electronic mechanization. Indeed it is quite likely that variable gain modifications to continuous steepest descent solutions are used in practice, although these modifications, if they exist, will probably be based on some heuristic reasoning rather than a knowledge of stochastic approximation[†]. We shall mention the topic of continuous-time estimation

[†]Although the theorist may not wish to admit it, often techniques developed with great mathematical rigour and effort have been used by a practitioner who has found by trial and error and a detailed knowledge of his problem that they work. His reasoning may have been completely heuristic and based totally on intuition, but the solution is not necessarily any the worse for that.

algorithms again in Chapter 10.

Search algorithms

If for some reason it is not possible to obtain the gradient of the realization $\nabla_{\hat{\underline{a}}} Q(\underline{x},\hat{\underline{a}})$, but sample values of $Q(\underline{x},\hat{\underline{a}})$ are available, then it is still possible to use SA by replacing the gradient in the algorithm by an estimate of the gradient obtained from the sample information; in fact this is the basis of the approach used by Kiefer and Wolfowitz (1952) mentioned earlier. How this estimate should be obtained is to a large degree open to the experimenter to choose; he may, for instance, choose one of the many methods that have been developed for deterministic hill climbing. To quote but one possible example, Tsypkin (1971) suggests perturbing $\hat{\underline{a}}$ in a certain manner and then computing the average gradient over the perturbation from the sample values of the realizations $Q(\underline{x},\hat{\underline{a}})$ obtained during the perturbation experiments.

Acceleration of convergence

We have commented little up to this point on the rate of convergence of SA algorithms. If designed properly they are guaranteed to converge "almost surely" as $k \to \infty$. But such asymptotic convergence may be intolerably slow. Here, as is often the case in the design of hill climbing algorithms, there are two approaches open to the designer: one heuristic, the other analytic. Of the first type there are many examples but we will mention here only one, namely the method of Kesten and Lapidus *et al.*, (see Wilde, 1964).

Kesten and later Lapidus *et al.*, note that the ideal SA scheme will take large steps when far from the optimum but should shorten its steps rapidly as it approaches the optimum. With this in mind, they monitor the gradient (actually the average gradient for they use a Kiefer-Wolfowitz scheme) and argue that, when far from the optimum, there will be a few reversals of the sign of the gradient but as the optimum is approached, the random effects of the noise will cause repeated reversals of sign. They then apply the simple rule that the SA gain will only be reduced when the direction of search changes, as indicated by changes of sign in the recursive residual: in this way there is a built-in "adaption" of the SA gain sequence and the scheme seems to be very effective in promoting more rapid convergence than that obtained using the standard SA gain sequence.

As far as analytic approaches are concerned, it would appear that if an optimal non-recursive analytic solution exists and is converted into an equivalent recursive solution, then the resulting implicitly generated SA gain sequence yields rapid convergence and it is not usually necessary to attempt any acceleration. Thus, for example, the RLS solution is nearly always superior in convergence to a scalar gain SA algorithm applied to the same problem (see e.g., Ho and Blaydon, 1966) simply because the implicity generated matrix gain sequence P_k contains within it information gleaned from the sampled data which allows it to adjust the estimates in a mutually dependent fashion and so achieve faster convergence. In fact, as we will see in the next Chapter, if the noise sequence e_{yi}, $i = 1,2,\ldots,k$ in the regression model (3.4) has the statistical properties of white noise as noted earlier, then P_k is directly proportional to the covariance of the estimation errors $P_k^* = E\{\underline{\tilde{a}}\underline{\tilde{a}}^T\}$, where $\underline{\tilde{a}} = \underline{\hat{a}} - \underline{a}$. The advantages of using such information in guiding the search are obvious. An attractive general approach to the analytic evaluation of the convergence properties of recursive algorithms has been suggested by Ljung (1977a,b), largely within a time-series context. But we shall refer to Ljung's work later in Chapter 9.

Some scalar gain SA algorithms can be exceptionally slow to converge and the potential user should, more than anything else, realize this. The point is, however, that heuristically designed SA algorithms are, just because of their simplicity, extremely flexible and widely applicable. And therein lies their attraction: better an algorithm which takes a long time to obtain an answer than no algorithm at all. After all there is at least the possibility with an SA algorithm that, having designed it and found it too slow to converge, the designer can modify it in some manner in order to improve its convergence properties.

3.4 Summary

We have seen in this chapter how the simple concepts introduced in the second Chapter can be extended to the multi-parameter case and how we can view all recursive algorithms of this general type from a stochastic approximation (SA) standpoint. This interpretation provides additional insight into the behaviour of the algorithms and provides a theoretical basis both for convergence analysis and the design of SA-like algorithms. Later, in Chapter 10, we will see how it can also help in the simplification of recursive algorithms in those situations where computational efficiency is important.

But stochastic approximation theory is not a panacea: it can be used to design algorithms with 'almost sure' convergence, but these may have rates of convergence which are intolerably slow. In this sense, naturally occurring recursive SA algorithms derived from optimal *en bloc* estimators or some approximation of the *en bloc* estimator are usually superior in performance.

Perhaps the most famous and useful recursive algorithm of this latter type is the recursive least squares (RLS) algorithm developed in this chapter. In the next two chapters, we will explore this RLS algorithm further from a statistical standpoint and see how it can be extended in various directions.

4 Recursive Least Squares Regression Analysis

The recursive least squares (RLS) algorithm II developed in the previous Chapter provides a general method of estimating the parameters in a multi-parameter regression model. But, as we have pointed out, the algorithm is a deterministic estimation procedure in the sense that it makes few assumptions about either the statistical nature of the signals or the noise and does not provide any statistical information on the nature of the estimates.

To illustrate how even the minimum of statistical information can help to improve this aspect of the algorithm, let us make certain mild assumptions about the nature of the measurement noise e_{yi} in equation (3.4).

In particular let us assume that:

(1) the e_{yi} are a zero mean sequence of random variables: i.e.

$$E\{e_{yi}\} = 0.$$

(2) the e_{yi} are serially uncorrelated and have a constant variance σ^2; i.e.

$$E\{e_{yi}e_{yj}\} = \sigma^2\delta_{ij}$$

where δ_{ij} is the Kronecker delta function:

$$\delta_{ij} = \begin{cases} 1 & i=j \\ 0 & i \neq j \end{cases}$$

(3) the e_{yi} are independent of the variables x_{ji} that compose the vector $\underline{x}_i$; i.e.

$$E\{\underline{x}_i e_{yj}\} = 0, \text{ for all } i, j.$$

In this situation, the model (3.4), i.e.

$$y_i = \underline{x}_i^T\underline{a} + e_{yi};\; i = 1,....,k$$

is known as the *General Linear Regression Model* and the Least Squares equations (3.5) are termed the *Normal Equations of Least Squares Regression Analysis* (Kendall and Stuart, 1961).

4.1 The RLS Algorithm for the General Linear Regression Model

To simplify the subsequent analysis, equation (3.4) can be written in the alternative vector-matrix form

$$\underline{y} = X\underline{a} + \underline{e}_y \tag{4.1}$$

where

$$\underline{y} = [y_1, \ldots, y_k]^T$$

$$\underline{e}_y = [e_{y1}, \ldots, e_{yk}]^T$$

and X is a k x n matrix with rows $\underline{x}_i^T$, i = 1,....,k. Consequently, the normal equations (3.5) can be written concisely as

$$X^T X \hat{\underline{a}} - X^T \underline{y} = 0 \tag{4.2}$$

so that

$$\hat{\underline{a}}_k = [X^T X]^{-1} X^T \underline{y} \tag{4.3}$$

Substituting for $\underline{y}$ from (4.1) we obtain

$$\hat{\underline{a}}_k = [X^T X]^{-1} X^T \{X\underline{a} + \underline{e}_y\}$$

or

$$\tilde{\underline{a}}_k = \hat{\underline{a}}_k - \underline{a} = [X^T X]^{-1} X^T \underline{e}_y \tag{4.4}$$

where $\tilde{\underline{a}}_k = \hat{\underline{a}}_k - \underline{a}$ is the *estimation error* after k samples.

It is now a simple matter to determine the statistics of the estimation error. The expected value of $\tilde{\underline{a}}_k$ is given by

$$E\{\tilde{\underline{a}}_k\} = E\{[X^T X]^{-1} X^T \underline{e}_y\}$$

But since $\underline{e}_y$ is independent of X from assumption (3) above and has zero mean from assumption (1), it is clear that this expression can be simplified in the following manner:

$$E\{\tilde{\underline{a}}_k\} = [X^T X]^{-1} X^T E\{\underline{e}_y\}$$

$\therefore$

$$E\{\tilde{\underline{a}}_k\} = 0 \quad \text{or} \quad E\{\hat{\underline{a}}\} = \underline{a} \tag{4.5}$$

and we see that the estimate $\hat{\underline{a}}_k$ is unbiased.

The covariance matrix of the estimation errors, P_k^*, is defined as

$$P_k^* = E\{\underline{\tilde{a}}_k \underline{\tilde{a}}_k^T\} = \begin{bmatrix} E\{\tilde{a}_1^2\} & E\{\tilde{a}_1\tilde{a}_2\},\ldots, & E\{\tilde{a}_1\tilde{a}_n\} \\ E\{\tilde{a}_2\tilde{a}_1\} & E\{\tilde{a}_2^2\},\ldots, & E\{\tilde{a}_2\tilde{a}_n\} \\ \vdots & \ddots & \vdots \\ E\{\tilde{a}_n\tilde{a}_1\} & E\{\tilde{a}_n\tilde{a}_2\},\ldots, & E\{\tilde{a}_n^2\} \end{bmatrix}$$

Now, by using the well known matrix product rule $[PQR]^T = R^TQ^TP^T$ and noting that $[X^TX]^{-1}$ is symmetric, we see that

$$P_k^* = E\{\underline{\tilde{a}}_k \underline{\tilde{a}}_k^T\} = E\{[X^TX]^{-1} X^T \underline{e}_y \underline{e}_y^T X[X^TX]^{-1}\}$$

Once again this expression can be simplified by noting assumption (3), to yield

$$P_k^* = [X^TX]^{-1} X^T E\{\underline{e}_y \underline{e}_y^T\} X[X^TX]^{-1}$$

But from assumption (2) the covariance matrix of the noise $\underline{e}_y$ i.e.,

$$E\{\underline{e}_y \underline{e}_y^T\} = \begin{bmatrix} E\{e_{y1}^2\} & E\{e_{y1}e_{y2}\},\ldots, & E\{e_{y1}e_{yk}\} \\ E\{e_{y2}e_{y1}\} & E\{e_{y2}^2\},\ldots, & E\{e_{y2}e_{yk}\} \\ \vdots & \ddots & \vdots \\ E\{e_{yk}e_{y1}\} & E\{e_{yk}e_{y2}\},\ldots, & E\{e_{yk}^2\} \end{bmatrix}$$

assumes the special form,

$$E\{\underline{e}_y \underline{e}_y^T\} = \sigma^2 I_k$$

where I_k is the kth order identity matrix. As a result,

$$P_k^* = \sigma^2 [X^TX]^{-1} X^TX [X^TX]^{-1}$$
$$= \sigma^2 [X^TX]^{-1}$$

i.e.

$$P_k^* = \sigma^2 P_k \qquad (4.6)$$

where $P_k = [X^TX]^{-1} = [\sum_{i=1}^{k} \underline{x}_i\underline{x}_i^T]^{-1}$, as before. In other words, the covariance matrix of the estimation errors is simply the product of the variance σ^2 and the matrix P_k obtained from the recursive least squares analysis via the matrix inversion lemma, II(1).

Finally, an estimate $\hat{\sigma}^2$ of σ^2 can be obtained by defining the *vector of least squares residuals* $\underline{e}$ as

$$\underline{e} = \underline{y} - X\underline{\hat{a}}_k; \quad \underline{e} = [e_1 \ldots\ldots, e_k]^T \tag{4.7}$$

From (4.1) and (4.3) we see that

$$\begin{aligned}\underline{e} &= X\underline{a} + \underline{e}_y - X\{[X^TX]^{-1} X^T[X\underline{a} + \underline{e}_y]\} \\ &= \underline{e}_y - X[X^TX]^{-1} X^T \underline{e}_y \\ &= [I_k - X[X^TX]^{-1} X^T]\underline{e}_y\end{aligned}$$

or

$$\underline{e} = B\underline{e}_y \tag{4.8}$$

where

$$B = [I_k - X[X^TX]^{-1} X^T]$$

is a symmetric *idempotent matrix*, i.e.,

$$B^T = B; \quad B^2 = B$$

Now referring to (4.8) we can write the sum of the squares of the least squares residuals as

$$\begin{aligned}\sum_{i=1}^{k} e_i^2 = \underline{e}^T\underline{e} &= \underline{e}_y^T B^T B\underline{e}_y \\ &= \underline{e}_y^T B\underline{e}_y\end{aligned}$$

Consequently,

$$\begin{aligned}E\{\underline{e}^T\underline{e}\} &= E\{\underline{e}_y^T [I_k - X[X^TX]^{-1} X^T]\underline{e}_y\} \\ &= E\{\underline{e}_y^T\underline{e}_y - \underline{e}_y^T X[X^TX]^{-1} X^T\underline{e}_y\}\end{aligned} \tag{4.9}$$

But from assumption (2)

$$E\{\underline{e}_y^T \underline{e}_y\} = E\{e_{y1}^2 + \dots + e_{yk}^2\} = \sigma^2 k$$

and

$$E\{\underline{e}_y^T X[X^T X]^{-1} X^T \underline{e}_y\} = \sigma^2 \text{ trace } \{X[X^T X]^{-1} X^T\}$$

where the trace of matrix is defined as the sum of its diagonal elements. Thus (4.9) can be simplified to the following,

$$E\{\underline{e}^T \underline{e}\} = \sigma^2 [k - \text{trace } \{[X^T X]^{-1} X^T X\}]$$

$$= \sigma^2 [k - \text{trace } I_n]$$

i.e.

$$E\{\underline{e}^T \underline{e}\} = \sigma^2 [k-n] \tag{4.10}$$

since $X^T X$ is of order n. From this result, we can now see that an unbiased estimator $\hat{\sigma}^2$ of σ^2 can be obtained from

$$\hat{\sigma}^2 = \frac{1}{k-n} \underline{e}^T \underline{e} = \frac{1}{k-n} \sum_{i=1}^{k} e_i^2 \tag{4.11}$$

The results (4.5), (4.6) and (4.11) are the basic results of least squares regression analysis and the reader is advised to refer to one of the standard texts (e.g. Kendall and Stuart, 1961; Johnston, 1963; Brownlee, 1965) for further detailed discussion on the subject. Noting the result (4.6), however, it is now a simple matter to obtain the *recursive least squares regression algorithm* by substituting $P_k^*/\hat{\sigma}^2$ for P_k in the basic RLS algorithm II, i.e.

$$\hat{\underline{a}}_k = \hat{\underline{a}}_{k-1} - P_{k-1}^* \underline{x}_k [\hat{\sigma}^2 + \underline{x}_k^T P_{k-1}^* \underline{x}_k]^{-1} \{\underline{x}_k^T \hat{\underline{a}}_{k-1} - y_k\} \quad (1)$$

or

$$\hat{\underline{a}}_k = \hat{\underline{a}}_{k-1} - \frac{P_k^*}{\hat{\sigma}^2} \{\underline{x}_k \underline{x}_k^T \hat{\underline{a}}_{k-1} - \underline{x}_k y_k\} \quad (2)$$

while

$$P_k^* = P_{k-1}^* - P_{k-1}^* \underline{x}_k [\hat{\sigma}^2 + \underline{x}_k^T P_{k-1}^* \underline{x}_k]^{-1} \underline{x}_k^T P_{k-1}^* \quad (3)$$

III

In this algorithm $\hat{\sigma}^2$ is either an *a priori* estimate of the noise variance σ^2 or an estimate obtained from the data. For example it could be obtained rather inconveniently from a recursive algorithm such as,

$$\hat{\sigma}_k^{\,2} = \hat{\sigma}^2_{k-1} - \frac{1}{k}(\hat{\sigma}^2_{k-1} - e_k^{\,2}) \qquad (4) \qquad \text{III}$$

For low values of k, the estimates obtained from III(4) will be biased because it does not account for the "degrees of freedom" adjustment in equation (4.11), where the sum of the squares of the residuals is divided by k-n rather than just k as in (2.15b). Adjustments can be introduced to compensate for this or, alternatively the *en bloc* solution (4.11) can be used either completely (since it is so simple) or until the effect of the degrees of freedom adjustment is minimal. III(4) is not, however, a particularly satisfactory estimate of σ^2 if e_k is replaced by $\hat{e}_k$, as we see later in Chapter 5 (page 64).

The algorithm III(1) to (3) is a very powerful and elegant one: it not only supplies the parameter estimates at each sampling instant but also an indication of the accuracy of these estimates through the error-covariance matrix P_k^*. It is not difficult to see that P_k^* behaves like P_k except for the scale factor $\hat{\sigma}^2$ and is a strictly decreasing function of the sample size. This property is one manifestation of the statistical property of consistency (see e.g. Johnston, 1963), since it demonstrates how the error-covariance is a strictly decreasing function of sample size and the accuracy of the estimate increases as more data are utilised.

It is well known that, with certain statistical assumptions about the noise and signal characteristics, least squares estimation can be interpreted from both the maximum likelihood and Bayesian standpoints (see e.g. Young, 1968a). These same interpretations can, of course, also be applied to the RLS regression algorithm III. For instance suppose we assume that, in addition to its white noise properties, $\underline{e}_y$ in (4.1) has Gaussian amplitude distribution: it is then possible to argue that the estimation errors $\tilde{\underline{a}}$ also have a Gaussian normal distribution of the form (see Appendix 1).

$$\tilde{\underline{a}}_k = N\,(\hat{\underline{a}}_k, P_k^*)$$

so that their statistical characteristics are completely determined by the first two statistical moments; namely the mean $\hat{\underline{a}}_k$ and the error covariance matrix P_k^*. In this case, algorithm III can be considered as a recursive maximum likelihood (RML) estimator.

If we go further and assume that prior information on the mean and covariance matrix is available in the form of the initial estimates $\hat{\underline{a}}_0$ and P_0^* then the algorithm can be considered as a Bayesian estimator since it can be obtained from application of the Bayes Rule linking *a priori* and *a posteriori* probability statements (see e.g. Appendix 1). Indeed the algorithm can be viewed as the physical embodiment of Bayesian estimation: at each recursive step from the first, the *a priori* estimates are modified with the help of the new information imparted by the latest data sample $\{y_k, \underline{x}_k\}$ and the resultant calculation of the kth innovation (recursive residual) to yield *a posteriori* estimates which, in turn, become the *a priori* estimates for the next recursive step.

In this latter sense, the rule of thumb introduced previously for initialising the P_0 matrix in the RLS algorithm II as a diagonal matrix with large diagonal elements makes obvious sense. If little is known about $\underline{a}$ and $\hat{\underline{a}}_0$ is set to zero, then the initial P_0 matrix, interpreted in these covariance terms, should reflect this uncertainty: the large diagonal elements indicate high initial error variance and little confidence in the initial estimate $\hat{\underline{a}}_0$ with the zero off-diagonal elements showing that there is no *a priori* information on the covariance properties and the best estimate (guess??) is that they are zero.

But in the case of algorithm III, we now have more flexibility since $\hat{\underline{a}}_0$ and P_0^* can, in general, be selected to reflect quantitatively our prior knowledge of the model parameters[†]. For instance, a previous estimation experiment could have supplied an estimate $\hat{\underline{a}}$ with covariance P^* and these could be used to define $\hat{\underline{a}}_0$ and P_0^* for any new experiment. Or alternatively, $\hat{\underline{a}}_0$ and P_0^* could be provided from other physical considerations, as when the coefficients are physically meaningful and are known to lie within certain definable confidence bounds.

An example of how algorithm III enhances the estimation results when compared with the RLS algorithm II is given in Figure 4.1. Here the standard error bounds computed as the square root of the diagonal elements of the P_k^* matrix, are plotted as dotted lines on the graph of the recursive parameter estimates for the moving body example considered in Chapter 3. Since the computations are off-line (in the

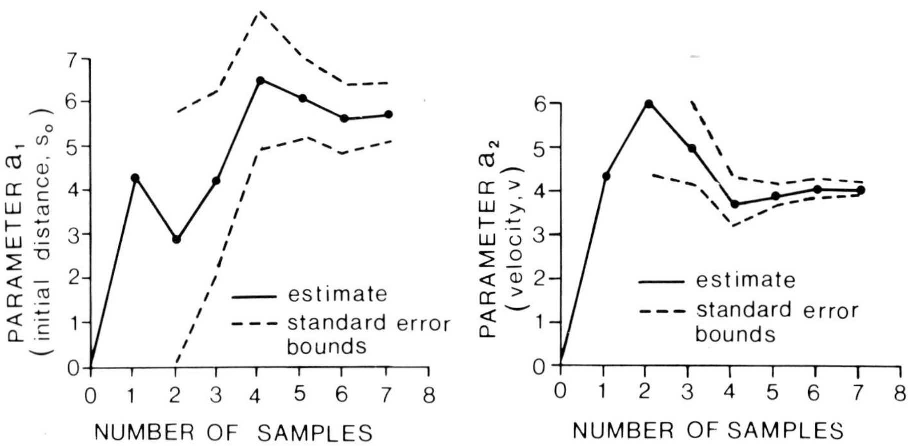

Figure 4.1 Statistical Estimation Results for Moving Body Example: Recursive Estimates and Standard Error Bounds.

† Appendix 2 shows how algorithm III can be obtained directly from a cost function including P_0^* and $\hat{\underline{a}}_0$.

sense that we have access to the whole data base prior to computing the recursive estimates and do not receive the samples sequentially), $\hat{\sigma}^2$ is estimated on the basis of the final estimates at t = 18 rather than being updated recursively. The convergence of the estimation errors with increasing sample size is clear from Figure 4.1. The elements of the P_k^* matrix are plotted in Figure 4.2: not surprisingly, given the nature of the problem, the estimate of the velocity ($\hat{a}_2$) is more accurate than the initial distance estimate ($\hat{a}_1$). Also, as we see from the final P_k^* matrix, i.e.

$$P_7^* = \begin{bmatrix} 0.62 & -0.05 \\ -0.05 & 0.0073 \end{bmatrix}$$

the covariance between the errors in $\hat{a}_1$ and $\hat{a}_2$ is negative at -0.05.

4.2 Some Cautionary Comments: Multiple Collinearity and Errors-in-Variables

Regression analysis is probably the most widely *used* yet, at the same time, the most widely *abused* statistical tool in both the physical and social sciences. Of the many caveats which should be issued to the unwary user of regression analysis, we will mention only two which are particularly pertinent to the subject of the present book, and become important in later chapters concerned with time-series analysis.

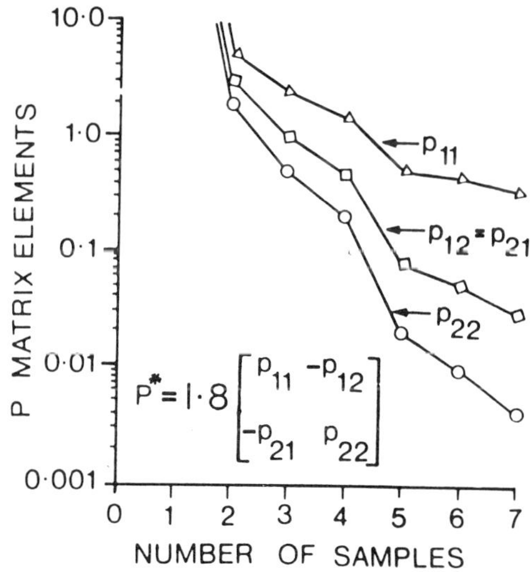

Figure 4.2 Statistical Estimation Results for Moving Body Example: Behaviour of Error-Covariance Matrix Elements.

Multiple collinearity

Although, as we have pointed out, the *regressors* or 'independent variables' x_{ji} that compose the vector $\underline{x}_i$ should be linearly independent, it will often be impossible to pose the estimation problem such that there is no correlation at all between the regressors. And if it happens that this correlation is high, so that there is a tendency towards at least 'partial linear dependence'[†] then it can be shown (e.g. Brownlee, 1965) that the estimates will have high estimation error variance, even though the sum of squares of the residuals e_i may be small and indicative of a good 'fit' or 'explanation' of the data. The reader should be wary of this problem, particularly in general practical applications of regression analysis, and should note that a check on the existence of problems of this type is available by using *Multiple Correlation Analysis* (e.g. Brownlee, 1965; Johnston, 1963).

We will have more to say on the topic of multiple collinearity in a time-series context later in Chapter 10. It is worth noting here, however, that the physical effect of multiple correlation problems in RLS analysis is for the various recursive estimates to apparently converge slowly and 'wander about' often in sympathy with each other. This is a consequence of the high correlation between regressors tending to drive the data covariance matrix $C_k = \sum^{k} \underline{x}_i\underline{x}_i^T$ towards singularity, so that the elements of its inverse P_k assume correspondingly high values. Given the relationship between P_k and the error covariance matrix P_k^*, we can see that this will imply high parametric estimation error variance (see Brownlee, 1965) and poorly defined parameter estimates.

From a geometric standpoint (see e.g. Young, 1965a) multiple collinearity can also be interpreted as a tendency towards the development of valley-like phenomena in the hypersurface associated with the cost function J_2 in the parameter-cost function hyperspace. Thus there is no clearly defined minimum in the surface and the estimates wander along the elongated, valley-like bottom in some partial relationship with each other: this leads to low residual errors but some ambiguity about the parameter values, as indicated by the high estimation error variance. In the case of exact linear dependence there will, of course, be an *exact* linear relationship between the regressors, a true valley in the cost function hypersurface and complete ambiguity. In theory, the C_k matrix will then be singular and P_k indeterminate: in practice, however, it turns out that the recursive matrix

[†]There are, of course, no relative measures of linear dependence and we use this phrase purely for illustrative purposes.

inversion lemma II(1) will yield an inverse with very large elements[†] indicative of the high level of uncertainty about the resulting parameter estimates.

Errors-in-variables and the structural model

Up to this point, we have only considered a particular form of linear estimation model, namely the *regression model*, in which the variables associated with the unknown parameters are *exactly known* quantities. In practice, a number of different estimation models are encountered and the techniques described here may need to be modified in some manner. This whole topic is discussed fully in a book by Graybill (1961), in which he treats all of the major possibilities in considerable detail. In the present context, it will suffice to mention one of the most important models met in practical situations, the *structural model*.

In the structural model, the basic relationship between the parameters is still in the form shown in (3.4) but the elements of $\underline{x}$ are no longer exactly known quantities and can only be observed in error. In other words, the observed value of $\underline{x}$ is $\underline{z}$ where

$$\underline{z} = \underline{x} + \underline{e}_z$$

and $\underline{e}_z$ is an n vector of measurement noise associated with the observation of $\underline{x}$. Once again, we will discuss the problems introduced by such "errors-in-variables" later in connection with time-series model estimation, where they are inherent and unavoidable. In the context of regression analysis, however, it is worth noting that the presence of errors-in-variables of the above type induces an asymptotic bias on the parameter estimates which is a function of the signal/noise ratio on the signals and is zero *only* when there is no noise, i.e., $\underline{e}_z \equiv 0$. In other words, no matter how many data are utilized to estimate the parameters in $\underline{a}$, the resultant estimates $\hat{\underline{a}}_k$ will always be biased away from the true values i.e.,

$$\underset{k \to \infty}{\text{p.lim.}}\ \hat{\underline{a}}_k \neq \underline{a}$$

where p.lim. is the probability in the limit (see Appendix 1).

The importance of the asymptotic bias on the parameter estimates will depend upon the application of the regression analysis. If it is being used merely for forecasting the "dependent variable" y_k on the basis of the noisy "independent variables" $\underline{z}_k$, then the results may be acceptable, since it can be shown that the forecast $y_k = \underline{z}_k^T\hat{\underline{a}}$ is unbiased in most situations, provided the noise is station-

---

[†]Note that this arises because the matrix inversion lemma can be considered as a psuedo-inverse in the sense of Penrose (1955).

ary and retains those statistical characteristics it possessed in the data used to obtain the estimate $\underline{\hat{a}}$ (see Johnston, 1963). But if the parameters or regression coefficients in $\underline{a}$ are important in their *own* right and, for example, have physical significance, then the problem can assume major proportions. Indeed one wonders how many exercises of regression analysis in the social sciences have, unknowingly, been affected by the problem of errors-in-variables, and how this in turn may have affected the conclusions of the analyst.

There are a number of ways of solving the structural model problem. If the noise statistics are known *a priori*, for example, then there is no real problem since it is possible to compensate directly for the bias (Levin, 1963), although the estimates can be highly sensitive to errors in the *a priori* assumptions. If we are prepared to resort to non-linear estimation, then it is possible to use more sophisiticated procedures such as maximum likelihood, which provide a good but usually more complicated general approach to the problem. Probably the simplest approach of this type is the "suboptimal" estimation procedure suggested by James, Souter and Dixon (1972) which involves the iterative (see Figure 2.2) application of the linear least squares solution, with modifications to allow for the estimation and removal of bias effects.

If we wish to retain the simplicity of linear least squares estimation, however, there seems to be only one real solution to the structural model problem: this is the method of *Instrumental Variables* (IV) (see e.g., Durbin, 1954; Kendall and Stuart, 1961). The IV method is an extremely simple technique which retains the single iteration, linear least squares-like solution and is particularly attractive *because it does not require detailed a priori information on the noise statistics to yield consistent, asymptotically unbiased estimates.* In relation to the linear regression problem, it involves first noting that, in a structural model situation, the normal equations (3.5) must be formulated as

$$[\sum_{i=1}^{k} \underline{z}_i \underline{z}_i^T]\underline{\hat{a}} - \sum_{i=1}^{k} \underline{z}_i y_i = 0 \tag{4.12}$$

since $\underline{x}_i$ has to be replaced by its noisily measured value $\underline{z}_i$. This solution is then modified, for the purposes of estimation, to yield the following IV normal equations

$$[\sum_{i=1}^{k} \underline{\hat{x}}_i \underline{z}_i^T]\,\underline{\hat{a}} - \sum_{i=1}^{k} \underline{\hat{x}}_i y_i = 0. \tag{4.13}$$

so that the IV estimate is obtained as

$$\underline{\hat{a}}_k = [\sum_{i=1}^{k} \underline{\hat{x}}_i \underline{z}_i^T]^{-1} \sum_{i=1}^{k} \underline{\hat{x}}_i y_i \tag{4.14}$$

or, in relation to equation (4.3)

$$\hat{\underline{a}}_k = [\hat{X}^T Z]^{-1}\, \hat{X}^T \underline{y} \tag{4.15}$$

where $\hat{X}$ is an k x n matrix with rows $\hat{\underline{x}}_i^T$ and Z a similarly dimensioned matrix with elements $\underline{z}_i^T$, i = 1,2,....,k.

In equation (4.14) $\hat{\underline{x}}_i$ is a vector of "instrumental variables" which are chosen to be as highly correlated as possible with the equivalent variables in the 'noise free' vector $\underline{x}_i$ but *totally* statistically independent of the noise $\underline{e}_z$, i.e.,

$$E\{\hat{\underline{x}}_i \underline{x}_j\} \neq 0; \quad \text{and } E\{\hat{\underline{x}}_i \underline{e}_{zj}\} = 0, \text{ for all i,j} \tag{4.16}$$

We shall see the reason why the introduction of the IV vector $\hat{\underline{x}}_i$ is so effective in combating the effects of the errors-in-variables in Chapter 7. For the moment, however, we will merely note that the *en bloc* solution (4.14) can be written in the form

$$\hat{\underline{a}}_k = \hat{P}_k \hat{\underline{b}}_k \tag{4.17}$$

where

$$\hat{P}_k = \hat{C}_k^{-1} = [\sum_{i=1}^{k} \hat{\underline{x}}_i \underline{z}_i^T]^{-1} \text{ and } \hat{\underline{b}}_k = \sum_{i=1}^{k} \hat{\underline{x}}_i y_i \tag{4.18}$$

As a result, it is a simple matter to obtain a recursive IV algorithm using an approach similar to that employed in the RLS case discussed in the previous Chapter. We will leave the reader to derive these equations, which take the following form,

$$\hat{\underline{a}}_k = \hat{\underline{a}}_{k-1} - \hat{P}_{k-1} \hat{\underline{x}}_k [1 + \underline{z}_k^T \hat{P}_{k-1} \hat{\underline{x}}_k]^{-1} \{\underline{z}_k^T \hat{\underline{a}}_{k-1} - y_k\} \quad (1)$$

or

$$\hat{\underline{a}}_k = \hat{\underline{a}}_{k\;1} - \hat{P}_k \{\hat{\underline{x}}_k \underline{z}_k^T \hat{\underline{a}}_{k\;1} - \hat{\underline{x}}_k y_k\} \quad (2)$$

where

$$\hat{P}_k = \hat{P}_{k-1} - \hat{P}_{k-1} \hat{\underline{x}}_k [1 + \underline{z}_k^T \hat{P}_{k-1} \hat{\underline{x}}_k]^{-1} \underline{z}_k^T \hat{P}_{k-1} \quad (3)$$

(equations (1)–(3): IV)

One problem with the general application of the IV algorithm is that of choosing the instrumental variables themselves: how can we obtain or generate variables with such specific statistical properties? The difficulty of answering this question in any general manner has acted as a strong deterrent to the widespread use of the IV approach in statistics (see Kendall and Stuart, 1961). But the question is surprisingly easy to answer in the time-series context considered in Chapters 6 and 7 and we will see there how the IV algorithm has assumed major significance in time-series analysis.

4.3 Summary

In this chapter we have seen how the deterministic, recursive least squares algorithm can easily be converted into a statistical form by making certain simple statistical assumptions about the nature of the observational errors. The resultant recursive least squares regression algorithm III is of wide general significance, providing a recursive method for estimating the parameters in the "general linear regression model" of statistics. But the algorithm, like its *en bloc* counterpart, can be abused if applied outside the range of restrictions inherent in its derivation. In this chapter, we have mentioned two such problem areas: multiple collinearity and errors-in-variables. In the next chapter, we will see how it is possible to resolve another potential problem; namely the difficulty introduced if the parameters, contrary to the assumptions, are non-stationary; i.e. they tend to vary significantly over the observation interval.

5 Recursive Estimation of Time-Variable Parameters in Regression Models

In time-series analysis, a series is said to be *stationary in the wide sense* or *stationary to second order* if the first and second statistical moments, i.e. the mean, variance and covariance, are not functions of time. In other words, for a series y_i, $i = -\infty,\ldots,-1,0,1,\ldots,\infty$

$$\left.\begin{aligned} &E\{y_i\} = a; \quad E\{(y_i - a)^2\} = \sigma^2 \\ &E\{(y_i - a)(y_j - a)\} = \mu(i - j) = \mu_\tau; \; \tau = i-j \end{aligned}\right\} \tag{5.1}$$

for all i,j, where μ is the covariance (Appendix 1). In this book, such a series will be termed simply *stationary* (although this term is often reserved for processes obeying stricter conditions than these).

A useful property of stationary series is that they are ergodic, in other words we can estimate the mean, variance and covariance by averages over time, as we did in the case of the Walgett rainfall data in Chapter 2, rather than by defining them in terms of ensemble averages over a number of random realizations of the series (see e.g. Brownlee, 1965). But if, as we suspect in the case of the Walgett data, the first two statistical moments are *not* constant and the series is *non-stationary*, then these ordinary temporal averages are not appropriate and alternative measures are required.

Non-stationarity in this general sense can arise for many different reasons and can be explained in many different ways. In standard statistical texts on time-series analysis however, discussion is often restricted to non-stationarity in the form of stochastically variable means or 'trends' in data. A well known example is the book by Box and Jenkins (1970), where such phenomena are ascribed to a form of non-stationarity described by Yaglom (1955) where the non-stationary characteristics are assumed to arise from the presence of one or more integrators in the stochastic part of signal generation process. We will discuss this in more detail later, but the simplest discrete-time example of such a process is the random walk sequence ξ_k, when

$$\xi_k = \xi_{k-1} + e_{k-1} \tag{5.2}$$

or in z transfer function terms (cf. Figure 2.3, Chapter 2, and Appendix 1)

$$\xi_k = \frac{z^{-1}}{1-z^{-1}} e_k = \frac{1}{z-1} e_k$$

where e_k is a zero mean, serially uncorrelated white noise sequence. The transfer function $^1/_{z-1}$ will be recognized from Figure 2.3 as a discrete summer or integrator. Clearly sequences such as ξ_k will 'wander' with stochastically variable mean value and models like (5.2), or similar ones with more than one integrator, are good candidates for the description of such behaviour in time-series. This is the motivation for the use of such models by Box and Jenkins (1970).

But Yaglom-type non-stationarity is not, in itself, an entirely appropriate description for all forms of non-stationary behaviour. One embodiment of non-stationarity particularly interesting and relevant to the control and systems analyst is *parametric time-variability*. This applies in those cases where the model of the time-series data can only be characterized adequately by parameters which vary over time in some significant manner. Apart from some notable exceptions (e.g. Granger and Hatanaka, 1964), the concept of parametric time-invariance is so embedded within the psyche of the statistician that parametric time-variability rarely receives the attention it deserves in texts on the subject.

Unlike his statistical counterpart, the control and systems analyst has always been worried by the possibility of parametric non-stationarity because he sees evidence of it in so many application areas. For example, linear, small perturbation, dynamic models are used very successfully in the design of autostabilization systems for aircraft, but often only if due allowance is made for systematic changes in the dynamic characteristics, as evidenced by variation in the model parameters (Young, 1979c). And in various industrial processes model parameters are known to vary because of factors such as catalyst ageing in chemical processes, changing carbon content in steel when using the basic oxygen furnace and changing characteristics in paper making plant.

From the theoretical standpoint, we might expect simple models of complex, nonlinear phenomena such as these to exhibit parametric non-stationarity: it is well known, for instance, that the linearization of nonlinear models often results in linear, time-varying model structures. Again the systems analyst is well aware of these problems and sees in the estimates of parametric variation possible information which could lead to the identification of the nature of the nonlinearities (Young, 1978).

These latter remarks apply particularly in the case of "badly defined" systems (Young, 1978; Young *et al.*, 1978; Kalman, 1979) such as those encountered in environmental and socio-economic systems analysis. In the socio-economic case, for example, the possibility of behavioural characteristics changing through time is often a topic of discussion; as, for instance, when it is suggested that the so-called "Phillips Curve" for wage inflation as a function of unemployment (Phillips, 1958) is not applicable during periods of strong union bargaining power. In such cases, the inference is that the (possibly nonlinear) relationships are themselves changing through the passage of time as a function of the changing socio-economic attitudes and conditions.

In all of these examples, the hypothesis of parametric non-stationarity (or its inverse) needs to be tested in some manner; and it is here that recursive estimation has a potentially important role to play. Conventional statistical hypothesis testing can be utilized, as we will discuss later. But the recursive parameter estimates themselves are clearly of primary utility in providing information not only on the existence of non-stationarity but also on the possible nature of the parametric variations (Young, 1975). In this chapter, therefore, we will explore how the recursive algorithms of the previous chapter can be extended in various ways to handle parametric variation and how they can, in this manner, provide a valuable additional weapon in the armoury of the statistician, econometrician and systems analyst.

5.1 Shaping the Memory of the Estimator

If a common factor $^1/_k$ is introduced into equation (3.5), we obtain the following least squares normal equation,

$$[\frac{1}{k}\sum_{i=1}^{k}\underline{x}_i\underline{x}_i^T]\hat{\underline{a}} - \frac{1}{k}\sum_{i=1}^{k}\underline{x}_i y_i = 0 \tag{5.3}$$

It is now clear that the data processing can be considered as the finite-time averaging of the various products and cross products in $\underline{x}_i\underline{x}_i^T$ and $\underline{x}_i y_i$. This means that, in arriving at the estimates, all data are weighted equally over the observation interval of k samples and so there is an implicit assumption that the parameters remain constant during this period. In order to allow for possible parametric variation, it is necessary to remove the effects of 'obsolete' data in some manner. If we consider that the conventional, time-invariant parameter solution (5.3) has an unrestricted memory of k samples, then such a procedure can be visualized as restricting or 'shaping' the memory of the estimator.

The two most obvious procedures for shaping the memory of an estimation scheme are: first, to base an estimation on only the most recent portion of the data, say S samples; and second to weight the data exponentially into the past with a particular exponential 'fading memory' of the form exp.$(-1/T_e)\tau$, where τ is time measured into the past and T_e is the time constant of the exponential decay. Both have been employed in conventional *en bloc* data processing and estimation, and can also be applied to recursive algorithms. They can best be visualized as moving window or weighting functions, as shown in Figure 5.1.

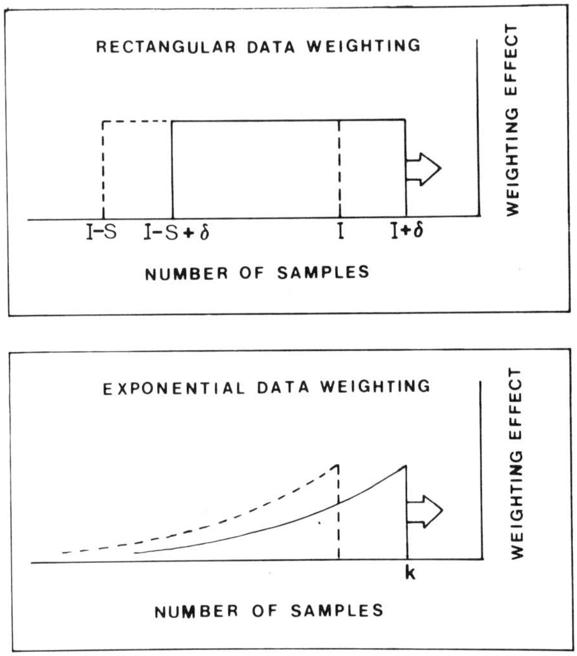

Figure 5.1 Rectangular and Exponential Data Weighting Functions

The moving rectangular window function

From Figure 5.1, we see that the procedure in this case is to choose an observation period of S samples which, on the basis of the conditions applying in the particular application under consideration, will provide estimates of the desired accuracy: typically this might require, for example, that the error covariance matrix obtained for S samples, i.e. $P_S^* = \sigma^2 P_S$, is acceptable in some sense. The estimation equations can then be solved for overlapping or adjoining intervals of this length. In other words, after each solution based on S samples is completed, δ new samples are taken, the δ oldest samples are removed and a new solution is obtained.

If we consider the least squares estimation problem in this manner then, after (j-1) estimation intervals each of S samples, the normal equations can be written (cf. equation 3.5)

$$[\sum_{i=I-S+1}^{I} \underline{x}_i \underline{x}_i^T]\ \hat{\underline{a}} = \sum_{i=I-S+1}^{I} \underline{x}_i \underline{y}_i \tag{5.4}$$

where I = S + (j-2)δ, the total number of samples up to the (j-1)th interval. After (j-1) estimation intervals, the solution to (5.4) can be written as

$$\hat{\underline{a}}_{j-1} = P_{j-1} \underline{b}_{j-1}$$

where

$$P_{j-1} = (C_{j-1})^{-1} = [\sum_{i=I-S+1}^{I} \underline{x}_i \underline{x}_i^T]^{-1}; \qquad \underline{b}_{j-1} = \sum_{i=I-S+1}^{I} \underline{x}_i y_i$$

Now when an additional set of δ samples are taken, the oldest set of δ samples must be removed and the estimate at the jth sampling instant is given by

$$\hat{\underline{a}}_j = P_j \underline{b}_j \tag{5.5}$$

where

$$P_j = [P_{j-1}^{-1} + \sum_{i=I+1}^{I+\delta} x_i x_i^T - \sum_{i=I-S+1}^{I-S+\delta} \underline{x}_i \underline{x}_i^T]^{-1} \tag{5.6}$$

$$\underline{b}_j = [\underline{b}_{j-1} + \sum_{i=I+1}^{I+\delta} \underline{x}_i y_i - \sum_{i=I-S+1}^{I-S+\delta} \underline{x}_i y_i]$$

In the most commonly useful case of this procedure, where only a single sample is added and removed at each new solution (δ=1), we obtain at the kth instant (i.e. j=k; I=k-1)

$$\left.\begin{aligned} P_k &= [P_{k-1} + \underline{x}_k\underline{x}_k^T - \underline{x}_{k-s}\underline{x}_{k-s}^T]^{-1} \\ \underline{b}_k &= [\underline{b}_{k-1} + \underline{x}_k y_k - \underline{x}_{k-s} y_{k-s}] \end{aligned}\right\} \qquad (5.7)$$

Employing a similar approach to that used in previous chapters, it is straightforward to obtain from these relationships and (5.5) a recursive, rectangularly - weighted - past (RWP) algorithm, which consists of the following two steps at each recursion,

receipt of new data at kth instant

$$\hat{\underline{a}}_k' = \hat{\underline{a}}_{k-1} - P_{k-1}\underline{x}_k[\underline{x}_k^T P_{k-1}\underline{x}_k + 1]^{-1}\{\underline{x}_k^T \underline{a}_{k-1} - y_k\} \qquad (1)$$

$$P_k' = P_{k-1} - P_{k-1}\underline{x}_k[\underline{x}_k^T P_{k-1}\underline{x}_k + 1]^{-1}\underline{x}_k^T P_{k-1} \qquad (2)$$

removal of data received at (k-S)th instant

$$\hat{\underline{a}}_k = \hat{\underline{a}}_k' - P_k'\underline{x}_{k-s}[\underline{x}_{k-s}^T P_k'\underline{x}_{k-s} - 1]^{-1}\{\underline{x}_{k-s}^T \hat{\underline{a}}_k' - y_{k-s}\} \qquad (3)$$

$$P_k = P_k' - P_k'\underline{x}_{k-s}[\underline{x}_{k-s}^T P_k'\underline{x}_{k-s} - 1]^{-1}\underline{x}_{k-s}^T P_k' \qquad (4)$$

V

The moving exponential window

Exponential-weighting-into-the-past (EWP) can be introduced into the least squares problem formulation by replacing the finite-time averaging operations in (5.3) by EWP averaging. This is equivalent to considering an EWP least squares cost function J_{EWP} of the form

$$J_{EWP} = \sum_{i=1}^{k} [\underline{x}_i^T\hat{\underline{a}} - y_i]^2\alpha^{k-i} \qquad (5.8)$$

where $0<\alpha<1.0$ is a constant related to the time constant T_e of the exponential weighting by the expression

$$\alpha = \exp.(-T_s/T_e)$$

where T_s is the sampling interval in time units appropriate to the application. Two typical exponential decay functions with $\alpha = 0.7$ ($T_e = 2.8\ T_s$) and $\alpha = 0.9$ ($T_e = 9.5\ T_s$) are shown in Figure 5.2. With $\alpha = 1.0$, J_{EWP} becomes, of course, the usual least squares cost function J_2 in (3.3).

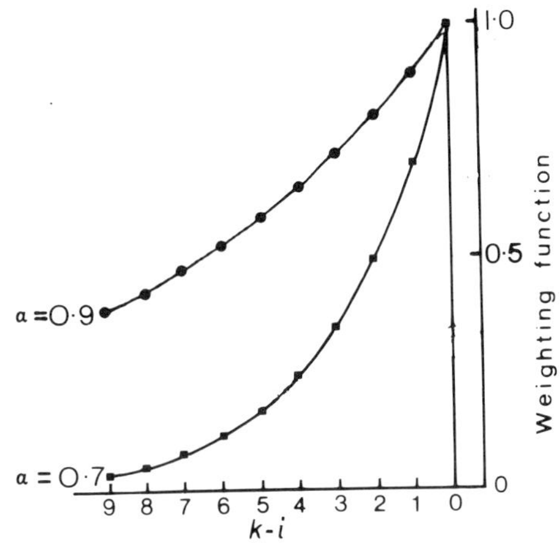

Figure 5.2 Typical EWP Weighting Functions

In a recursive context, the easiest way of incorporating these modifications is to note that a first order, discrete-time, dynamic system has an exponential unit impulse response and, therefore, an exponential weighting effect on past inputs. This can be demonstrated quite simply by considering the following discrete-time 'filter' with input u_k and output x_k

$$x_k = \alpha x_{k-1} + u_k \tag{5.9}$$

In z transform transfer function terms (Chapter 2 and Appendix 1) this can be written,

$$x_k = \frac{1}{1-\alpha z^{-1}} u_k$$

which, on division of 1 by $1-\alpha z^{-1}$, yields

$$x_k = (1 + \alpha z^{-1} + \alpha^2 z^{-2} + \ldots + \alpha^{\infty} z^{-\infty}) u_k$$

or†

$$x_k = u_k + \alpha u_{k-1} + \alpha^2 u_{k-2} + \ldots + \alpha^i u_{k-i} + \ldots + \alpha^{\infty} u_{k-\infty}$$

With this in mind, we can achieve the minimization of (5.8) by simply solving the following EWP normal equations

$$\langle \underline{x}\underline{x}^T \rangle_k \, \hat{\underline{a}} = \langle \underline{x}y \rangle_k \tag{5.10}$$

where $\langle \cdot \rangle$ denotes EWP averaging of the enclosed variables, and $\langle \underline{x}\underline{x}^T \rangle_k$ and $\langle \underline{x}y \rangle_k$ are obtained from the following recursive relationships (cf. equation 5.9)

$$\langle \underline{x}\underline{x}^T \rangle_k = \alpha \langle \underline{x}\underline{x}^T \rangle_{k-1} + \underline{x}_k \underline{x}_k^T$$

and (5.11)

$$\langle \underline{x}y \rangle_k = \alpha \langle \underline{x}y \rangle_{k-1} + \underline{x}_k y_k$$

Using a nomenclature similar to that used previously, it is then simple to see that

$$\hat{\underline{a}}_k = P_k \underline{b}_k \tag{5.12}$$

where

$$\left.\begin{aligned} P_k &= [\alpha P_{k-1}^{-1} + \underline{x}_k \underline{x}_k^T]^{-1} \\ \underline{b}_k &= [\alpha \underline{b}_{k-1} + \underline{x}_k y_k] \end{aligned}\right\} \tag{5.13}$$

---

† if u_k is a unit impulse, i.e. $u_k = 1.0$ for $k = o$; $u_k = o$ for $k > o$, then it is clear that $x_o = 1.0$; $x_1 = \alpha^2$;....; $x_i = \alpha^i$;....; $x_{\infty} = \alpha^{\infty}$, as required

and

$$P_k = [\langle \underline{x}\underline{x}^T \rangle_k]^{-1}; \quad \underline{b}_k = \langle \underline{x}y \rangle_k \tag{5.14}$$

The relationships (5.12) and (5.13) can be converted into the fully recursive form using an identical approach to that employed for the RLS algorithm II of Chapter 3, but taking care to carry the scalar α through the calculations. We will leave the reader to derive this EWP algorithm but the final recursive equations can be written in the form,

$$\left.\begin{aligned} \hat{\underline{a}}_k &= \hat{\underline{a}}_{k-1} - P_{k-1}\underline{x}_k[\alpha + \underline{x}_k^T P_{k-1}\underline{x}_k]^{-1}\{\underline{x}_k^T\hat{\underline{a}}_{k-1} - y_k\} & (1) \\ \hat{\underline{a}}_k &= \hat{\underline{a}}_{k-1} - P_k\{\underline{x}_k\underline{x}_k^T\,\hat{\underline{a}}_{k-1} - \underline{x}_k y_k\} & (2) \\ P_k &= \frac{1}{\alpha}\{P_{k-1} - P_{k-1}\underline{x}_k[\alpha + \underline{x}_k^T\, P_{k-1}\underline{x}_k]^{-1}\underline{x}_k^T P_{k-1}\} & (3) \end{aligned}\right\} \quad \text{VI}$$

Algorithm VI is one form of the recursive EWP least squares algorithm; other forms are possible depending upon the exact nature of the filter algorithm (5.9)[†] but, while superficially dissimilar, these other forms tend to perform in a similar manner to VI. In practice, it has been found advisable to replace the constant α by a variable α_k which is initially smaller than α but approaches it asymptotically. In this manner, the exponential decay time constant and, therefore, the fading memory is initially small and grows towards the specified length $T_e = -T_s/\log_e\alpha$ as time progresses and more samples are processed. This aids initial convergence by ensuring that the effects of initial conditions on the filters (5.11) are quickly removed: when α is constant, these effects themselves decay with a time constant T_e and this can slow convergence considerably.

One possible definition of α_k is the following,

$$\alpha_k = \lambda_o\alpha_{k-1} + (1 - \lambda_o)\alpha \tag{5.15}$$

with typical values of α_o and λ_o being 0.95 and 0.99 respectively. This is simply a 'unity steady state gain' filter with time constant (see Appendix 1) $T_i = -T_s/\log_e\lambda_o$ and input α: the reader can verify that the output α_k rises

---

[†] For example, the 'unity steady state gain' filter, $x_k = \alpha x_{k-1} + (1-\alpha)u_k$ or its equivalent is sometimes preferred (Bray *et al.*, 1965; Young, 1969a, c).

exponentially with a time constant T_i from the initial condition α_o to a steady state level of α (the solution of (5.15) when $\alpha_k = \alpha_{k-1}$). An alternative is to use the following algorithm for α_k,

$$\left.\begin{aligned} \alpha_k &= 1 - \gamma_k \\ \text{where} \quad \gamma_k &= \frac{1-\alpha}{1-(\alpha)^{k+1}} \end{aligned}\right\} \quad (5.16)$$

This is a derivation of the scheme used successfully by Bray et al. (1965) and Young (1969c) and is rather simpler to use than (5.15) since it does not require the specification of any additional parameters, such as α_o and λ_o in (5.15)†. α_k in this case approaches α with increasing k starting from an initial condition $\alpha/1+\alpha$.

It is easy to see that EWP algorithms with time varying exponential weighting, such as those obtained with the definitions (5.15) and (5.16) for α_k, can be very useful in practice. A case in question is the recursive estimation of the observational error variance $\hat{\sigma}^2$, as given in equation III(4) of Chapter 4. If this algorithm is utilized with e_k defined as the recursive residual or innovation $\hat{e}_k = \underline{x}_k^T \hat{\underline{a}}_{k-1} - y_k$, then it is clear that, since the initial $\hat{e}_k$ values will be influenced by the relatively poor estimates obtained when little data have been processed, the estimate may not be particularly good in comparison with the 'equivalent' estimate obtained using the *a posteriori* residuals calculated from equation (4.7). But if the following EWP algorithm is used in place of III(4), with α_k defined by (5.15), for example, then the 'poor' early estimates $\hat{e}_k$, for small k, will be exponentially forgotten,

$$\hat{\sigma}_k^2 = \hat{\sigma}_{k-1}^2 - p_k[\hat{\sigma}_{k-1}^2 - \hat{e}_k^{\,2}]$$

$$p_k = \frac{1}{\alpha_k}[p_{k-1} - \frac{p_{k-1}^2}{\alpha_k + p_{k-1}}]$$

And if α is chosen as unity, then $\lim_{k\to\infty} \alpha_k = 1.0$ and so this algorithm will forget the poor initial estimates $\hat{e}_k$ but progressively utilize later data with equal weighting; i.e. it will progressively convert to the normal least squares algorithm ($\alpha = 1.0$). In this case, we would not expect the estimate $\hat{\sigma}_k^{\,2}$ to be as good as that obtained off-line, but we would at least have a reasonable on-line estimator. We will see other examples where this kind of device proves useful in the later Chapters on time-series analysis.

† although it approaches α rather slowly

It is interesting to investigate the effect of the fading memory modifications in (a) and (b) above on the RLS estimator. This is most easily accomplished by looking again at the mean value estimator considered in Chapter 2. In the case of rectangular data weighting, the effect is immediately obvious: from equation (2.13) we see that p_k is strictly decreasing during the period when the first S samples are being processed, but thereafter it remains constant since the additions to p_k from V(3) and V(4) compensate exactly for the reductions arising from V(1) and V(2); indeed in this simple case, V(3) and V(4) are redundant since V(1) and V(2) could be used for the first S samples and then p_k could be frozen at its p_S value thereafter.

In the EWP situation, we see from VI(3) that the recursive relationship for p_k is now of the form

$$p_k = \frac{1}{\alpha}\{p_{k-1} - \frac{p_{k-1}^2}{\alpha+p_{k-1}}\} \tag{5.17}$$

Once again p_k is no longer a strictly decreasing function of sample size: in the steady state ($p_k = p_{k-1} = p_\infty$) the solution of (5.17) is given by

$$p_\infty = \frac{1}{\alpha}\{p_\infty - \frac{p_\infty^2}{\alpha+p_\infty}\}$$

so that

$$p_\infty\{p_\infty + (\alpha-1)\} = 0$$

and the non zero solution is $p_\infty = (1-\alpha)$. In other words, for $p_0>1-\alpha$, p_k reduces from its initial condition to a steady state solution of $(1-\alpha)$ which, as we might expect, is zero only for $\alpha=1.0$ corresponding to the 'infinite' time constant and unlimited memory. For $1-\alpha>p_0>0$, p_k increases slowly to again reach a steady state value of $(1-\alpha)$. Plots of p_k from (5.17) for $\alpha=0.9$ (T_e = 9.5 sampling intervals) and $p_0 = 10^4$, p_0 = 0.05 are given in Figure 5.3. Also shown on the plot is the p_k for the RWP algorithm with S = 10: both algorithms have the same steady state gain of p_∞ = 0.1 but the RWP reaches this at k = 10 and is constant thereafter while, in the EWP case, the steady state value is approached asymptotically.

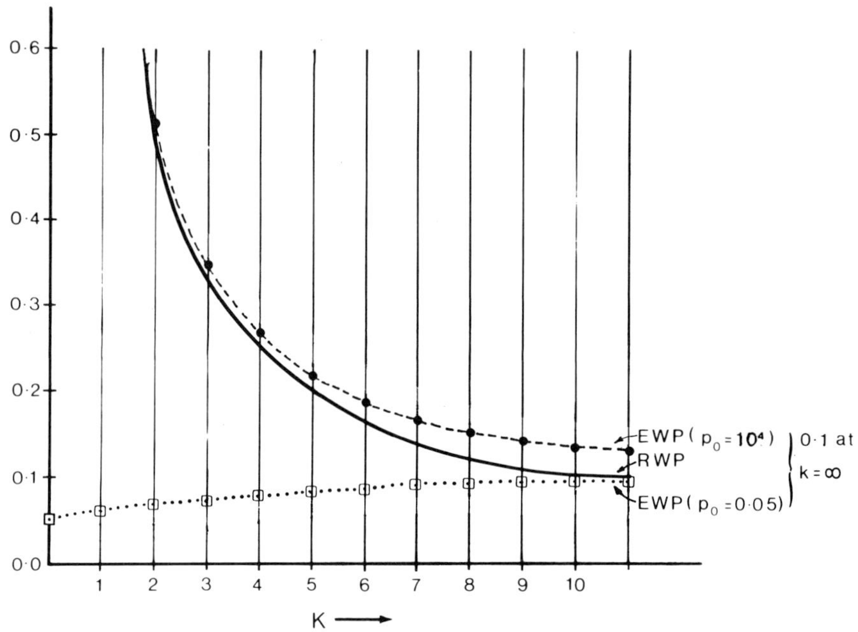

Figure 5.3 Variation of p_k for EWP and RWP Estimation of Mean Value

5.2 Modelling the Parameter Variations

In practice, shaping the memory of the estimation algorithm can prove to be a somewhat inflexible approach to the problem of estimating time variable parameters. In algorithms V and VI, for example, the analyst can only specify a single parameter (the memory length S or the exponential time-constant T_e, respectively) in order to 'match' the algorithm to the specific time-variable parameter model under investigation. If there happened to be *a priori* information available which suggested that certain of the parameters varied at different rates or, indeed, if it was known that only some of the parameters varied while others were constant, then it would be difficult for the analyst to make use of this information.

The statistical interpretation of the least squares equations discussed earlier suggests a more flexible alternative to the RWP and EWP approaches to the problem of detecting parameter variations. Suppose that instead of assuming parametric invariance, i.e.

$$\underline{a}_k = \underline{a}_{k-1} \tag{5.18}$$

we assume that the parameters vary in a manner that can be described by a Gauss-Markov stochastic difference equation of the following form (see Appendix 1 and Bryson and Ho, 1969)

$$\underline{a}_k = \Phi\underline{a}_{k-1} + \Gamma\underline{q}_{k-1} \tag{5.19}$$

where $\Phi = \Phi(k,k-1) = \Phi_k$ is an nxn transition matrix,

$\Gamma = \Gamma(k,k-1) = \Gamma_k$ is a nxm input matrix

both of which may be time variable; and $\underline{q}_{k-1}$ is an mx1 white noise vector of serially-independent random variables with zero mean and covariance matrix Q_p, i.e.

$$E\{\underline{q}_k\} = 0; \quad E\{\underline{q}_k\underline{q}_j^T\} = Q_p\delta_{kj}$$

where δ_{kj} is the Kronecker delta function, $\delta_{kj} = \begin{cases} 1; & k=j \\ 0; & k\neq j \end{cases}$

For the moment, we will assume that Φ and Γ are time-invariant and omit the subscript k in the subsequent analysis. However, as we will see in Section 5.4, similar results could be obtained with Φ, Γ and, indeed, Q_p defined as time-variable matrices.

The simplest examples of the model (5.19) are the multivariable random walk (R W) similar to that discussed earlier in equation (5.2), i.e. in the present context,

$$\underline{a}_k = \underline{a}_{k-1} + \underline{q}_{k-1} \tag{5.20}$$

and the first order Gauss-Markov model for n variables, i.e.

$$(\underline{a}_k - \underline{\bar{a}}) = \beta(\underline{a}_{k-1} - \underline{\bar{a}}) + \underline{q}_{k-1} \tag{5.21}$$

where

$$\beta = \begin{bmatrix} \beta_1 & & \mathbf{0} \\ & \beta_2 & \\ \mathbf{0} & & \ddots \\ & & & \beta_n \end{bmatrix}$$

with, for stability, $0<\beta_i<1$ for $i = 1,2,\ldots,n$; and $\underline{a}$ is the mean value about which the parameter is assumed to vary.

If a parameter variation model of this general form (5.19) is available, it clearly provides additional *a priori* information which can be used to advantage when attempting to estimate the parameter vector $\underline{a}_k$. The purpose of this section is to demonstrate how this *a priori* information may be used to construct dynamic least squares algorithms capable of tracking parametric variations. We will discuss later the problem of obtaining prior information in this form.

Let us consider the problem of estimating the n parameter vector $\underline{a}_k$ in a *pure linear regression model* of the form

$$y_k = \underline{x}_k^T \underline{a}_k + e_{yk} \tag{5.22}$$

where now $\underline{a}_k$ is a time variable parameter vector which can be described by the stochastic difference equation (5.19). By assuming this model for the parameter variations, we have introduced additional prior information into the estimation problem that allows us to make *a priori* updates $\hat{a}_{k/k-1}$ and $P^*_{k/k-1}$ to the estimate $\hat{a}_{k-1}$ and covariance matrix P^*_{k-1} obtained at the previous (k-1)th instant from the recursive regression algorithm III of Chapter 4. Here, as before, the subscript k/k-1 denotes the *a priori* value at the kth instant given information only to the (k-1)th instant.

To show how these *a priori* updates can be obtained we first note that, because of the statistical properties of $\underline{q}_k$,

$$E\{\underline{a}_k\} = \Phi \underline{a}_{k-1}$$

Although heuristic, it seems intuitively reasonable to assume that, given a least squares estimate $\hat{\underline{a}}_{k-1}$ at the (k-1)th instant, the best *a priori* estimate of $\underline{a}_k$ at the kth instant (i.e. prior to the receipt of new data at the kth instant) can be generated by

$$\hat{\underline{a}}_{k/k-1} = \Phi \hat{\underline{a}}_{k-1} \tag{5.23}$$

where the subscript k/k-1 indicates the estimate at k based on the previous estimate at k-1 and *a priori* knowledge of the parameters variation law (5.19). Now, if the *a priori* estimation error is defined as

$$\underline{\tilde{a}}_{k/k-1} = \underline{\hat{a}}_{k/k-1} - \underline{a}_k$$

then from (5.19) and (5.23)

$$\underline{\tilde{a}}_{k/k-1} = -\Phi\underline{a}_{k-1} - \Gamma\underline{q}_{k-1} + \Phi\underline{\hat{a}}_{k-1}$$

$$= \Phi\underline{\tilde{a}}_{k-1} - \Gamma\underline{q}_{k-1}$$

where $\underline{\tilde{a}}_{k-1} \triangleq \underline{\hat{a}}_{k-1} - \underline{a}_{k-1}$ is the *estimation error* at the (k-1)th instant.

The covariance matrix $P^{\star}_{k/k-1}$ of the *a priori* estimation errors $\tilde{a}_{k/k-1}$, can now be obtained from

$$P^{\star}_{k/k-1} = E\{\underline{\tilde{a}}_{k/k-1}\underline{\tilde{a}}^T_{k/k-1}\} = E\{[\Phi\underline{\tilde{a}}_{k-1} - \Gamma\underline{q}_{k-1}][\Phi\underline{\tilde{a}}_{k-1} - \Gamma\underline{q}_{k-1}]^T\}$$

$$= \Phi E\{\underline{\tilde{a}}_{k-1}\underline{\tilde{a}}^T_{k-1}\}\Phi^T + \Gamma E\{\underline{q}_{k-1}\underline{q}^T_{k-1}\}\Gamma^T - \Phi E\{\underline{\tilde{a}}_{k-1}\underline{q}^T_{k-1}\}\Gamma^T$$

$$- \Gamma E\{\underline{q}_{k-1}\underline{\tilde{a}}^T_{k-1}\}\Phi^T \tag{5.24}$$

From equation (5.20) we see that $\underline{q}_{k-1}$ only affects $\underline{a}_k$, so that

$$E\{\underline{\tilde{a}}_{k-1}\underline{q}^T_{k-1}\} = 0; \quad E\{\underline{q}_{k-1}\underline{\tilde{a}}^T_{k-1}\} = 0.$$

Consequently, equation (5.24) can be reduced to

$$P^{\star}_{k/k-1} = \Phi P^{\star}_{k-1}\Phi^T + \Gamma Q_p \Gamma^T \tag{5.25}$$

where $P^{\star}_{k-1} = E\{\underline{\tilde{a}}_{k-1}\underline{\tilde{a}}^T_{k-1}\}$ and Q_p is the covariance matrix of the parameter variation disturbance vector in equation (5.19).

Equations (5.23) and (5.25) provide the required *a priori predictions* $\hat{\underline{a}}_{k/k-1}$ and $P^*_{k/k-1}$ of $\hat{\underline{a}}_{k-1}$ and P^*_{k-1}, respectively. Bearing in mind the basic recursive least squares regression algorithm in the *time invariant parameter* case, these equations can be used to construct the following least squares prediction-correction algorithm for estimating the *time variable parameter* $\underline{a}_k$ defined by equation (5.19) i.e.

prediction:

$$\hat{\underline{a}}_{k/k-1} = \Phi\hat{\underline{a}}_{k-1} \quad (1)$$

$$P^*_{k/k-1} = \Phi P^*_{k-1}\Phi^T + \Gamma Q_p \Gamma^T \quad (2)$$

correction:

$$\hat{\underline{a}}_k = \hat{\underline{a}}_{k/k-1} - P^*_{k/k-1}\underline{x}_k[\hat{\sigma}^2 + \underline{x}_k^T P^*_{k/k-1}\underline{x}_k]^{-1}\{\underline{x}_k^T\hat{\underline{a}}_{k/k-1} - y_k\} \quad (3)$$

$$P^*_k = P^*_{k/k-1} - P^*_{k/k-1}\underline{x}_k[\hat{\sigma}^2 + \underline{x}_k^T P^*_{k/k-1}\underline{x}_k]^{-1}\underline{x}_k^T P^*_{k/k-1} \quad (4)$$

(equations (1)–(4): VII)

At this point, the reader may well argue that, while the Time Variable Parameter (TVP) algorithm (VII) has theoretical appeal, it may be exceedingly difficult to specify a parameter variation model such as (5.19). Fortunately, such understandable pessimism is not fully justified in this case. For example, consider the situation when the random walk (R W) model (5.20) is used to describe parameter variations which are themselves non-stationary in a Yaglom (1955) sense. In this case $\Phi = \Gamma = I$ so that the prediction equations become simply

$$\hat{\underline{a}}_{k/k-1} = \hat{\underline{a}}_{k-1} \quad (5.26)$$

and

$$P^*_{k/k-1} = P^*_{k-1} + Q_p \quad (5.27)$$

In order to utilize this simplified algorithm, it is only necessary to specify the covariance matrix Q_p of the random parameter variations between samples. In practice, this is a relatively straightforward task; experimental evidence suggests (e.g. Young, 1969b,c; 1970) that the algorithm (VII) with (1) and (2) in the form of equations (5.26) and (5.27) provides a reasonable approach to the estimation of slowly variable parameters (i.e. parameters whose percentage change per sampling instant is small, say < 5%). In such cases the Q_p matrix can be chosen to be diagonal in form (so implying a simplifying *a priori* assumption that the parameter variations are not correlated) with the diagonal elements selected to reflect the expected rate of variation between samples. For example, if the ith element a_i of $\underline{a}$ is expected to have a variance $\sigma_{a_i}^2$ between samples, then the ith diagonal element $[q_p]_{ii}$ of Q_p should be set to $\sigma_{a_i}^2$. This is inherently more flexible than the equivalent EWP approach because different expected rates of change can be specified for different parameters; for instance, any known time-invariant parameters can be accommodated simply by setting the appropriate diagonal element of Q_p to zero.

In the single parameter case, we can see that the algorithm VII with the RW model performs quite similarly to the EWP algorithm. For example in the case of the mean value estimation problem, the recursive relationship for p_k is obtained from VII (4) and (5.27) as

$$p_k^* = p_{k/k-1}^* - p_{k/k-1}^{*2} / (1 + p_{k/k-1}^*) \tag{5.28}$$

$$p_{k/k-1}^* = p_{k-1}^* + q_p$$

where $\hat{\sigma}^2$ is assumed to be unity to facilitate comparison with the EWP equivalent equation (5.17). The steady state solution of these equations is obtained when $p_k^* = p_{k-1}^* = p_\infty$ and is given by the solution of

$$p_\infty^2 + p_\infty q_p - q_p = 0$$

i.e.

$$p_\infty = \frac{-q_p}{2} \pm \frac{\sqrt{(q_p^2 + 4q_p)}}{2} \tag{5.29}$$

Equating p_∞ in (5.29) with p_∞ in (5.17), we find that, for similar steady state gains

$$q_p = \frac{1}{\alpha}(1 - 2\alpha + \alpha^2) \tag{5.30}$$

When $\alpha = 0.9$, therefore, $q_p = 0.0111$. A plot of p_k^* from (5.28) with q_p set to this value is given in Figure 5.4: as we can see by comparing Figures 5.3 and 5.4, the time variable gain effect in this simple case lies somewhere between that obtained with the RWP and EWP algorithms.

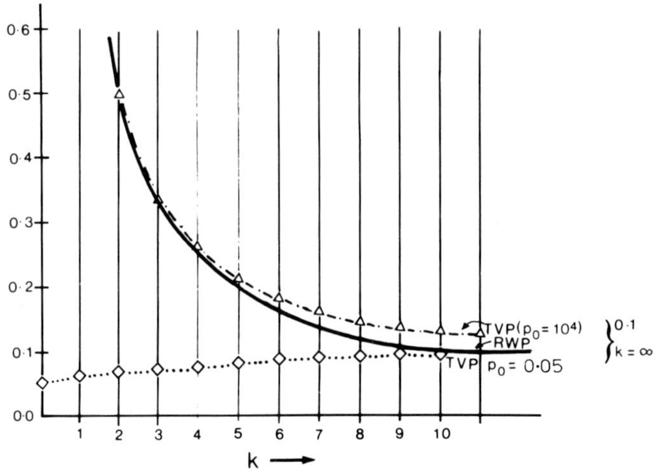

Figure 5.4 Variation of p_k^* for T.V.P. Estimation of Mean Value as RWP Model

A typical example of the gain variations in a multi-parameter example is given in Figure 5.5 (Young, 1969c). Here the data were generated by a deterministic linear dynamic system with a single time-variable coefficient and three time-invariant parameters. The Figure shows the variation in the diagonal element of the P_k matrix associated with the time variable parameter in the TVP case . Also shown is the same element in the situation where the algorithm was given (incorrectly in this case) an unlimited memory (i.e. $\alpha = 1.0$, $Q_p \equiv 0$).

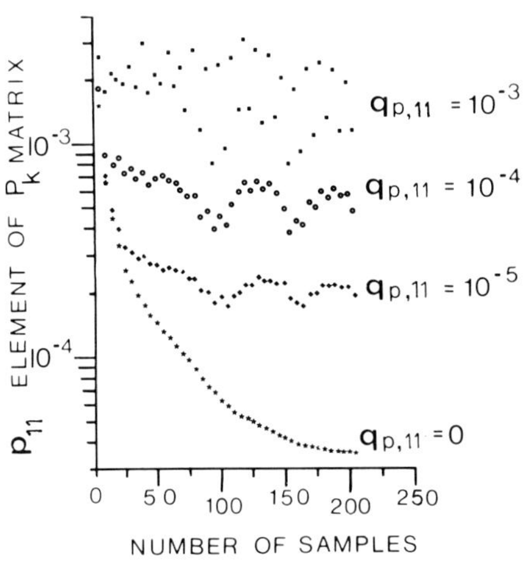

Figure 5.5 Typical Variation of P_k Matrix Elements in EWP and TVP Estimation

The first order Gauss-Markov model (5.21) can also be useful in time variable parameter estimation, but it is clearly only appropriate if the parameter variations are around a given constant mean value $\bar{\underline{a}}$. The smoothness of the expected variation can be specified by the selection of the β_i: by analogy with (5.9), the time constant of the smoothing for the ith parameter, T_{a_i}, is defined by

$$T_{a_i} = - T_s/\log_e\beta_i$$

so that the closer β_i is to unity, the greater the smoothing effect.

Other models that have been found useful in those cases where the parameter variations are themselves non-stationary in the Yaglom sense are the 'Integrated Random Walk' (I R W) and the 'Smoothed Random Walk' (S R W) i.e.

Integrated Random Walk:

$$\left.\begin{aligned} \underline{a}_k &= \underline{a}_{k-1} + \underline{s}_{k-1} \\ \underline{s}_k &= \underline{s}_{k-1} + \underline{q}_{k-1} \end{aligned}\right\} \quad (5.31)$$

Smoothed Random Walk:

$$\left.\begin{aligned} \underline{a}_k &= \beta\underline{a}_{k-1} + \underline{s}_{k-1} \\ \underline{s}_k &= \underline{s}_{k-1} + \underline{q}_{k-1} \end{aligned}\right\} \quad (5.32)$$

Intuitively, we would expect equation (5.31) to be useful when there are expected to be large and smooth variations in the parameters (Norton, 1975); while (5.32) is a compromise between (5.20) and(5.31) which requires specification of the degree of smoothing via the matrix β, defined in a similar manner to β in (5.21). It can be considered, in fact, as the equivalent of (5.21) when the mean value of the parameter is time variable. The general model (5.19) is likely to be less useful in practical terms since it requires much more *a priori* information than the simpler random walk models. Nevertheless, it can still be used in certain practical situations, as we shall see subsequently in Section 5.5.

In order to obtain a better idea of the way in which the assumptions (5.20), (5.31) and (5.32) affect the performance of the estimator, it is instructive to consider the nature of the prediction equations VII(1) for the various simple random walk processes. In the RW case, we have seen already that the *a priori* prediction $\hat{\underline{a}}_{k/k-1}$ is simply the estimate obtained at the previous recursive instant.

For the IRW and SRW model, however, an additional unknown parameter vector $\underline{s}_k$ is introduced and estimated.

To see the implications of this on the prediction equations, it is convenient to consider the scalar situation: then in both the IRW and SRW cases, we can substitute from the s_k equation into the a_k equation and obtain the following results

$$\text{IRW:}\quad \hat{a}_{k/k-1} = \hat{a}_{k-1} + (\hat{a}_{k-1} - \hat{a}_{k-2}) = \hat{a}_{k-1} + \Delta\hat{a}_{k-1}$$

$$\text{SRW:}\quad \hat{a}_{k/k-1} = \hat{a}_{k-1} + \beta(\hat{a}_{k-1} - \hat{a}_{k-2}) = \hat{a}_{k-1} + \beta\Delta\hat{a}_{k-1}$$

where $\Delta\hat{a}_{k-i} = \hat{a}_{k-i} - \hat{a}_{k-i-1}$ is the rate of change of the estimate between the $(k-i-1)$ and $k-i$ recursions. In other words, the *a priori* prediction is obtained by projecting the *a priori* estimate obtained at the previous (k-1)th recursion in proportion to its immediate past rate of change: in the IRW case the projection is equal to the past rate of change; in the SRW case it is made a fraction (β) of this rate of change.

If we consider higher order integrated random walks then it is clear that more past information on rate of change would be used for *a priori* prediction. In the case of the double integrated random walk (DIRW), for example, we have

$$a_k = a_{k-1} + s_{k-1}$$

$$s_k = s_{k-1} + r_{k-1}$$

$$r_k = r_{k-1} + q_{k-1}$$

so that,

$$\hat{a}_{k/k-1} = \hat{a}_{k-1} + \Delta\hat{a}_{k-1} + (\Delta\hat{a}_{k-1} - \Delta\hat{a}_{k-2})$$

Simple illustrative examples of the RW, SRW, IRW and DIRW *a priori* predictions are given in Fig. 5.6 for different past $\hat{a}_k$ histories. It is interesting to note that in the IRW case, this procedure is analogous to that used in a hybrid (analog-digital) estimation algorithm by Young (1965b), where the past rates of change of the estimates were generated using purely analog equipment. It is also clear from the examples in Fig. 5.6 that, although the applicability of the different random walk models will depend to some extent upon the nature of the parametric variation, all of the models (except perhaps the DIRW) seem to provide *reasonable* extrapolations when it is remembered that these extrapolations are automatically corrected by the update equations VII(3) on receipt of the next data sample.

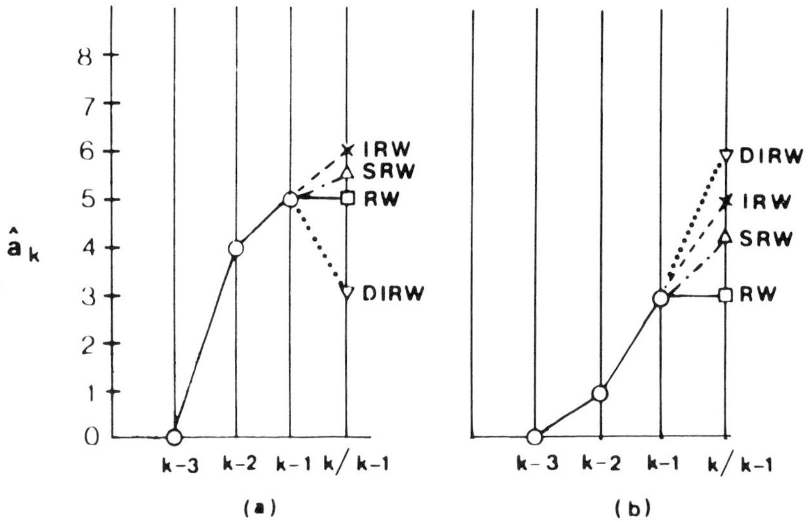

Figure 5.6 Typical Predictions for Random Walk Models

5.3 Vector Measurements and the Kalman Filter-Estimation Algorithm

The analysis in Chapters 3, 4 and Section 5.2 can be extended[†] to the case where the equation (5.22) is in terms of a p vector of observations $\underline{y}_k$ at each sampling instant, i.e.

$$\underline{y}_k = H_k \underline{a}_k + \underline{e}_{yk} \tag{5.33}$$

where $\underline{y}_k = [y_1, y_2, \ldots, y_p]_k^T$ and $\underline{e}_{yk}$ is a vector of zero mean observational errors whose elements are assumed to be serially uncorrelated white noise disturbances but which may, however, be *instantaneously* correlated with covariance matrix R_N, i.e.

[†]Clearly the RWP and EWP algorithms could also be extended quite straightforwardly in a similar fashion.

$$E\{\underline{e}_{yk}\} = 0;\ E\{\underline{e}_{y_k}\underline{e}_{y_j}^T\} = R_N\delta_{kj} \tag{5.34}$$

Finally Π_k is a p x n matrix of exactly known and possibly time-variable 'regressors' assumed to have the same properties as the analogous regression variables in the scalar observation case.

In order to develop a recursive least squares algorithm in this vector case, it is first necessary to define a suitable cost function. Clearly a simple least squares function such as (3.3) with its scalar norm $e_i^2 = [\underline{x}_k^T\hat{\underline{a}} - y_i]^2$ is no longer appropriate and we must consider a vector norm of some kind (see Appendix 1). The most general quadratic norm in the present context is the L_2 norm, $||\underline{e}_i||_W^2$,

where $\underline{e}_i = \Pi_i\hat{\underline{a}}_i - \underline{y}_i$ and

$$||\underline{e}_i||_W^2 = \underline{e}_i^T W\underline{e}_i = [\Pi_i\ \hat{\underline{a}}_i - \underline{y}_i]^T W[\Pi_i\hat{\underline{a}}_i - \underline{y}_i] \tag{5.35}$$

in which W is a p x p, symmetric, positive definite weighting matrix with elements w_{kj}. By expansion of (5.35), the reader can verify that it is equivalent to the following summation

$$\underline{e}_i^T W\underline{e}_i = \sum_{k,j=1}^{p} w_{kj}\, e_{ki}\, e_{ji} \tag{5.36}$$

where e_{ki}, k = 1,2,...,p are the elements of $\underline{e}_i$. Thus the L_2 norm (5.35) is a general quadratic form in the elements of the error vector with weightings w_{kj} to be specified by the analyst.

We shall consider how the weighting matrix should be specified subsequently but, for the moment, let us pose the estimation problem in terms of the norm (5.35) as one of minimizing the cost function J_{2W} with respect to $\hat{\underline{a}}$, where

$$J_{2W} = \sum_{i=1}^{k} ||\underline{e}_i||_W^2 \tag{5.37}$$

In the case of constant parameters, i.e. $\underline{a}_k = \underline{a}$ for all k, this becomes

$$J_{2W} = \sum_{i=1}^{k} [\Pi_i\hat{\underline{a}} - \underline{y}_i]^T W[\Pi_i\hat{\underline{a}} - \underline{y}_i] \tag{5.38}$$

so that, setting the gradient of J_{2W} to zero in the normal manner, we obtain the following normal equations (see Appendix 1)

$$[\sum_{i=1}^{k} \Pi_i^T W \Pi_i]\hat{\underline{a}}_k = \sum_{i=1}^{k} \Pi_i^T W \underline{y}_i \tag{5.39}$$

As a result, the recursive expressions (3.7) and (3.8) in the scalar error case now take the form,

$$\begin{aligned} P_k^{-1} &= P_{k-1}^{-1} + \Pi_k^T W \Pi_k \\ \underline{b}_k &= \underline{b}_{k-1} + \Pi_k^T W \underline{y}_k \end{aligned} \tag{5.40}$$

and the reader will find that the recursive stimation algorithms for $\hat{\underline{a}}_k$ and P_k can be derived quite straightforwardly in the following form,

$$\left.\begin{aligned} \hat{\underline{a}}_k &= \hat{\underline{a}}_{k-1} - P_{k-1}\Pi_k^T[W^{-1} + \Pi_k P_{k-1}\Pi_k^T]^{-1}\{\Pi_k\hat{\underline{a}}_{k-1} - \underline{y}_k\} \quad (1) \\ P_k &= P_{k-1} - P_{k-1}\Pi_k^T[W^{-1} + \Pi_k P_{k-1}\Pi_k^T]^{-1}\Pi_k P_{k-1} \quad (2) \end{aligned}\right\} \text{VIII}$$

Before proceeding to develop the time-variable parameter equivalent of algorithm VIII, it is first necessary to consider how the weighting matrix should be chosen and what the statistical properties of the parameter estimates will be with this choice of W. In relation to the choice of weighting matrix, it is well known (Johnston, 1963; Young, 1968a) that if the observational errors $\underline{e}_{yk}$ have a Gaussian amplitude distribution (Appendix 1) i.e.

$$\underline{e}_{yk} = N(\underline{0}, R_N) \tag{5.41}$$

where $N(\underline{a},B)$ denotes that the distribution is Gaussian normal with mean $\underline{a}$ and covariance matrix B defined by equation (5.34), then the cost function (5.38) is equivalent to the maximum likelihood cost function when $W = R_N^{-1}$. And in this situation, it can be shown (Johnston, 1963), by an argument similar to that used in Chapter 4, that the estimates obtained from VIII are zero mean and have an error covariance matrix P_k^* defined by

$$P_k^* = [\sum_{i=1}^{k} \Pi_i^T R_N^{-1} \Pi_i]^{-1} \tag{5.42}$$

Given this information and pursuing the same, partially heuristic approach used to develop algorithm VII, it is possible to derive the following prediction-correction algorithm for the case where $\underline{a}_k$ is now stochastically time-variable and can be modeled by equation (5.19),

prediction:

$$\hat{\underline{a}}_{k/k-1} = \Phi\hat{\underline{a}}_{k-1} \quad (1)$$

$$P^*_{k/k-1} = \Phi P^*_{k-1}\,\Phi^T + \Gamma Q_p \Gamma^T \quad (2)$$

correction: IX

$$\hat{\underline{a}}_k = \hat{\underline{a}}_{k/k-1} - P^*_{k/k-1}\,\Pi_k^T[R_N + \Pi_k P^*_{k/k-1}\,\Pi_k^T]^{-1}\{\Pi_k\hat{\underline{a}}_{k/k-1} - \underline{y}_k\} \quad (3)$$

$$P_k = P^*_{k/k-1} - P^*_{k/k-1}\,\Pi_k^T[R_N + \Pi_k P^*_{k/k-1}\,\Pi_k^T]^{-1}\Pi_k P^*_{k/k-1} \quad (4)$$

It is reasonable to surmize that, given the assumptions about the observational errors and the stochastic nature of the parameter variation model, the TVP algorithm IX will be optimal in a maximum likelihood sense. And further that, given *a priori* information on the initial estimate $\hat{\underline{a}}_0$ and its associated initial covariance matrix P_0^*, the estimates will also be optimal in a Bayesian sense. Such conjecture is borne out by more rigorous analysis: indeed algorithm IX is algorithmically identical to the optimal filter-estimation algorithm first proposed by R.E. Kalman (1960) within a stochastic state-space context (see e.g. Bryson and Ho, 1969).

Kalman considered the problem of estimating a state vector $\underline{x}_k$ associated with a stochastic dynamic system modelled by†

$$\underline{x}_k = A_k\underline{x}_{k-1} + D_k\underline{\eta}_{k-1} \quad (1) \quad (5.43)$$

† We use $\underline{x}$ to denote the state for simplicity of nomenclature in later chapters; it should not be confused with the regressor vector $\underline{x}$ used previously.

from vector observations $\underline{y}_k$ related linearly to $\underline{x}_k$ by the measurement equation[†]

$$\underline{y}_k = C_k \underline{x}_k + \underline{\xi}_k \qquad (2) \quad (5.43)$$

where $\underline{\eta}_k$ and $\underline{\xi}_k$ are zero mean, statistically independent, white noise disturbance vectors with possibly time-variable covariance matrices Q_k and R_k respectively, i.e.

$$E\{\eta_k\} = E\{\xi_k\} = 0; \quad E\{\underline{\eta}_k \underline{\xi}_j^T\} = 0, \text{ for all } k,j \qquad (5.44)$$

$$E\{\underline{\eta}_k \underline{\eta}_j^T\} = Q_k \delta_{kj}; \quad E\{\underline{\xi}_k \underline{\xi}_j^T\} = R_k \delta_{kj}$$

By an argument based on orthogonal projection, Kalman showed that the optimal estimate $\hat{\underline{x}}_k$ of $\underline{x}_k$ in (5.43) can be obtained from the algorithm

prediction:

$$\hat{\underline{x}}_{k/k-1} = A_k \hat{\underline{x}}_{k-1} \qquad (1)$$

$$P^*_{k/k-1} = A_k P^*_{k-1} A_k^T + D_k Q_k D_k^T \qquad (2)$$

correction:

$$\hat{\underline{x}}_k = \hat{\underline{x}}_{k/k-1} - P^*_{k/k-1} C_k^T [R_k + C_k P^*_{k/k-1} C_k^T]^{-1} \{C_k \hat{\underline{x}}_{k/k-1} - \underline{y}_k\} \qquad (3)$$

$$P_k^* = P^*_{k/k-1} - P^*_{k/k-1} C_k^T [R_k + C_k P^*_{k/k-1} C_k^T]^{-1} C_k P^*_{k/k-1} \qquad (4)$$

(equations (2)–(4) bracketed as X)

which is, of course, identical to algorithm IX except that here the model matrices A_k, D_k, C_k and the covariance matrices Q_k, R_k are considered as possibly time-variable; in other words, not only are the basic state variables $\underline{x}_k$ time-variable but they can also be generated from a non-stationary dynamic system excited by non-stationary stochastic disturbances! Bearing this result in mind, it is clearly reasonable to extend the parameter estimation algorithm IX in a similar manner, if required. We will discuss an example where this extension proves useful in the next section.

[†] Actually, in his original paper Kalman (1960) considers $\underline{\xi}_k = 0$ and measurement noise enters via the state equations.

Note that the Kalman filter is often derived in terms of a model with a vector of external deterministic inputs or exogenous inputs $\underline{u}_k$ affecting the state equations, i.e. equation (5.43), (1) becomes

$$\underline{x}_k = A_k\underline{x}_{k-1} + B_k\underline{u}_{k-1} + D_k\underline{\eta}_{k-1} \tag{5.45}$$

The only effect of this modification is to change the prediction equation X(1) to the form

$$\hat{\underline{x}}_{k/k-1} = A_k\hat{\underline{x}}_{k-1} + B_k\underline{u}_{k-1} \qquad \text{(1A)} \qquad \text{X}$$

The covariance prediction equation remains unchanged since the elements of $\underline{u}_k$ are assumed purely deterministic variables. The reader can verify that these equations make sense by introducing a similar vector of deterministic input disturbance into the Gauss Markov equation (5.19) and repeating the previous analysis. We shall refer to the resulting TVP algorithm subsequently in Section 5.4.

Many different derivations of the Kalman filter algorithm have been suggested since the appearance of Kalman's seminal paper in 1960 (e.g. Rauch, Tung and Striebel, 1965; Kailath and Frost, 1968; Duncan and Horn, 1972). The treatment of Bryson and Ho (1969) is, however, of particular merit because of its simple derivation and clarity in statistical terms. Using a Bayesian argument under the assumption that the probability densities $p(\underline{x}_0)$, $p(\underline{\xi}_k)$ and $p(\underline{\eta}_k)$, $k = 1,2,\dots,T$ are all Gaussian, they point out that the conditional probability $p(\underline{x}_0|Y_k)$ is also Gaussian (where Y_k denotes all data up to sample instant k) and is thus characterized completely by its mean and covariance matrix, with the mean giving the maximum of the density. They then derive equations for the conditional mean $\hat{\underline{x}}_k$ and covariance P_k^* of the densities $p(\underline{x}_k, Y_k)$, which are precisely the Kalman filter equations. Rauch, Tung and Streibel (1965) use a similar approach from a maximum likelihood standpoint, but concentrate on the "smoothing solution" (i.e. the estimation of $\hat{x}_{l/k}$ where $l<k$) which we consider later in Section 5.6.

At this point it is interesting to note that, from a starting point of time-invariant parameter estimation, we have progressed via algorithms for slowly-variable parameter estimation to procedures for estimating the possibly rapid variations in the state of a stochastic dynamic system. This link between state and parameter estimation, which was first pointed out by R.C.K. Lee (1964), raises certain philosophical questions regarding the relationship between state variables and parameters in dynamic system models. Although it is inappropriate for us to enter this kind of discussion in the present context, it is interesting to note how the similarities between states and parameters in dynamic systems are emphasized in the simple moving body example discussed earlier. Here, the unknown 'parameters' can actually be interpreted as the states of a dynamic system; we are able to consider them as stationary parameters simply because they do not vary over the observation interval. In the light of the above discussion, the problem could equally well have been considered as one of state estimation and solved by application of the Kalman filter-estimation algorithm X. In the case of the single unknown velocity problem, for example, the reader may easily verify that the state equations are as follows,

$$\begin{bmatrix} x_1 \\ x_2 \end{bmatrix}_k = \begin{bmatrix} 1 & a_{12} \\ 0 & 1 \end{bmatrix} \begin{bmatrix} x_1 \\ x_2 \end{bmatrix}_{k-1}$$

$$y_k = [\,1 \quad 0\,] \begin{bmatrix} x_1 \\ x_2 \end{bmatrix} + \xi_k$$

where $x_1 = s - s_0$; $x_2 = v$ and a_{12} is the sampling interval. Consequently, estimates $\hat{x}_1$ and $\hat{x}_2$ of x_1 and x_2, respectively, could be obtained from algorithm X simply by making the following substitutions:

$$A_k = \begin{bmatrix} 1 & a_{12} \\ 0 & 1 \end{bmatrix}; \quad D_k = [0]; \quad Q = [0]; \quad C_k = [1 \quad 0]; \quad R = \sigma^2$$

The results obtained in this manner would be directly equivalent to those obtained via linear regression analysis.

Finally in relation to the Kalman filter, it should be noted that although the two algorithms IX and X are algorithmically identical, we feel that it is a mistake to refer to IX as a Kalman filter algorithm (although this often occurs in the literature). The reason is that, whereas in practice the observation matrix C_k in the Kalman filter is an assumed known matrix which simply explains the relationship between the observation y_k and the state $\underline{x}_k$, in the general regression problems Π_k may be a matrix of *measurements* which happen to be free from error. Recognizing this, the analyst will be careful to check for the presence of, for example, errors-in-variables (see Chapters 4 and 6); whereas if he considers the algorithm as a Kalman filter, he may well omit to do this and incorrectly use the algorithm in a *structural* model situation. This subtle point becomes clearer when we consider structural time-series models in subsequent chapters: in such cases, it is necessary to modify the equations to allow for the error-in-variables and to avoid asymptotic bias on the estimates. They are then, however, significantly different from the Kalman filter equations.

A simple scalar example which can be classified as either a recursive parameter or state estimation problem is the estimation of the mean value for the Walgett rainfall considered in Chapter 2 and it is instructive to look at the results obtained when the mean is assumed to vary according to the IRW and SRW models. The estimation results are given in Figure 5.7. In the IRW and SRW cases, the initial estimates were set at $a_o = 0$, $s_o = 0$ and the initial P matrix was set to $P_o = 10^6$ diagonal. This choice of initialisation values implies little confidence in the initial estimates and, as expected, there is considerable variance in the recursive estimates for the first 25 samples. Thereafter, however, the SRW and IRW estimates for lower q_p values (2.0 and 0.002 respectively) settle down to relatively small changes of about 1 to 2 inches with a slow 'drift' downwards of about one inch in rainfall. This continues until about 68 samples (the year 1948) when a quite sharp change in behaviour is noted, with an upward movement of some 4-5 inches, which tends to reduce again after 86 samples (1966).

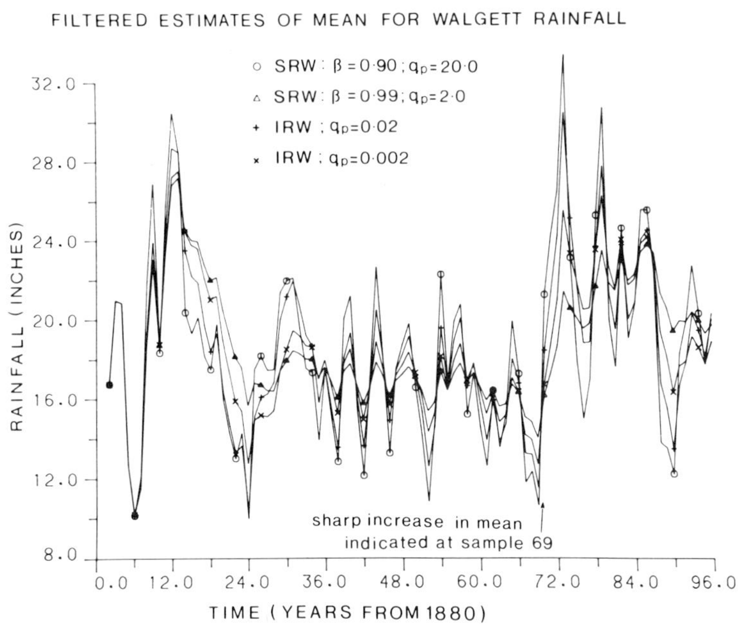

Figure 5.7 Recursive Estimation of Mean for Walgett Rainfall

The results in Figure 5.7 are shown for various settings of the model parameters. In the IRW case, these parameters are the assumed variance of the parameters between samples, as reflected in the choice of q_p (Q_p is a 2 x 2 matrix with only one non-zero lower diagonal element in the IRW case). In the SRW situation two such 'program parameters' need to be specified; q_p again and also β the exponential smoothing constant. Clearly the effect of different choices of these program parameters can be considerable and, in these examples, our choice is completely heuristic. In all cases, however, the sharp upward change in 1948 is a notable characteristic and we might conclude that 'something' affected the rainfall at around this date and resulted in a predominantly higher mean measured rainfall over the period 1948 to 1975 than from 1905 to 1948. We can say little of the behaviour between 1880 and 1905 from these results, however, because of the high estimation variance at low sample size for $k = 0 \rightarrow 25$. This could be improved, however, by utilizing a recursive smoothing algorithm, as discussed in Section 5.6.

Similar results are obtained from analyzing other rainfall data in NSW over this period, with the upward movement being apparently larger in some cases than in others. In the absence of any evidence that the measurement procedures (via rain gauges) were changed around 1948, it is possible to hypothesize that some 'significant' climatic variation affected N.S.W. in the immediate post-second world war period and was sustained for some 30 years. Whether this is an example of climatic 'variation' or more permanent 'change' is arguable, but it is clear that the recursive estimation results are useful in initially assessing the data for the possibility of non-stationarity. Additional research is, of course, required in order to reach any more detailed conclusions in this case and the work of Pittock (1975), which suggests that the increase in mean is due primarily to changes in two specific climatic factors, is a step in this direction.

But can we say anything further simply on the basis of the recursive estimation results? Can we for example assess the standard error on the estimates using the recursive covariance equations VII (4)? Or, continuing in this manner, is it possible to test whether the apparent change in mean is statistically significant? We will consider the second possibility in Section 5.7. As far as the standard error evaluation is concerned, the applicability of result (4.6) in Chapter 4 and, therefore, the interpretation of P_k^* in algorithm VII depends upon whether the e_{yk} random variables satisfy the requirements (1) to (3) on page 42.

In this example, where there is high variance in the series (reflecting the high variability of rainfall in Australia), it is difficult to be sure on this point: first the variance of e_y itself may be varying; second, the sample auto-correlation function of the least squares residuals obtained in the time-variable parameter estimation case does not yield any clear indication as to whether or not they are serially uncorrelated. In such a situation it is, perhaps, best not to quote the standard errors obtained by interpreting P_k^* as a covariance matrix but merely to note the qualitative behaviour of the estimates for *all* assumed models of parameter variation. The initial indication of parameter variation obtained in this manner could then be tested further by more detailed time-series analysis in which the rainfall series is assumed to be described by a more complex time-series model than (2.2).

5.4 The Estimation of Rapidly Varying Parameters

The simple random walk models (5.20), (5.31) and (5.32) are useful in practice simply because they require the specification of so little *a priori* information on the nature of parameter variations. But if it is likely that the parameters will vary rapidly between samples, then we cannot expect algorithms based on these simple parameter variation laws to necessarily function satisfactorily: substantial estimation lags are likely between the occurrence of rapid parameter change and its estimation by the algorithm. In this situation, we need to look for procedures whereby additional information on the nature of the parameter variations can, if available, be incorporated into the algorithm to aid its 'parameter tracking' performance; in other words, procedures which utilize this information to give the algorithm prior warning of the direction and magnitude of the parametric change.

The most obvious approach to the problem is to make use of the more complicated Gauss-Markov model of parameter variation given in (5.19). But, as we have pointed out, this would require the specification of the entire Φ and Γ matrices; a somewhat daunting task given that they contain $n^2 + nm$ elements! As a result there is no general procedure for using this model: its efficacy in practice will depend upon the problem at hand and the nature of the *a priori* knowledge available on parameter variation.[†] We can, however, consider *special* situations where the *a priori* information may be available and is of a form which allows Φ and Γ to be specified without too much difficulty.

Simple parameter decomposition

The simplest examples where the Gauss-Markov model can be useful are encountered in situations where the unknown variation in the parameters can be characterized by a vector of constant or *slowly* variable coefficients denoted by $\underline{a}_k^*$ but where the *major*

[†]Needless to say, if the parameters are known to vary in a particular fashion, e.g. sinusoidally with known frequency ω_n then Φ and Γ can be chosen accordingly to incorporate this knowledge; and if this variation is purely deterministic then, of course, $\Gamma = 0$ and only Φ has to be specified.

variations derive from a vector $\underline{t}_k$ of measured (or otherwise *known*) variables with which the variation in $\underline{a}_k$ can be associated, either on the basis of *a priori* knowledge of the system or via some physically meaningful argument of the kind in the missile example discussed later, i.e.

$$\underline{a}_k = \underline{f}(\underline{t}_k, \underline{a}_k^*) \tag{5.46}$$

In (5.46) $\underline{a}_k^*$ might be constant, in which case,

$$\underline{a}_k^* = \underline{a}_{k-1}^* \tag{5.47}$$

or, more likely in practice, it will be slowly variable, in which case we will assume that its variations are not known *a priori* but can be modeled approximately by the random walk model (5.20) introduced in the last section, i.e. in the present situation,

$$\underline{a}_k^* = a_{k-1}^* + \underline{q}_{k-1}; \qquad E\{\underline{q}_k \underline{q}_j^T\} = Q_p \delta_{kj} \tag{5.48}$$

A simple example of this kind of parameter decomposition is the case where the functional relationship $\underline{f}$ is linear-in-the-parameters (Young, 1969a, b, c; 1970), i.e.

$$\underline{a}_k = T_k \, \underline{a}_k^* \tag{5.49}$$

where T_k is an n x n matrix whose elements are *known* functions of the elements of the $\underline{t}_k$ vector. In this case, we see that by substituting from (5.48) into (5.49), we obtain

$$\underline{a}_k = T_k \underline{a}_{k-1}^* + T_k q_{k-1}$$

which, from (5.49) can be written as

$$\underline{a}_k = T_k T_{k-1}^{-1} \underline{a}_{k-1} + T_k \underline{q}_{k-1} \tag{5.50}$$

provided T_k is chosen to be non-singular.

Equation (5.50) is, of course, a Gauss-Markov model of the form (5.19) with $\Phi = T_k T_{k-1}^{-1}$ and $\Gamma = T_k$ and so algorithm VII can be used directly to estimate $\underline{a}_k$. Alternatively, we could simply estimate $\underline{a}_k^*$: this becomes clear if we note that, by substituting from (5.49) into time variable parameter regression relationship (5.22), we obtain

$$y_k = \underline{x}_k^T T_k \underline{a}_k^* + e_{yk}$$

or

$$y_k = \underline{\pi}^T \underline{a}_k^* + e_{yk} \tag{5.51}$$

where $\underline{\pi}^T = \underline{x}_k^T T_k$ represents the vector of regressors $\underline{x}_k^T$ transformed by the *known* matrix T_k. Equation (5.51) is then simply a regression model in the constant or slowly variable parameters composing the $\underline{a}_k^*$ vector. And these parameters can be estimated by using either the constant parameter recursive regression algorithm III of Chapter 4 or the time variable parameter estimation algorithm VII with $\Phi = \Gamma = I$, whichever is applicable.† Because the prediction equations (5.26) and (5.27) are so simple in this second case, it is probably preferable in computational terms to the alternative of estimating $\underline{a}_k$ directly using the model (5.50). And of course $\hat{\underline{a}}_k$ can easily be recovered from $\hat{\underline{a}}_k^*$ by using equation (5.49).

Except in special circumstances, a_k^* in the above formulation should be considered slowly variable rather than constant. Constancy of $\underline{a}_k^*$ would imply complete determinism, which is unlikely in most real problems, particularly in the present situation where we are assuming a simple linear-in-the-parameters relationship such as (5.49). Also it is interesting to observe how the presence of the random vector $\underline{q}_k$ in (5.48) and (5.50) adds a useful statistical degree of freedom to the problem and not only allows for residual slow variations in the parameter vector $\underline{a}_k^*$, which will usually be required because of the approximate nature of the relationship (5.49), but also allows for uncertainty in the measurement of the variables in T_k. Note, however, that in this latter situation the relationship (5.51) would then change from a regression to a structural model, with its attendant problem of errors-in-variables (Chapter 4). But this problem can sometimes be obviated, as we shall see in the later Chapters on time-series analysis.

The approach described above has been applied to the problem of estimating the parameters of a simple differential equation model for a missile (Young and Yancey, 1971; Young, 1979c). The missile was simulated over an 80 second mission involving rocket boost and considerable changes in altitude and speed. The recursively estimated parameters were required to continuously update the control

---

† i.e. with $\underline{x}_k = \underline{\pi}$

gains in a self-adaptive autostabilization system. The model equations in this case can, under certain conditions of low-measurement noise (but allowing for stochastic gust disturbances), be considered as two *regression-like* models which relate the missile state variables and are of the form

$$y_1 = \underline{x}_1^T \underline{a}_1 + e_{y1} \qquad \text{(i)}$$

and

$$y_2 = \underline{x}_2^T \underline{a}_2 + e_{y2} \qquad \text{(ii)} \qquad (5.52)$$

where,

$$y_1 = \dot{q};\ \underline{x}_1^T = [a_z\ q\ \delta];\ \underline{a}_1 = [a_1\ a_2\ a_3]^T$$

and

$$y_2 = a_z;\ \underline{x}_2^T = [\alpha_T\ \delta];\ \underline{a}_2 = [a_4\ a_5]^T$$

Here, the variables e_{y1} and e_{y2} represent "equation errors" which arise more because the simple linear relationships cannot completely describe the vehicle dynamics than because of measurement errors on y_1 and y_2 (see Young 1979c). The vehicle state variables in these relationships are the pitch rate q, as measured by a pitch rate gyro; the normal acceleration a_z, measured by an accelerometer; the actuator deflection δ, measured by an angular position sensor and the angle of attack α_T obtained either from a strap down inertial guidance system or a vane sensor. In 5.52 (i) $\dot{q}$ represents the time-derivative of q and is not normally measurable. However, this difficulty can be circumvented by performing special "state variable filtering" operations on q and the other state variables; these filtering procedures are, however, mentioned in Chapter 10 and the reader can, for the moment, assume that $\dot{q}$ is available directly for measurement (see also Young, 1979c).

With the above assumptions, the parameter vectors $\underline{a}_1$ and $\underline{a}_2$ can be estimated using the deterministic recursive least squares method and a TVP algorithm. All the variables in (5.52) were sampled at 0.05 second intervals and the $\underline{a}_{1k}$ and $\underline{a}_{2k}$ parameter vectors were decomposed using transformation matrices T_{1k} and T_{2k} of the diagonal form

$$T_{1k} = \begin{bmatrix} c_1\frac{m_k}{M_k} & 0 & 0 \\ 0 & c_2 q_{ok} & 0 \\ 0 & 0 & c_3 q_{ok} \end{bmatrix};\qquad T_{2k} = \begin{bmatrix} c_4\frac{q_{ok}}{m_k} & 0 \\ 0 & c_5 q_{ok} \end{bmatrix} \qquad (5.53)$$

where c_i, i = 1, 2, ..., 5, are normalization coefficients selected to normalize the $\underline{a}_{1k}^*$ and $\underline{a}_{2k}^*$ parameter values about unity and q_{ok}, m_k and M_k are, respectively, the changes in dynamic pressure, mass and Mach number associated with the mission. It is well known that such "air data" variables lead to variations in the coefficients of a simple linear dynamic model of an airborne vehicle (Blakelock, 1965); and the particular relationships in T_{1k} and T_{2k} were selected on the basis of this theory and then adjusted by trial and error (Young and Yancey, 1971). The transformed parameters $\underline{a}_{1k}^*$ and $\underline{a}_{2k}^*$, respectively, were then estimated using TVP algorithms primed with the random walk model for parameter variations (5.20). In other words, the "regression" relations (5.52) in $\underline{a}_1$ and $\underline{a}_2$ were transformed into relationships in $\underline{a}_1^*$ and $\underline{a}_2^*$, as shown in (5.51), and then two separate algorithms of the form VII with $\Phi = \Gamma = I$ were used to estimate $\underline{a}_1^*$ and $\underline{a}_2^*$. The covariance matrices Q_{p1} and Q_{p2} required to implement the prediction equations (5.26) and (5.27) were selected by *experiment* to provide good estimation performance.

Additional details of this example are given in Young (1979c) which discusses adaptive control applications. For present purposes, it will suffice to note that, with the considerable changes in the parameters occurring over the rocket boost period, it proved necessary to make the Q_{p1} and Q_{p2} covariance matrices themselves functions of time over this period, so giving the estimation algorithm greater flexibility to track the extremely rapid variations during rocket boost. A possible alternative would have been to utilize the integrated random walk model (5.31) rather than the pure random walk model (5.20).

Some of the results obtained during simulation of the above example are shown in Figures 5.8 to 5.13. Figure 5.8 gives the normal acceleration commands and the resulting response of the adaptively controlled missile over the 80 second mission, while Figures 5.9 to 5.13 compare the estimated parameter $\hat{\underline{a}}_i$, i = 1, 2, ..., 5 with the true parameter variations; here the vectors $\hat{\underline{a}}_1$ and $\hat{\underline{a}}_2$ were obtained continuously from $\hat{\underline{a}}_1^*$ and $\hat{\underline{a}}_2^*$ by the use of the transformation equation (5.49).

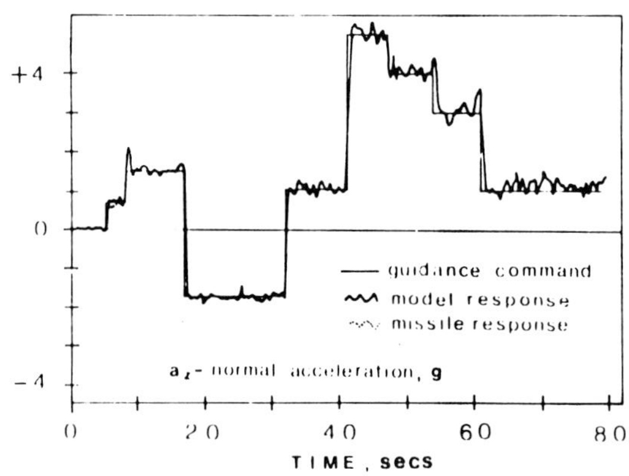

Figure 5.8 Missile Response Over 80 Second Mission

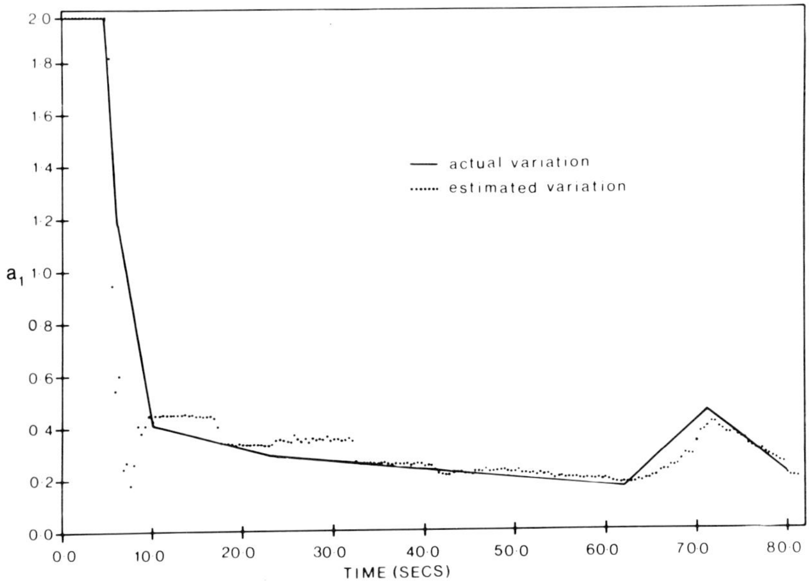

Figure 5.9 Recursive Estimate of Missile Model Parameter a_1

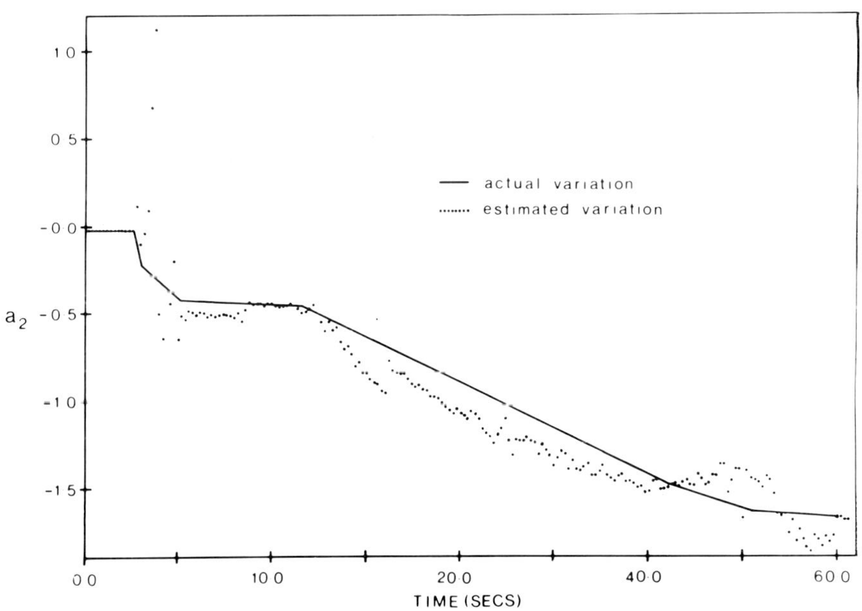

Figure 5.10 Recursive Estimate of Missile Model Parameter a_2

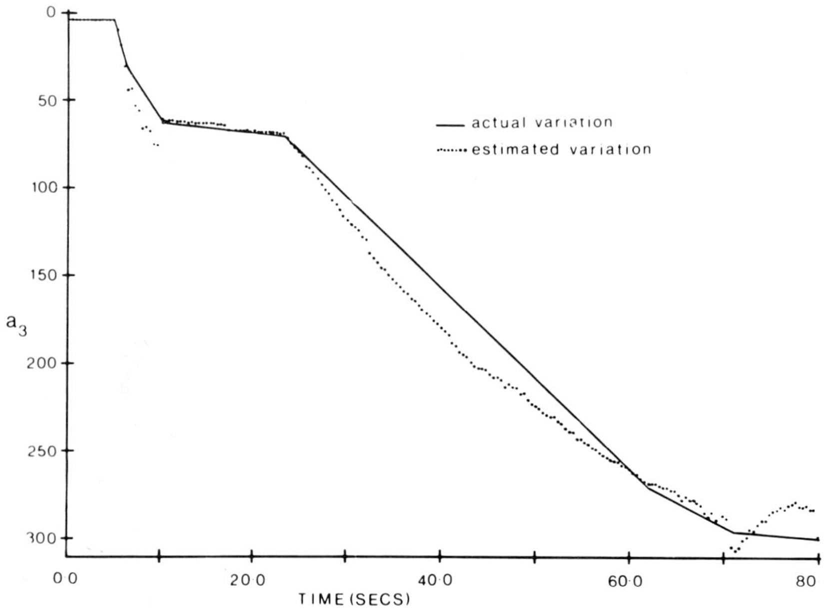

Figure 5.11 Recursive Estimate of Missile Model Parameter a_3

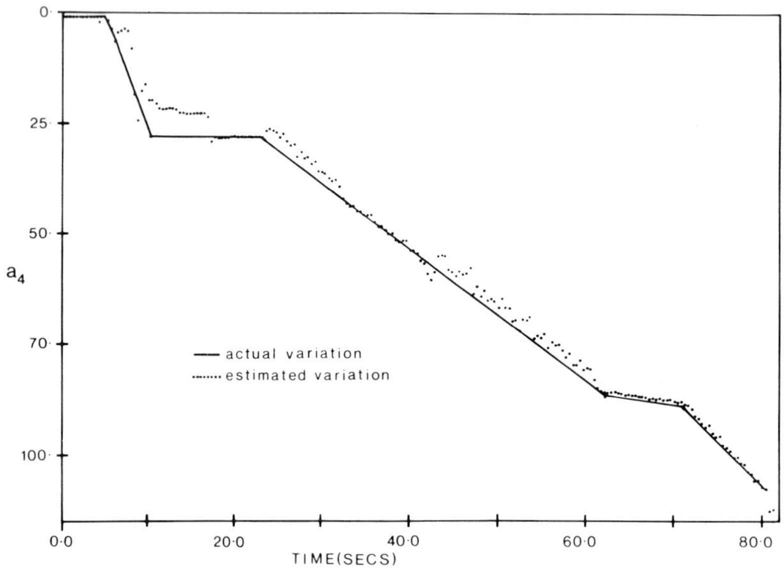

Figure 5.12 Recursive Estimate of Missile Model Parameter a_4

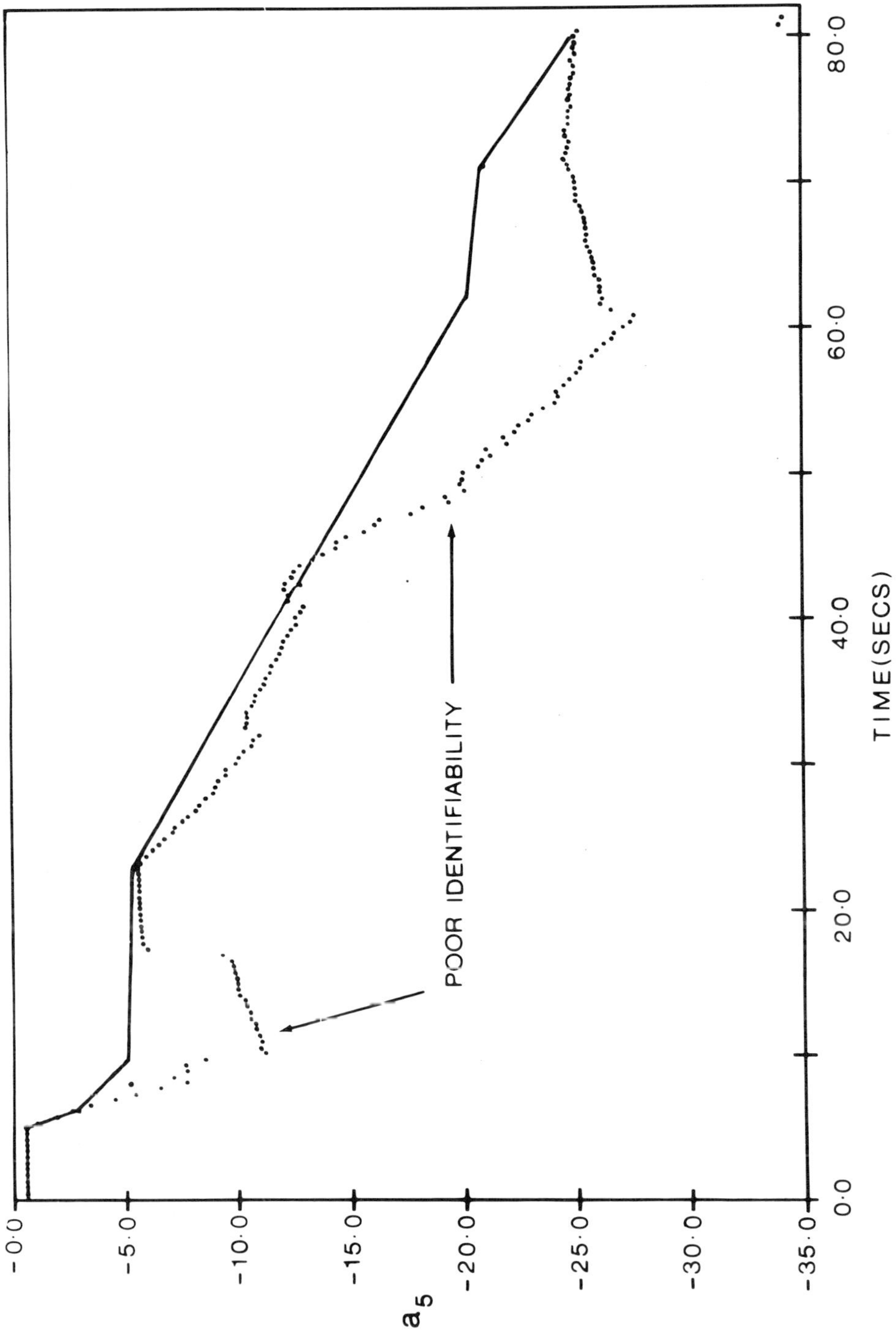

Figure 5.13 Recursive Estimate of Missile Model Parameter a_5

It will be noted that all parameters change particularly rapidly over the rocket boost period (5-10 seconds), when the Mach number and dynamic pressure are changing considerably (e.g. a_2 changes by a factor of 57.9). Also we see that, at various times over the mission, the parameter a_5 is estimated rather poorly: this arises because a_5 becomes relatively unimportant during these periods, having little effect on the dynamic behaviour of the missile. As a result the sensitivity of the measured state variables to a_5 is small so that the information on a_5 contained in the measured perturbations of the state variables is correspondingly limited.

This latter observation demonstrates how model parameters can sometimes become poorly "identifiable", in the sense that they either have exceptionally high error variance or, in the completely "unidentifiable" case, cannot be uniquely estimated at all. Identifiability in the present regression context is linked to the conditioning of the regressor covariance matrix X^TX in the normal equations (equation (4.2) of Chapter 4). Lack of identifiability corresponds to the complete singularity of this matrix and can depend both upon the chosen structure of the model and the nature of the observed data. The most common identifiability problem encountered in regression problems is probably that which arises from collinearity, as mentioned previously in Chapter 4. But identifiability is perhaps more interesting in a time-series context and so we will delay further discussion on this topic until we consider time-series models in the next Chapter.

More complex methods

The simple method of parameter decomposition discussed above may not always be suitable, although it appears to have quite wide application potential. We can however, conceive of more complicated approaches, but these will naturally result in more complex algorithmic forms for parameter estimation. For example, as in the case of the Kalman Filter, we might extend the Gauss-Markov model (5.19) to include a vector of deterministic inputs $\underline{u}_k = [u_{1k}, u_{2k}, \ldots u_{nk}]^T$ i.e.

$$\underline{a}_k = \Phi\underline{a}_{k-1} + \Omega\underline{u}_{k-1} + \Gamma\underline{q}_{k-1} \qquad (5.54)$$

where $\underline{u}_k$ could be a function of the additional information available about the parameter variations; for instance, in the missile example it could be composed of functions of the air data variables. As in the Kalman filter, the prediction equations of the general TVP algorithm VII would then need to be modified to incorporate this new term in the parameter variation equation, i.e.

$$\hat{\underline{a}}_{k/k-1} = \Phi\hat{\underline{a}}_{k-1} + \Omega\underline{u}_{k-1} \qquad (1A)$$

$$P^*_{k/k-1} = \Phi P^*_{k-1}\Phi^T + \Gamma Q_p\Gamma^T \qquad (2A)$$

VII

where, as in the equivalent Kalman filter case, the second covariance matrix equation is unchanged because $\underline{u}_k$ is assumed completely deterministic. Of course, this approach still leaves the analyst to specify the Φ, Γ and Q_p matrices, which is clearly not a simple task in general terms.

One simplifying possibility in this case would be to select Φ as a diagonal matrix with all diagonal elements ϕ_i less than unity, while Ω could also be chosen to be diagonal with elements w_i selected by experiment to fulfil a similar role to the normalization coefficients c_i in the missile experiment. For $\Gamma = I$, this would be equivalent to assuming that the a_{ik} parameters composing the vector $\underline{a}_k$ each vary according to a parameter variation law of the following type

$$a_{ik} = \phi_i a_{ik-1} + w_i u_{ik-1} + q_{k-1};\ i = 1,2,\ldots,n \qquad (5.57)$$

In other words, each of the coefficients a_{ik} are assumed to be related to the variations in u_{ik} via a stable first order dynamic process (low pass smoothing filter) with time constant $\tau = -T_s/\log\phi_i$ and steady-state gain $w_i/(1-\phi_i)$; once again a statistical degree of freedom is added to this relationship by incorporating the random variable vector $\underline{q}_{k-1}$ as an input to the equation (see Appendix 1).

This is certainly a simpler problem than that posed by the general equation (5.54), but it still requires the specification of at least 3n algorithmic parameters (ϕ_i, w_i and the diagonal elements of Q_p). In practice, the analyst might attempt to specify these parameters by first carrying out initial estimation using one of the simple random walk models (5.20), (5.31) and (5.32) for parameter variation. He could then attempt to estimate the parameters ϕ_i and w_i via *separate* estimation experiments on equations (5.57) using these estimates and the u_i variables. Alternatively he might, as in the missile example, have additional *a priori* information available which would enable him to pre-specify ϕ_i and w_i. It has even been suggested that recursive algorithms such as those discussed in this chapter should be made "adaptive", in the sense that the unknown coefficients of a parameter variation law such as (5.19) should be estimated *simultaneously* with the basic unknown regression parameters in $\underline{a}_k$. This is clearly possible but can add considerable complexity to the problem. One simple exercise of this general type has been described by Bennett (1976, 1977), who employs an adaptive method for estimating the noise covariance matrix Q_p. This approach is based on a procedure developed previously for use with Kalman filter algorithms by Jazwinski (1969, 1970) and Sage and Husa (1969). Bennett also discusses other adaptive procedures in his recent book (Bennett, 1979). At this point in time it is difficult to assess the potential of such adaptive procedures although it seems likely that their complexity may often make them unattractive in practical terms.

But we hope that, by this stage, the reader will have realized that the various possibilities are so numerous and so problem dependent that it serves little useful purpose to proceed any further with this kind of general speculation. We hope merely that by mentioning a small number of the many alternative approaches to this problem, we may have alerted the reader to the great flexibility of the TVP regression algorithms and that this will stimulate him to consider their use in data analysis. However two caveats are necessary in this regard; first, care should be taken to ensure either that the model is *indeed* of a regression form (or at least that the consequences of it *not* being a regression model have been fully considered); second, it has been said that the synthesis of recursive estimation algorithms, even in the constant parameter case, is a "fiddler's paradise" (Astrom and Eykhoff, 1971) - the reader should note this carefully and avoid the pitfall of getting more interested in "fiddling" with ever more sophisticated algorithmic mechanisms, rather than solving the real data processing problems with which he is confronted.

5.5 Simplified 'Constant Gain' Algorithms for Time Variable Parameter Estimation

During the past twenty years, a considerable amount of research has been devoted to the estimation of parameters in deterministic dynamic systems using continuous time recursive algorithms with constant gains of the type referred to in section 3.3 of Chapter 3 (see e.g. Young, 1979a; Narendra, 1976). The difficulties introduced by stochastic disturbances are rarely discussed in publications on this topic and it is only recently that their performance in stochastic terms has received any attention at all (see Young, 1979 a). In their discrete-time form, these algorithms can, however, be considered directly within the general framework of time variable parameter estimation discussed in this Chapter. The continuous-time situation is then directly analogous and can be interpreted in the same manner.

If we consider the nature of the algorithmic gains p_k in the scalar case as shown in Figures 5.3 and 5.4, we see that they approach a constant value asymptotically. And in this asymptotic situation, all three algorithms (RWP, EWP and TVP) behave in an identical and very simple manner. Indeed, given the filtering interpretation of recursive mean value estimation in section 2.1 of

Chapter 2, we can consider that all the algorithms in this situation are constant gain first order filters governed by the equation

$$\hat{a}_k = \hat{a}_{k-1} + (1-\alpha)\ [y_k - \hat{a}_{k-1}] = \alpha\hat{a}_{k-1} + (1-\alpha)\ y_k \qquad (5.58)$$

This is simply a low pass filter with unity steady state gain and time constant $-T_s/\log_e\alpha$; i.e. $\hat{a}_k$ is the output of a first order, linear, discrete time filter with equivalent continuous-time band width 0 to $-\log_e\alpha/T_s$ rad/sec which progressively attenuates all frequencies greater than $-\log_e\alpha/T_s$ rad/sec at 6 db's/octave (see a control text; e.g. Takahashi *et al.*, 1970).[†] In the purely deterministic case when e_{yi} in (2.2) is zero, then y_k in (5.40) is simply $\bar{a}$ and, from its initial condition $\hat{a}_0$, $\hat{a}_k$ simply approaches $\bar{a}$ exponentially for increasing k, eventually becoming identical to $\bar{a}$.

This deterministic case is not, of course, very realistic since with $e_{yi} \equiv 0$ we already know $\bar{a}$ and the algorithm (5.40) is redundant! But it does provide useful insight into the nature of the algorithm in the stochastic case: if y_k is a random variable then $\hat{a}_k$, being the output of a linear filter, will also be a random variable (e.g. Box and Jenkins 1970; Truxal, 1965) whose characteristics will depend upon the passband of the filter. And this in turn is governed by the asymptotic value of the algorithmic gain $(1-\alpha)$. For example, if $(1-\alpha) = 0.1$, as in our examples, and $T_s = 0.01$ second then the band width of the 'equivalent' continuous time system is 0 to 10.54 rad/sec.[†] The attenuation of the noise effects can best be seen by evaluating the mean square value $\sigma_{\hat{a}}^2$ of $\hat{a}_k$. The reader can verify that $\sigma_{\hat{a}}^2$ is given, in this case, by

$$\sigma_{\hat{a}}^2 = \frac{(1-\alpha)^2}{1-\alpha^2}\,\sigma^2\ ; \qquad \text{Hint:}\quad E\{\hat{a}_k^{\ 2}\} = E\{[\alpha\hat{a}_{k-1} + (1-\alpha)y_k]^2\} \text{ and } E\{\hat{a}_{k-1}y_k\} = 0$$

where σ^2 is the variance of y_k. For $\alpha = 0.9$, therefore, $\sigma_{\hat{a}}^2 = .053\sigma^2$ and the residual variance of the estimate $\hat{a}_k$ about the mean $\bar{a}$ (the true value) is quite small in relation to the 'observation' noise variance σ^2. And at the same time the convergence from an initial large error in the estimate $\tilde{a}_0 = \hat{a}_0 - \bar{a}$ will take place with a reasonable time constant of $-1/\log_e \alpha = 9.49$ sampling intervals (0.09 secs)

Of course, by passing some of the noise in this manner the algorithm is also able to detect variations in $\bar{a}$. And these variations should be apparent to the analyst provided they are of a different nature to and larger than the noise effects. In general this will be the case, with the noise tending to have wide

---

[†]Note that here, for simplicity, we are referring to the frequency response characteristics of the 'equivalent' continuous-time system. This assumes that the input (y_k here) is constant between samples (Appendix 1), which may not be the case. But the overall argument is not affected by this simplification.

band characteristics with power at the higher frequencies while the changes in $\bar{a}$ will usually be in the form of low frequency 'trends' or slow variation,as in the Walgett rainfall case (Section 5.4).

The situation is, of course, much more complicated in the multi-parameter case with a time varying vector of regressors $\underline{x}_k$ (or matrix Π_k). It is still reasonable to assume, however, that a constant gain matrix P can be selected which, in some sense, 'approximates' the P_k matrix for large k and which yields asymptotic results similar to those of the algorithm with time varying P_k. In one important case, namely the situation where the $\underline{x}_k$ (or Π_k) are constant and it is possible to assume a random walk model (5.17) for the parameter (state?) variations, the asymptotic P matrix is constant. Here it would be possible to compute the asymptotic gain matrix, as in the simple example above, and use this in the implementation of the constant-gain algorithm.

The main disadvantage of constant gain algorithms lies in their initial transient performance. As we have seen, the transient behaviour of recursive algorithms is a function of the recursive gain and if this is small, as it may well be when asymptotically equivalent values are utilized, then the transient convergence could be quite slow in certain circumstances. Thus, if the initial estimate $\hat{\underline{a}}_0$ is in error and far from the true value $\underline{a}$, then the initial convergence could be deliteriously affected, since the low gains in P would be unable to correct the errors in $\hat{\underline{a}}_k$ very quickly. In this situation, we would need to revert to time variable gains either as in the RWP, EWP and TVP situations, or via some form of scalar SA like gain.

The main advantage of the constant gain algorithms is their simplicity and computational efficiency, and they clearly have a potential role to play where low storage capacity microprocessors are being used in on-line applications, e.g. adaptive control. Here the user would need to ensure that the asymptotically equivalent gain matrix does not lead to problems: either by guaranteeing that the initial error $\tilde{\underline{a}}_0 = \hat{\underline{a}}_0 - \underline{a}$ is small enough to be corrected adequately in a reasonable time; or by employing some simple variable gain mechanism, similar to that used in the EWP case of equations (5.12) and (5.13).

5.6 Recursive Smoothing Algorithms

At any instant k, the recursive estimate $\hat{\underline{a}}_k$ obtained from algorithm VII is based on the data up to and including the kth sample, but it is clearly *not* a function of the future data k+1, k+2, etc. In the situation where the data are being processed 'on-line' as they are received (for example from some measurement device or system) then this is reasonable since the future data are not yet available. But if we are recursively processing a block of T samples collected during some prior experiment or monitoring exercise, then it is tempting to consider how we might generate an 'off-line' recursive estimate at sample k which is a function of *all* the available data, i.e. using our previous terminology, the estimate of $\hat{\underline{a}}_k$ given *all* the data, or $\hat{\underline{a}}_{k/T}$. Such an estimate is termed a "smoothed" recursive estimate and the corresponding estimation algorithm is referred to as a "recursive smoothing algorithm". Note that such a smoothed recursive estimate is only relevant in the time-variable parameter (or state) estimation context: clearly if the parameters are constant, then $\hat{\underline{a}}_{k/T} \equiv \hat{\underline{a}}_T$ for all k.

Much has been written on smoothing algorithms, particularly in relation to the Kalman Filter and we will not deal with the subject in detail in these Lecture Notes. A good introductory treatment is given in Gelb et al (1974) who point out that the smoothed estimate at time k can be considered as the optimal combination of the recursive estimate obtained up to k, starting at the beginning of the data (k=0), and the estimate obtained working backwards through the data from T to k.

An alternative method of derivation is to consider the problem from a Bayesian viewpoint (Rauch, Tung and Streibel, 1965; Bryson and Ho 1969). We have stressed previously that recursive estimation can be considered as the embodiment of Bayesian estimation and that the Kalman filter, for example, can be obtained in a rigorous fashion from Bayesian concepts. In the smoothing case, however, the solution is not so well defined as in the previous recursive (or filtering) context. Norton (1975) has discussed this matter and compared various solutions to the problem which differ in computational and numerical stability terms. He concludes that the most useful form of the smoothing algorithm in the TVP regression case is that derived by Bryson and Ho (1969) under the condition that the probability densities $p(\underline{a}_0)$, $p(e_k)$ $p(q_k)$ are all Gaussian. In this case, the conditional density of the 'filtered' estimate $\hat{\underline{a}}_k$ is also Gaussian and characterised completely by its mean and covariance matrix. And it is this conditional mean which is used as the 'filtering' estimate and is generated by either algorithm VII or the Kalman filter 'equivalent' algorithm X.

In the smoothing case, the problem can be posed in optimization terms and, in the solution of this optimization problem, the smoothed estimates are obtained from the results of an initial recursive (filtering) run through the data, by the equations

$$\hat{\underline{a}}_{k/T} = \hat{\underline{a}}_k - P_k^* \Phi^T \underline{\lambda}_k \qquad (1)$$

or

$$\hat{\underline{a}}_{k/T} = \Phi^{-1}(\hat{\underline{a}}_{k+1/T} + \Gamma Q_p \Gamma^T \underline{\lambda}_k) \qquad (2)$$

XI

The associated error covariance matrix $P^*_{k/T}$ can be generated in a number of ways;

for example Rauch *et al.*(1965) use

$$P^*_{k/T} = P^*_k + P^*_k \Phi P^{*\,-1}_{k+1/k} (P^*_{k+1/T} - P^*_{k+1/k}) P^{*\,-1}_{k+1/k} \Phi P^*_k \qquad (3) \qquad \text{XI}$$

where P^*_k and $P^*_{k+1/k}$ are the covariance matrices generated by the basic recursive algorithm VII (where strictly, using this nomenclature, P^*_k should be denoted by $P^*_{k/k}$ to avoid ambiguity). The vector $\underline{\lambda}_k$ in algorithm XI is a Lagrange multiplier term required in the solution of the optimization problem in order to ensure that the estimate $\hat{\underline{a}}_{k/T}$ is obtained under the constraint that the parameter vector evolves according to the Gauss-Markov equation (5.19); it is obtained recursively from the equations

$$\underline{\lambda}_{k-1} = (I_n - P^*_k \frac{\underline{x}_k \underline{x}_k^T}{\hat{\sigma}^2})^T (\Phi^T \underline{\lambda}_k - \frac{\underline{x}_k}{\hat{\sigma}^2} (y_k - \underline{x}_k^T \Phi \hat{\underline{a}}_{k-1})) \qquad (4) \qquad \text{XI}$$

$$\underline{\lambda}_T = 0$$

Other derivations are considered by Norton but are not favoured because of numerical and computational aspects.

In contrast to the forward filtering pass, the covariance matrix equation XI(3) is not essential to the computation of the estimate, as can be seen from the above algorithm, and need only be computed if required to assess the error covariance properties. The fact that the $P^*_{k/T}$ will always be less than P^*_k may be sufficient to put bounds on the smoothed estimates which are satisfactory for practical purposes; if so, this will introduce considerable computational savings. Alternatively it is possible to employ a slightly lengthier algorithm in place of XI(3) which avoids matrix inversion and seems computationally more attractive (Bryson and Ho, 1969).

But these are all clearly problem dependent factors. They serve to emphasize, however, that the recursive smoothing algorithms involve considerable additional computational burden in comparison to their filtering relatives and that they should be used with discretion where they will be of maximum assistance (see e.g. Jakeman and Young, 1980) Fig. 5.14 shows the recursive smoothing results obtained in the above manner for the Walgett rainfall data (c.f. Figure 5.7).

5.7 Statistical Methods for Detecting the Presence of Parameter Variation

En-bloc procedures for testing whether statistically significant parameter variation occurs over a set of observations have received a considerable amount of

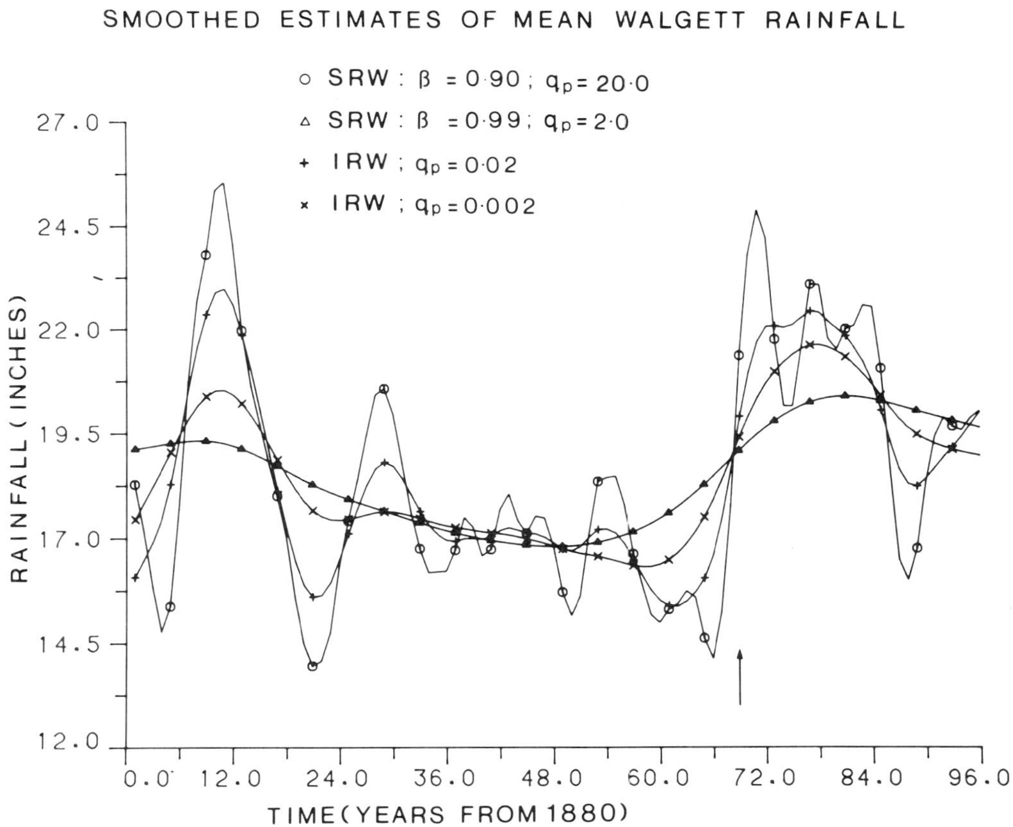

Figure 5.14 Recursive Smoothed Estimates of Mean Value for Walgett Rainfall

attention. Usually the data set is split into subsections and tests are applied to see if there is a significant difference in the estimates obtained from the different subsections. There appears to be three main sets of test statistics available: those based on the 't' test of residual means (e.g. Quandt, 1960); those based on analysis of variance (e.g. the 'F' test of Quandt, (1960) or its extension by Chow (1960)) as a test of so-called "structural change" - see also Harvey (1976); and those based on analysis of variance in the frequency domain using "evolutionary spectra" (e.g. Priestley and Rao, 1969).

Recursive methods for testing for parameter variation are of more recent origin and do not have such a rich history. Here we will consider only one typical procedure, as suggested by Brown *et al.* (1975), which will help to exemplify the kind of approach which is possible in the recursive situation.

As we pointed out in Chapter 3, the recursive residuals or innovation process $\hat{e}_k$ should, if the assumptions regarding the regression problem are satisfied, be a zero mean, serially uncorrelated sequence of random variables with changing variance. Furthermore, if it is assumed that the errors of observation e_{yk} have a normal Gaussian amplitude distribution $N(0, \sigma^2)$ then the innovations will also be Gaussian. Brown *et al.* show that the following normalized innovations sequence $\hat{e}_{Nk}$ where

$$\hat{e}_{Nk} = [1 + \underline{x}_k^T P_{k-1}\underline{x}_k]^{-\frac{1}{2}} \{y_k - \underline{x}_k^T\hat{\underline{a}}_{k-1}\}$$

is, under the assumption of parametric invariance, distributed as $N(0, \sigma^2)$. As a result, they suggest that a test for detecting a departure from the zero expectation can function as a test for parametric change. Such a test can be derived from the cumulative sum or CUSUM, W_k, where

$$W_k = \frac{1}{\hat{\sigma}} \sum_{i=h+1}^{k} \hat{\underline{e}}_{Ni} \quad \text{with } \hat{\sigma}^2 = \frac{1}{T-h} \sum_{i=h+1}^{T} \hat{e}_{Ni}^2 = \frac{S_T}{T-h}$$

where e_{Ni} are obtained from the normal constant parameter recursive least squares algorithm III (Chapter 4) for each point i, and h is the number of estimated parameters. As might be expected, a plot of W_k against k will yield a graph which varies about zero if the parameters are indeed time-invariant, but which moves away from zero if permanent parametric change occurs. The significance of these movements is tested by the application of a procedure in which the probability limits are given by the lines joining the points $\{h, \pm a(T-h)^{\frac{1}{2}}\}$ and $\{T, \pm 3a(T-h)^{\frac{1}{2}}\}$, where 'a' is a coefficient given by Brown et al for various significance levels α, as shown below

$\alpha = 0.01, \quad a = 1.143$

$\alpha = 0.05, \quad a = 0.948$

$\alpha = 0.10, \quad a = 0.850$

Brown *et al.* also suggest a CUSUM - squared test (as well as other tests based on likelihood ratios which we will not consider here). In this case, it is argued that the cumulative sum of squares s_k , where

$$s_k = \frac{1}{S_T} \sum_{i=h+1}^{k} \hat{e}^2_{Ni}$$

is distributed as a beta distribution in the time-invariant parameter case, with a mean of (k-h)/(T-h) for h parameters. This follows a straight line on the plot of s_k against k and probability limits are given by drawing lines parallel to the mean value line at distances $\pm c_0 + (k-h)/(T-h)$, where c_0 is a parameter given in Table 5.1 for different significance levels α and sample sizes T.

Table 5.1

Probability of Crossing CUSUM-squared Test Lines at Significance Levels α (after Durbin, 1969; see Brown et al., 1975)

T	SIGNIFICANCE LEVEL α		
	0.1	0.05	0.01
10	0.259	0.302	0.385
15	0.224	0.261	0.331
20	0.201	0.233	0.295
25	0.184	0.213	0.269
30	0.170	0.197	0.249
40	0.151	0.174	0.219
50	0.134	0.158	0.198
60	0.126	0.145	0.182
70	0.118	0.135	0.170
100	0.100	0.115	0.144

In practice the CUSUM and CUSUM-squared graphs are plotted with associated significance bands, chosen in accordance with the requirements of the analyst. If the test statistic crosses or comes close to the probability limits, then the hypothesis of parametric invariance is rejected and the possibility of parametric change needs to be investigated further. Brown et al point out that the CUSUM-squared test is likely to be more sensitive to noise effects and changes in variance. They also stress that the significance bands are only approximate guides since the innovations are estimated model residuals,and tests such as those of Kolmogorov-Smirnov may be affected by this factor.

Of course, the failure of the above tests to indicate the presence of parametric non-stationarity does not mean that parameter variation has not occurred; nor does the availability of such tests negate the need for evaluating parametric change by the application of the TVP algorithms. As Brown *et al.* agreed in discussion on their paper (see Young,1975), the recursive estimates of the parameters are themselves valuable indicators of parameter change and should always be inspected. What is called for, however, is more theoretical research on statistical procedures for evaluating the significance of estimated changes on the *individual* parameters, as well as in relation to the regression function as a whole. And it might be better in this case to develop tests applied to the results obtained from the TVP algorithm where the constraint of a strictly decreasing P_k matrix is removed and deviations in parameter estimates are made more easily observable. This is, however, a difficult problem to formulate in statistical terms and there does not yet appear to have been any research on such an approach.

If we consider the application of the CUSUM and CUSUM - squared tests to the Walgett data then the results are a little ambiguous. Even with $\alpha = 0.1$ the CUSUM plot does not transgress the significance band; on the other hand the CUSUM-squared plot does come very close to the lower boundary for $\alpha = 0.1$ at $k = 65 \rightarrow 68$ (1946 - 1948) indicating that the change in mean is significant at about the 90% confidence level. Note, however, that the procedure is not strictly applicable in this case since the least squares residuals are not clearly white (serially uncorrelated), as required by the assumptions.

5.8 Summary

In this long chapter, we have explored the many aspects of time variable parameter estimation from the simple 'fading memory' approach,to the optimal state variable estimation procedures of R.E. Kalman. The object has not just been simply to derive algorithms but more to provide the reader with an insight into *why* and *how* they work. In this manner, it is hoped that both the synthesis and the use of the various algorithms will become more transparent.

But the discussion has been limited in one important practical respect: we have restricted the analysis to the 'general linear regression model' and have not addressed the important problem of errors-in-variables mentioned in Chapter 4. This will be corrected later in Chapters 7 and 8, which show how it is possible to modify the regression algorithms to handle the very real errors-in-variables problems that arise naturally when attempting to estimate parameters in certain important time-series models of stochastic dynamic systems.

Part 2
Recursive Time-Series Analysis

6 The Time-Series Estimation Problem

In the first part of this book, we have considered in some detail the recursive estimation of parameters in linear regression models. Such models are, however, primarily utilized in the evaluation of static relationships between variables and, while they may well figure in some aspects of systems analysis (for example the characterization of equilibrium (steady state) behaviour) they are not generally suitable for use in dynamic systems analysis. Of course, dynamic systems can be modeled in various ways and so consideration of methods for estimating the parameters of dynamic system models is dependent to some extent at least on the nature of the model chosen to characterize the system. In this Chapter we will consider a specific class of models for linear stochastic dynamic systems and show how the recursive methods of estimation discussed in the previous chapter can be modified to handle the estimation of the parameters which characterize these models.

In the classical statistical literature, the models we shall consider have been termed 'time-series' representations since they attempt to describe a time-series of data, i.e. data arranged in temporal order (see e.g. Box and Jenkins 1970). While our treatment of these models will be similar to that used in statistical texts, its emphasis will be rather different: the classical approach tends to concentrate on the stochastic aspects of the time-series model, while here we emphasize the relationship between the assumed deterministic input or "exogenous" variables and that part of the measured output of the stochastic dynamic system that can be "explained" in relation to these inputs. This emphasis arises because we are concerned here with the mathematical characterization of stochastic dynamic systems for the purposes of designing control and management schemes; and it contrasts with much of the statistical literature which concentrates on the problem of forecasting or predicting the output variables.

The reader has already been introduced to one type of linear stochastic dynamic model in the last chapter: namely the discrete time Gauss-Markov or state-space model used both as the vehicle for the development of the TVP estimation algorithm IX and the Kalman state variable filter algorithm X. This seems, therefore, an appropriate starting point for our present discussion. If we consider the description (5.43) used in relation to the Kalman filter, then this will serve to emphasize that we are concerned here with the modeling of a general, linear, stochastic dynamic system; rather than considering the model simply as a device for representing the variations in parameters. In this manner, we can retain the dynamically equivalent but symbolically different representation (5.19) for the latter purpose and consider its use in time-variable parameter estimation in later Chapters.

6.1 The Time-Series Model in the Observation Space

To begin with, we will assume that the dynamic system is stable and stationary so that the relevant state-space description is (5.43) with A, C and D all matrices composed of time-invariant real numbers and with the eigenvalues of the characteristic equation

$$\det[I - Az^{-1}] \triangleq 1 + a_1 z^{-1} + \dots + a_n z^{-n} = 0 \tag{6.1}$$

all lying outside the unit circle in the complex plane.[†] In other words, if we consider that the system has deterministic exogenous inputs $\underline{u}_k$ (see (5.45), page 80), then the model takes the form

$$\left.\begin{aligned} \underline{x}_k &= A\,\underline{x}_{k-1} + B\underline{u}_{k-1} + D\underline{\eta}_{k-1} && \text{(i)} \\ \underline{y}_k &= C\,\underline{x}_k + \underline{\xi}_k && \text{(ii)} \end{aligned}\right\} \tag{6.2}$$

Nominally, the problem of parameter estimation for this model (6.2) is to infer, in some statistical manner, the parameters of the A, B, C and D matrices, on the basis of the measurements $\underline{u}_k$ and $\underline{y}_k$ over some observational interval, say $k = 1,2,\dots,T$. In his 1960 paper, Kalman recognized this as a difficult and, at that time, unsolved problem when he noted that his approach to state variable estimation was limited by its assumption that the model parameters were known *a priori*. He concluded, "to be sure the two problems (state *and* parameter estimation) should be optimized jointly if possible; the author is not aware, however, of any study of the joint optimization problem".

Methods for joint recursive estimation of the states and parameters in equations such as 6.2 have since become available. The Extended Kalman Filter (EKF), for example, is one relatively straightforward approach in which the state vector $\underline{x}$ is augmented to include an unknown parameter vector $\underline{a}$ which includes all the unknown elements in the model matrices A, B, C and D; estimation then proceeds using a Kalman filter-like algorithm obtained by linearizing the now non-linear relationships (arising from the products between the unknown parameters and the unknown states) about the current estimates at each recursive step.

[†] i.e. equivalent to the roots of $z^n + a_1 z^{n-1} + \dots + a_n = 0$ lying *inside* the unit circle in the complex plane.

The EKF approach originated in the work of Kopp and Orford (1963) and has since been applied relatively successfully to the modeling of various real dynamic systems. But it does not always work satisfactorily and is known to have poor statistical efficiency, in the sense that its parameter estimates can have rather high error variance when compared with the theoretically possible minimum variance. Superior results in this sense can be obtained, albeit at some considerable additional cost in complexity, by considering the problem from a maximum-likelihood (ML) standpoint, as suggested by Mehra and Tyler (1973) and Astrom and Kallstrom (1973). Such an approach is nominally recursive, but it also involves complex hill climbing procedures and cannot be considered as realistic for on-line use. We will describe both the ML and EKF methods in a little more detail when we discuss other recursive estimation techniques in Chapter 9. For the moment, however, we will evaluate alternative and somewhat simpler approaches to the problem.

The major difficulty with the estimation of parameters in (6.2) arises from its relative complexity as a description when viewed in parametric estimation terms. Because the model explicitly includes the unmeasured states comprising the state vector $\underline{x}_k$, the problem is clearly nonlinear from an estimation standpoint with the nonlinearity, as we have remarked above, arising from the product terms involving the unknown states *and* parameters. What we would prefer is a representation in which the unknown parameters are, wherever possible, associated with *measured* variables in a linear model. But is such a representation possible; or is the problem inherently nonlinear? To answer the question, it is necessary to attempt some simplification of the model (6.2).

One simplification arises if all the state variables are either directly measureable, or if n independent linear combinations of the state variables are measurable, i.e., if the observation matrix C in equation (6.2)(ii) is either the n x n identity matrix I_n, or is an n x n non-singular matrix. In either case, C^{-1} exists so that we can then substitute from (ii) into (i) and obtain, in the general case,

$$\underline{y}_k = CAC^{-1}\underline{y}_{k-1} + CB\underline{u}_{k-1} + \underline{\rho}_k \tag{6.3}$$

where

$$\underline{\rho}_k = (Iz - CAC^{-1})\underline{\xi}_{k-1} + CD\underline{\eta}_{k-1} \tag{6.4}$$

Now, if we define $A_1 \triangleq CAC^{-1}$ and $B_1 = CB$ (note $A_1 = A$ and $B_1 = B$ if $C = I_n$), then it is possible to consider the problem of estimating the coefficientis in A_1 and B_1 by linear least squares methods since they are now associated directly with the measured variables in $\underline{y}_k$ and $\underline{u}_k$. The problem is still not trivial, however, since it is clear that the noise term ρ_k is quite complex in statistical terms being both serially correlated, and also correlated with the measurement $\underline{y}_k$. In other words, the problem is still rather difficult in the sense that the least squares estimates of the parameters in A_1 and B_1, although easily obtainable by a normal regression approach, will be asymptotically biased and inconsistent. Consistent estimation will require either simultaneous estimation of the noise model (6.4), which involves both unknown parameters and associated *unmeasurable* noise variables $\underline{\xi}_k$ and $\underline{\eta}_k$ (clearly a nonlinear problem) or the use of some device, such as the instrumental variable (IV) method mentioned in Chapter 4, which yields a linear-like solution to the problem (Young and Whitehead, 1977). The first approach is clearly complicated; while discussion of the second, using instrumental variables, is best deferred until later in Chapter 7.

But in any case, the model (6.3) obviously presents a very special case and it is unlikely in practice that such fortuitous circumstances will often be encountered. It seems advisable, therefore, to consider further the more general case where the dimension of $\underline{y}_k$ is *less* than the dimension of the state $\underline{x}_k$. In this connection, the model (6.3) gives us one clue for further progress, since it is a model which directly relates the *measured* variables: it is, in other words, a model in the 'observation space' rather than the 'state space', with the model dimension happening to coincide with the dimension of the state-space simply because of our rather unreal assumption that dim. $\underline{y}_k \equiv$ dim $\underline{x}_k$.

If the dimension of $\underline{y}_k$ is less than $\underline{x}_k$, however, we can still obtain an observation space description. This is most easily accomplished by introducing the z^{-1} operator and converting (6.2) into operator notation form. In the simplest single input, single output (SISO) case, where $\underline{y}_k$, $\underline{u}_k$ and $\underline{\eta}_k$ are scalars rather than vectors, we obtain the following representation:

$$\left.\begin{aligned} [I - Az^{-1}]\underline{x}_k &= \underline{b}z^{-1}u_k + \underline{d}z^{-1}\eta_k && \text{(i)} \\ y_k &= \underline{c}^T \underline{x}_k + \xi_k && \text{(ii)} \end{aligned}\right\} \quad (6.5)$$

where $\underline{b}$, $\underline{d}$ and $\underline{c}$ are vectors which replace the matrices B, D and C in this scalar or SISO situation. Rearranging (6.5)(i) and substituting from (ii), we then obtain

$$y_k = \underline{c}^T[I - Az^{-1}]^{-1} \underline{b}z^{-1}u_k + \underline{c}^T[I - Az^{-1}]^{-1} \underline{d}z^{-1}\eta_k + \xi_k \quad (6.6)$$

which, upon evaluation of the matrix inverse $[I - Az^{-1}]^{-1}$ (Appendix 1), yields

$$y_k = \frac{1}{A(z^{-1})}\underline{c}^T \text{Adj}[I - Az^{-1}]\underline{b}z^{-1}u_k + \frac{1}{A(z^{-1})}\underline{c}^T \text{Adj}[I - Az^{-1}]\underline{d}z^{-1}\eta_k + \xi_k \quad (6.7)$$

where $A(z^{-1}) = \det[I-Az^{-1}]$ is the characteristic polynomial in z^{-1} which will have the following general form

$$A(z^{-1}) = 1 + a_1z^{-1} + a_2z^{-2} + \ldots\ldots + a_nz^{-n} \quad (6.8)$$

with a_i, $i = 1,2,\ldots,n$ defined accordingly, while Adj [.] denotes the Adjoint of the matrix.

Whilst superficially complicated, equation (6.7) is really quite simple. We can see this by noting that the state space description is not unique and that there are, therefore, an infinite number of different representations for A, $\underline{b}$, $\underline{d}$ and $\underline{c}$ which can provide identical input-output behaviour (see Ogata, 1967). As a result, we can choose a specific,simple but general 'canonical' (minimally parameterized and uniquely identifiable) form for these matrices and then evaluate (6.7) directly using these definitions. Numerous such canonical forms are available but the most convenient in the present context is the following (see e.g. Young, 1979b):

$$A = \begin{bmatrix} -a_1 & 1 & 0 & & 1 \\ -a_2 & 0 & 1 & & 0 \\ \cdot & & & \ddots & \cdot \\ \cdot & & & & 1 \\ -a_n & 0 & \cdot & \cdot & 0 \end{bmatrix} \qquad \underline{b} = \begin{bmatrix} b_1 \\ b_2 \\ \vdots \\ b_n \end{bmatrix} \quad (6.9)$$

$$\underline{d} = \begin{bmatrix} d_1 \\ d_2 \\ \vdots \\ d_n \end{bmatrix} \qquad \underline{c}^T = [1\ 0\ldots\ldots0]$$

The reader can verify by evaluation of (6.7) with these definitions that

$$y_k = \frac{B(z^{-1})}{A(z^{-1})}u_k + \frac{D(z^{-1})}{A(z^{-1})}\eta_k + \xi_k \quad (6.10)$$

where $B(z^{-1})$ and $D(z^{-1})$ are polynomials in the z^{-1} operator of the form

$$B(z^{-1}) = b_1 z^{-1}+.....+b_n z^{-n} \tag{6.11}$$

$$D(z^{-1}) = d_1 z^{-1}+.....+d_n z^{-n}$$

From (6.10) we see that

$$A(z^{-1})y_k = B(z^{-1})u_k + D(z^{-1})\eta_k + A(z^{-1})\xi_k \tag{6.12}$$

so that, by substituting from (6.8) and (6.11) and then expanding in discrete-time terms, we obtain

$$y_k + a_1 y_{k-1} +.....+a_n y_{k-n} = b_1 u_{k-1}+....+b_n u_{k-n} + \mu_k \tag{6.13}$$

where $\mu_k = d_1\eta_{k-1} +....+d_n\eta_{k-n} + \xi_k + a_1\xi_{k-1} + ... + a_n\xi_{k-n}$.

Equation (6.13) represents a simple linear relationship between the measured variables y_{k-i} and u_{k-j}, $i = 0,1,2, ..n$, $j = 1,2,....,n$ and the unknown system parameters a_i, $i = 1,2,....,n$ and b_j, $j = 1,2,...,n$. But once again, as in the case of equation (6.3), the noise term μ_k is not simple in statistical terms, being a linear function of the *two* white noise source terms η_k and ξ_k.

6.2 Various Observation Space Model Forms

At this stage, it seems unlikely that we can avoid the presence of statistically complicated noise terms if we are to obtain a general linear representation of the system in terms of the basic system parameters a_i and b_i. But we can note that μ_k in (6.13) is the sum of two independent 'moving average' (MA) noise processes, i.e.

$$\mu_k = \mu_{1k} + \mu_{2k} \tag{6.14}$$

where $\mu_{1k} = d_1\eta_{k-1}+....+d_n\eta_{k-n}$ and $\mu_{2k} = \xi_k + a_1\xi_{k-1}+....+a_n\xi_{k-n}$. Here the term 'moving average' has arisen in the statistical literature on time-series because μ_{1k} and μ_{2k} are, respectively, composed of weighted sums of the random white noise variables η_{k-i} and ξ_{k-i}, i.e. μ_{1k} and μ_{2k} are both linearly dependent upon a (usually) finite number of previous white-noise terms. As Box and Jenkins (1970) point out, the term "moving average" is rather misleading, although in very common use, since the weighting coefficients $d_1,\ldots,d_n$ and $a_1,\ldots,a_n$ need not sum to unity, nor need they be positive, as would be the case if the expressions were true moving "averages". But like Box-Jenkins, we will bow to convention and continue to use the expression.

Since it is well known (e.g. Granger and Morris, 1976) that the sum of any number of independent moving average processes is itself a moving average process, then we can simplify the expression (6.13) somewhat by replacing μ_k by a moving average in a *single* white noise source e_k which is different from both η_k and ξ_k but is still zero mean and serially uncorrelated. Such a representation can, however, be obtained in an alternative manner which yields greater insight in statistical terms and provides yet another link with our previous discussion of the Kalman filter.

In his original derivation, Kalman did not derive the filter exactly in the form of algorithm X; rather he obtained the estimate $\hat{\underline{x}}_k$ given information *up to but not including* the kth observation (see Sage (1968), page 273). If we consider this form of the Kalman filter algorithm for the model system (6.9) then

$$\hat{\underline{x}}_k = A\hat{\underline{x}}_{k-1} + \underline{b}\, u_{k-1} + \underline{g}_k(y_{k-1} - \underline{c}^T \hat{\underline{x}}_{k-1}) \tag{6.15}$$

where $\underline{g}_k$ is the Kalman gain vector given by

$$\underline{g}_k = P^*_{k-1}\, \underline{c}[r + \underline{c}^T P^*_{k-1}\, \underline{c}]^{-1} \tag{6.16}$$

with r the variance of scalar noise affecting y_k, and P^*_{k-1} the covariance matrix associated with $\hat{\underline{x}}_{k-1}$. If we now define the recursive residual or innovations process e_k as

$$e_k = y_k - \underline{c}^T\hat{\underline{x}}_k$$

then it is clear that,

$$\hat{\underline{x}}_k = A\hat{\underline{x}}_{k-1} + \underline{b}u_{k-1} + \underline{g}_k e_{k-1} \qquad \text{(i)}$$

and

$$y_k = \underline{c}^T\hat{\underline{x}}_k + e_k \qquad \text{(ii)} \tag{6.17}$$

Equations (6.17) constitute an alternative state-space representation of the system with the new states defined as the *estimates* or 'conditional expectations' of the states comprising $\underline{x}$ rather than the original state variables themselves. Here the term 'conditional expectations' arises because $\hat{\underline{x}}_k$ can be considered as the best estimate of the random process $\underline{x}_k$ conditional on all the time-series data up to sampling instant k-1, i.e. $\hat{\underline{x}}_k = E\{\underline{x}_k/Y_{k-1}\}$, where Y_{k-1} represents all of the time-series data available up to k-1.

But now we see that (6.17) is rather simpler than the original description in stochastic terms since there is only one noise term, namely the innovations process e_k. For this reason, it is usually termed the 'innovations' or 'Kalman filter' description (see e.g. Astrom, 1970). The innovations process e_k is known to possess white noise properties but, whilst serially uncorrelated it does have changing variance because of the variable nature of the P_k^* covariance matrix. However, if the system and noise parameters are stationary, P_k^* and hence $\underline{g}_k$ will achieve a constant asymptotic value $P_k^* = P_\infty^*$ and $\underline{g} = \underline{g}_\infty$, respectively, for k tending to infinity and will be virtually constant for moderate k values. Consequently, in this condition, e_k will be a zero mean, white noise sequence with constant variance, $r = \sigma^2$ i.e.

$$E\{e_j e_k\} = \sigma^2 \delta_{jk}$$

where $\delta_{jk} = \{\begin{smallmatrix} 1, & j=k \\ 0, & j \neq k \end{smallmatrix}\}$ is the Kronecker delta function.

It is now quite straightforward to derive the observation-space transfer function description of (6.17) by introducing the z^{-1} operator as before. The reader can verify that this yields a model of the form

$$y_k = \frac{B(z^{-1})}{A(z^{-1})} u_k + \frac{D(z^{-1})}{A(z^{-1})} e_k \qquad \text{(i)}$$

or

$$A(z^{-1}) y_k = B(z^{-1}) u_k + D(z^{-1}) e_k \qquad \text{(ii)} \qquad (6.18)$$

where $A(z^{-1})$ and $B(z^{-1})$ are defined as before, while $D(z^{-1})$ is now of the form

$$D(z^{-1}) = 1 + d_1 z^{-1} + \ldots + d_n z^{-n} \qquad (6.19)$$

and

$$d_i = a_i + g_i \qquad (6.20)$$

where g_i are the elements of the Kalman gain vector $\underline{g}$, i.e.

$$\underline{g}^T = [g_1,g_2,\dots,g_n]$$

Converting (6.18)(ii) into discrete-time model form, we obtain

$$y_k = -a_1 y_{k-1} \dots -a_n y_{k-n} + b_1 u_{k-1} + \dots + b_n u_{k-n} + \eta_k \tag{6.21}$$

where

$$\eta_k = e_k + d_1 e_{k-1} + \dots + d_n e_{k-n} \tag{6.22}$$

This is the required observation-space description which is linear in the a_i and b_i parameters and characterized by a noise disturbance η_k defined in terms of a 'moving average' of the present and past values of the white noise source, e_k.

Equation (6.18) is well known in statistical texts (Hannan, 1970) as the 'autoregressive - moving average - exogenous variables' (ARMAX) description. This terminology arises because the noise effect on y_k appears in (6.18) as white noise e_k passed through a linear dynamic system or filter, with transfer function $D(z^{-1})/A(z^{-1})$. In other words, if we denote this noise term by ξ_k (note $\neq \eta_k$), then

$$\xi_k = \frac{D(z^{-1})}{A(z^{-1})} e_k \tag{6.23}$$

or, in discrete-time equation form,

$$\xi_k = -a_1 \xi_{k-1} \dots -a_n \xi_{k-n} + e_k + d_1 e_{k-1} + \dots + d_n e_{k-n} \tag{6.24}$$

Here the ξ_k at the latest kth sampling instant is a linear function of the n previous past values of itself and a 'moving average' of the present and n previous past values of the white innovations variable e_k. Because of algebraic similarity with a linear regression relationship, the linear function of the ξ_k is termed an 'auto-regression' while the overall relationship (6.24) or its equivalent operator notation form (6.23) are considered as 'autoregressive-moving average' (ARMA) processes. The 'X' is then appended to this description to denote the presence of the 'exogenous' or input variables u_k, which affect the output via the transfer function $B(z^{-1})/A(z^{-1})$ i.e.

$$x_k = \frac{B(z^{-1})}{A(z^{-1})} u_k \tag{6.25}$$

or
$$x_k = -a_1 x_{k-1} -, \ldots, -a_n x_{k-n} + b_1 u_{k-1} + \ldots + b_n u_{k-n}$$

where x_k can be considered as a hypothetical and unmeasurable 'noise-free' output of the system. With this nomenclature (6.18)(i) can be written as

$$y_k = x_k + \xi_k \tag{6.26}$$

with the deterministic output x_k and the 'coloured' noise ξ_k defined by equations (6.23) and (6.25) respectively.

The term coloured noise arises because the 'white' noise input e_k, which is serially uncorrelated, is shaped by its passage through the ARMA filter and so possesses serial correlation properties, the nature of which is defined by the ARMA model coefficients a_i and d_i. Sometimes the reader may also see ξ_k referred to as noise with 'rational spectral density': in simple terms, this merely means that it is derived from white noise via a 'rational' transfer function such as (6.23) which is the ratio of two polynomials in z^{-1}. Clearly the autoregressive (AR) process, i.e.

$$\xi_k = -a_1 \xi_{k-1} - \ldots - a_n \xi_{k-n} + e_k \tag{6.27}$$

and the moving average (MA) process

$$\xi_k = e_k + d_1 e_{k-2} + \ldots + d_n e_{k-n} \tag{6.28}$$

are special cases of (6.23). A detailed discussion on the statistical properties of AR, MA and joint ARMA processes is given in Box and Jenkins (1970) and Chatfield (1975), and the reader is urged to consult these texts for further background on the subject.

Figure 6.1(a) shows a block diagram of the ARMAX system (6.18). Whilst the input signal u_k and the white noise e_k enter the system via different pathways, it is worth noting that the system transfer function $B(z^{-1})/A(z^{-1})$ and the ARMA noise transfer function $D(z^{-1})/A(z^{-1})$ are interrelated, since they have a common denominator polynomial $A(z^{-1})$. On the basis of this observation, we might consider the alternate representation in Figure 6.1(b) to be more "general" in some sense: here the noise transfer function is chosen as $D(z^{-1})/C(z^{-1})$ and is *independent* of the system transfer function $B(z^{-1})/A(z^{-1})$. In searching for greater generality we ought also allow for the output to be affected *instantaneously* by the input u_k: this can be achieved simply by introducing an additional term b_o in the definition of $B(z^{-1})$ i.e.

$$B(z^{-1}) = b_o + b_1 z^{-1} + \ldots\ldots + b_n z^{-n}$$

so that (6.25) now yields a relationship between x_k and u_k of the form

$$x_k = -a_1 x_{k-1} - \ldots\ldots - a_n x_{k-n} + b_o u_k + \ldots + b_n u_{k-n}$$

It is clear that x_k at the kth instant now has a component $b_o u_k$ which is dependent upon the value of u_k *at the same instant.*

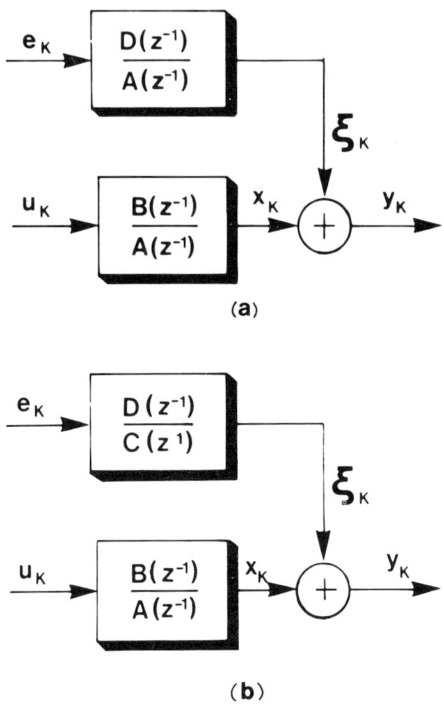

Figure 6.1 Linear Time-series Models

The time-series model obtained with these two modifications is usually referred to as the transfer function (TF) or Box-Jenkins model, since it is the principal model considered by Box and Jenkins (1970) in their monograph on time-series analysis. In its deterministic form, this model is the classical 'transfer function' representation of control systems analysis, and it has been discussed in stochastic terms within a control and systems framework by Young *et al.*(1971), Young (1974) and Bohlin (1970). Clearly, this model can be written as

$$y_k = \frac{B(z^{-1})}{A(z^{-1})} u_k + \frac{D(z^{-1})}{C(z^{-1})} e_k \tag{6.29}$$

or, decomposing it into its constituent parts,

$$\left.\begin{aligned} &\text{system model : } x_k = \frac{B(z^{-1})}{A(z^{-1})} u_k && \text{(i)} \\ &\text{noise model : } \xi_k = \frac{D(z^{-1})}{C(z^{-1})} e_k; \quad E\{e_j e_k\} = \sigma^2 \delta_{jk} && \text{(ii)} \\ &\text{observation equation : } y_k = x_k + \xi_k && \text{(iii)} \end{aligned}\right\} \tag{6.30}$$

Here, to summarize, $A(z^{-1})$, $B(z^{-1})$, $C(z^{-1})$ and $D(z^{-1})$ are defined as follows

$$A(z^{-1}) = 1 + a_1 z^{-1} + \ldots + a_n z^{-n}$$

$$B(z^{-1}) = b_o + b_1 z^{-1} + \ldots + b_n z^{-n}$$

$$C(z^{-1}) = 1 + c_1 z^{-1} + \ldots + c_n z^{-n}$$

$$D(z^{-1}) = 1 + d_1 z^{-1} + \ldots + d_n z^{-n}$$

and, to be consistent with our assumptions of stationarity in the original state-space form, we assume that the roots of the characteristic equation of the noise model, i.e. the roots of

$$1 + c_1 z^{-1} + \ldots + c_n z^{-n} = 0$$

all lie completely outside the unit circle in the complex plane. Note that Box and Jenkins refer to non-stationary situations where certain roots may lie *on* the circle, so that there are integrators (or summers) on the noise process, as in the random walk model (5.2) considered previously. The reader is referred to Box and Jenkins for discussion on this topic but is warned that their suggested device for removing such non-stationarity, namely differencing the data, can be dangerous in practice and can lead to high noise amplification (see Chapter 10).

In relation to the TF model (6.30), note that, for complete generality, we might let the polynomials A, B, C and D be of different orders, rather than specifying them all as being nth order. However, since we can introduce such modifications simply by assuming that certain of the coefficients in the polynomials may be zero, it is convenient to retain the present nomenclature and so reduce the complexity of the mathematical statements. Note also that *both* the D and C polynomials are chosen with a leading coefficient of unity (i.e. they are 'monic' polynomials). This is made possible because the variance of e_k is specified separately as σ^2: in effect, the use of unity leading coefficients together with the definition of e_k as white noise with variance σ^2 completely specifies the statistical characteristics of the noise ξ_k as an ARMA process. This is the conventional method of defining models for noise with rational spectral density (see e.g., Box and Jenkins, 1970) and it can be contrasted with the deterministic model transfer function $B(z^{-1})/A(z^{-1})$, where only $A(z^{-1})$ has a unity leading coefficient. $B(z^{-1})$, on the other hand, has a leading coefficient b_0 which is to be estimated and which allows the input u_k to have arbitrary and unspecified magnitude characteristics.

It is clear that the TF model (6.30) is more general than the ARMAX model in the sense that it separates out the deterministic and stochastic effects completely; a factor which, as we shall see later, has some important theoretical and algorithmic consequences. But it is not really any more *general* as a stochastic description of the relationship between the input and output data, as we can readily see if we consider a TF model

$$y_k = \frac{B}{A} u_k + \frac{D}{C} e_k \tag{6.31}$$

where, for convenience here and subsequently, we omit the argument and let $A = A(z^{-1})$, $B = B(z^{-1})$ etc. Multiplying through (6.31) by AC then yields

$$ACy_k = BCu_k + DAe_k \tag{6.32}$$

which is, of course, an ARMAX model of the form (6.18)(ii) with A, B, and D replaced by A^*, B^* and D^*, where

$$A^* = AC; \quad B^* = BC; \quad \text{and } D^* = DA \tag{6.33}$$

Similarly, if we have an ARMAX model

$$Ay_k = Bu_k + De_k \tag{6.34}$$

then this corresponds to a TF model of the form (6.29) with

$$A = A; \quad B = B; \quad C = A; \quad \text{and } D = D \tag{6.35}$$

It is clear from these examples that the ARMAX and TF models can both be used as the description of the same stochastic dynamic system and, as such, are interchangeable descriptions. Note, however, that the ARMAX description of the TF model involves definitions of the A, B and D polynomials in (6.33) which specify *common factors* between any two polynomials. This implies that estimation of this model based on data generated by a TF model may need to involve some constraints on the nature of the polynomials. Jakeman and Young (1981a and b) show that this is indeed the case and that estimation problems can occur when such constraints are not imposed.

The reader will probably have realized by now that the ARMAX and TF model forms are not the only ones which could be considered. For instance, we can construct models which are combinations of the two but which, in terms of the TF model, involve special noise structures. For example the moving average noise in the ARMAX model could be replaced by ARMA noise i.e.

$$A\, y_k = B\, u_k + \frac{D}{C}\, e_k \tag{6.36}$$

which, in TF model format, becomes

$$y_k = \frac{B}{A}\, u_k + \frac{D}{CA}\, e_k \tag{6.37}$$

Or, alternatively, the MA noise could be replaced by autoregressive noise, i.e.

$$A\, y_k = B\, u_k + \frac{1}{C}\, e_k \tag{6.38}$$

which corresponds to a TF model

$$y_k = \frac{B}{A}\, u_k + \frac{1}{CA}\, e_k \tag{6.39}$$

where the noise is also autoregressive but with a polynomial with two factors, one of which is dependent upon the system transfer function denominator A. The model (6.38) has been quite popular in the literature on recursive time-series modeling and is often termed the 'dynamic adjustment' (DA) model with autogressive residuals (it has been discussed by Clarke, 1967; and Hastings-James and Sage, 1969). The model (6.36), which seems less popular, is simply the DA model with ARMA residuals.

The reader may well feel that, with so many different OS model forms available, the position as regards model structure selection in any particular application is going to be rather confusing. On the surface this is certainly the case: there is little to choose between the various OS model forms (6.18), (6.29), (6.36) and (6.38). All could be seen as candidate model structures in the observation space, and the selection of which format is most appropriate will depend, for example, upon the specific objectives of the analysis, as well as on the statistical properties of the time-series. But we shall have much more to say on this problem of model structure 'identification' in Chapter 10. For the moment, we will continue the discussion in relation to the estimation of parameters in the TF model and refer subsequently to the estimation of the other alternative model forms.

6.3 Least Squares Estimation: Its Advantages and Limitations

Considering the TF model (6.30), we see from (i) that

$$[1 + a_1 z^{-1} +....+ a_n z^{-n}]x_k = [b_o + b_1 z^{-1}+....+b_n z^{-n}]u_k$$

or

$$x_k + a_1 x_{k-1} + a_2 x_{k-2} +....+ a_n x_{k-n} = b_o u_k + b_1 u_{k-1} +....+ b_n u_{k-n}$$

so that the deterministic (noise free) output of the system is given by

$$x_k = -a_1 x_{k-1} -....- a_n x_{k-n} + b_o u_k + b_1 u_{k-1} +....+ b_n u_{k-n} \tag{6.40}$$

in other words, x_k is a function of past values of x_{k-i} $i = 1,2,....,n$ and present and past values of u_{k-i}, $i = 0,1,....,n$. If the observation equation (6.30)(iii) is now substituted into (6.40) then we obtain the following relationship between the measures variables y_k and u_k

$$\left.\begin{aligned} y_k &= -a_1 y_{k-1} -....-a_n y_{k-n} + b_o u_k +....+ b_n u_{k-n} + \eta_k \\ &\text{where} \\ \eta_k &= \xi_k + a_1\, \xi_{k-1} +....+ a_n\, \xi_{k-n} \end{aligned}\right\} \tag{6.41}$$

Equation (6.41) can be written in simple vector terms (c.f. equation (3.4)) as

$$y_k = \underline{z}_k^T \underline{a} + \eta_k \tag{6.42}$$

where

$$\underline{z}_k^T = [-y_{k-1}, -y_{k-2}, \ldots, -y_{k-n}, u_k, \ldots, u_{k-n}] \tag{6.43}$$

and,

$$\underline{a} = [a_1, a_2, \ldots, a_n, b_o, \ldots, b_n]^T \tag{6.44}$$

In a similar manner, the coloured noise term ξ_k is generated by

$$\xi_k = -c_1 \xi_{k-1} - \ldots - c_n \xi_{k-n} + d_1 e_{k-1} + \ldots + d_n e_{k-n} + e_k$$

or in vector terms

$$\xi_k = \underline{n}_k^T \underline{c} + e_k \tag{6.45}$$

where

$$\underline{n}_k^T = [-\xi_{k-1}, -\xi_{k-2}, \ldots, -\xi_{k-n}, e_{k-1}, \ldots, e_{k-n}]$$

and, (6.46)

$$\underline{c} = [c_1, c_2, \ldots, c_n, d_1, \ldots, d_n]^T$$

In other words, both the system and noise model can be written in linear terms which superficially resemble the regression relationships discussed in previous Chapters.

But this superficial resemblance is misleading. Equation (6.42) is obviously *not* a regression relationship since it is derived in a manner which demonstrates its error-in-variables or structural form (Chapter 4) with each of the output variables y_{k-i}, $i = 1,2,\ldots,n$ in the vector $\underline{z}_k$ contaminated by noise ξ_{k-i} and with the noise term η_k, in consequence, being both correlated with the y_{k-i} variables and serially correlated in time. And again, while the noise expression (6.45) is apparently nearer to the regression form - with the residual e_k serially uncorrelated and independent (by definition) of the ξ_{k-i} and e_{k-i} variables in $\underline{n}_k^T$, it cannot be considered directly in regression terms because both ξ_{k-i} and e_{k-i}, $i = 0,1,\ldots,n$ are unavailable for measurement!

With these observations in mind, it is clear that least squares analysis is not generally applicable to the estimation of the parameters in either of the models (6.42) and (6.45). In the case of (6.42) estimation is possible but we would expect from previous discussion in Chapter 4 that such analysis will normally yield estimates which are asymptotically biased and inconsistent. In order to demonstrate that this is indeed the case, let us write down the least squares 'normal' equations for the model (6.42) where at least we have access to all the variables in the $\underline{z}_k$ vector.

After k samples, these equations are of the form (c.f. equation 4.12)

$$[\sum_{i=1}^{k} \underline{z}_i\underline{z}_i^T]\hat{\underline{a}}_k = \sum_{i=1}^{k} \underline{z}_i y_i \tag{6.47}$$

If we now define a vector $\underline{x}_k$ of 'noise free' variables as

$$\underline{x}_k = [-x_{k-1},\ldots, -x_{k-n}, u_k,\ldots,u_{k-n}]^T \tag{6.48}$$

then

$$\underline{z}_k = \underline{x}_k + \underline{\xi}_k \tag{6.49}$$

where the noise vector $\underline{\xi}_k$ is defined as

$$\underline{\xi}_k = [-\xi_{k-1},\ldots,-\xi_{k-n}, 0,\ldots,0]^T \tag{6.50}$$

With this nomenclature and multiplying through equation (6.47) by $^1/_k$ we obtain

$$\{\frac{1}{k}\sum_{i=1}^{k}[\underline{x}_i + \underline{\xi}_i][\underline{x}_i + \underline{\xi}_i]^T\}\hat{\underline{a}}_k = \frac{1}{k}\sum_{i=1}^{k}[\underline{x}_i + \underline{\xi}_i]y_i \tag{6.51}$$

or, noting that $y_i = \underline{z}_i^T\underline{a} + \eta_i$ from (6.42),

$$\{\frac{1}{k}\Sigma[\underline{x}_i + \underline{\xi}_i][\underline{x}_i + \underline{\xi}_i]^T\}\hat{\underline{a}}_k = \frac{1}{k}\Sigma[\underline{x}_i + \underline{\xi}_i][\underline{x}_i + \underline{\xi}_i]^T\underline{a} + \frac{1}{k}\Sigma[\underline{x}_i + \underline{\xi}_i]\eta_i \tag{6.52}$$

As a result,

$$\frac{1}{k}\Sigma[\underline{x}_i + \underline{\xi}_i][\underline{x}_i + \underline{\xi}_i]^T\{\hat{\underline{a}}_k - \underline{a}\} = \frac{1}{k}[\Sigma\, \underline{x}_i + \underline{\xi}_i]\eta_i$$

or

$$\tilde{\underline{a}}_k = \hat{\underline{a}}_k - \underline{a} = [\frac{1}{k}\Sigma\ [\underline{x}_i + \underline{\xi}_i][\underline{x}_i + \underline{\xi}_i]^T]^{-1}\ \frac{1}{k}\ \Sigma[\underline{x}_i + \underline{\xi}_i]\eta_i \quad (6.53)$$

In order to examine the asymptotic situation we can now take the probability in the limit as $k\to\infty$ i.e.

$$\text{p.lim}\ \tilde{\underline{a}}_k = \text{p.lim}\{\frac{1}{k}C_k^{-1}\ \frac{1}{k}\underline{b}_k\} \quad (6.54)$$

where

$$\left.\begin{aligned} C_k &= \sum_{i=1}^{k} [\underline{x}_i + \underline{\xi}_i][\underline{x}_i + \underline{\xi}_i]^T && \text{(i)} \\ \underline{b}_k &= \sum_{i=1}^{k} [\underline{x}_i + \underline{\xi}_i]\eta_i && \text{(ii)} \end{aligned}\right\} \quad (6.55)$$

Now by Slutsky's theorem (Elgerd, 1967)

$$\text{p.lim}\{\frac{1}{k}C_k^{-1}\ \frac{1}{k}\underline{b}_k\} = \text{p.lim}\frac{1}{k}C_k^{-1}\ \text{p.lim}\frac{1}{k}\underline{b}_k \quad (6.56)$$

provided the elements of $\frac{1}{k}C_k$ and $\frac{1}{k}\underline{b}_k$ converge in probability, as they do in this case because of the stationarity and stability assumptions. Now

$$\text{p.lim}\ \frac{1}{k}\ C_k = \text{p.lim}\ \frac{1}{k}\ \Sigma\underline{x}_i\underline{x}_i^T + \text{p.lim}\ \frac{1}{k}\Sigma\ \underline{\xi}_i\underline{\xi}_i^T$$

$$+\ \text{p.lim}\ \frac{1}{k}\ \Sigma\underline{x}_i\underline{\xi}_i^T + \text{p.lim}\ \frac{1}{k}\Sigma\ \underline{\xi}_i\underline{x}_i^T$$

but, since $\underline{x}_i$ and $\underline{\xi}_i$ are statistically independent by definition, then the last two terms on the RHS are identically zero so that

$$\text{p.lim}\ \frac{1}{k}\ C_k = R_{\underline{x},\underline{x}} + R_{\underline{\xi},\underline{\xi}} \quad (6.57)$$

where $R_{\underline{x},\underline{x}}$ and $R_{\underline{\xi},\underline{\xi}}$ are the covariance matrices (Appendix 1) of $\underline{x}$ and $\underline{\xi}$ respectively.[†] Similarly, because $\underline{x}_i$ and $\underline{\xi}_i$ are statistically independent,

---

[†]Assuming for convenience that $\underline{x}_i$ is zero mean.

$$\text{p.lim}\ \frac{1}{k}\, b_k = \text{p.lim}\ \frac{1}{k}\, \Sigma[\underline{x}_i + \underline{\xi}_i]\eta_i = R_{\underline{\xi},\eta} \tag{6.58}$$

where $R_{\underline{\xi},\eta}$ is a 2n+1 covariance vector whose first n elements are defined as the covariances of ξ_{i-j} and η_i, j = 1,2,....,n; and whose final (n+1) elements are zero.

Substituting from (6.58) and (6.57) in (6.54) and noting (6.56) we now find that

$$\text{p.lim}\ \underline{\tilde{a}}_k = [R_{\underline{x},\underline{x}} + R_{\underline{\xi},\underline{\xi}}]^{-1}\, R_{\underline{\xi},\eta} \tag{6.59}$$

Provided that the input u_k continues to perturb or 'excite' the process and, at the same time, the coefficients in the polynomial $B(z^{-1})$ are not all zero, then the output x_i will exist for all i over the observation interval. As a result, $R_{\underline{x},\underline{x}}$ will also be non-zero and its inverse will exist.† Thus the only way in which the asymptotic error can be zero is for $R_{\underline{\xi},\eta}$ to be identically zero. But from the definition of η_i in (6.39) i.e. $\eta_i = \xi_i + a_1\xi_{i-1} + \ldots + a_n\xi_{i-n}$, we know that, in general, ξ_i and η_i are correlated and $R_{\underline{\xi},\eta}$ will *only* be zero if ξ_i is zero for all i. In other words, the asymptotic error will normally be zero and the least squares estimates will be unbiased asymptotically *only* if there is no noise on the data. And if there is noise on the data, then the estimates will be biased asymptotically to an extent determined by the relative ratio of noise to signal variance.

Nominally then, least squares estimation is not appropriate for estimating the parameters in the TF model since the estimates will be inconsistent and the bias error will remain *no matter how much data we use in the estimation*. We should remember, however, that the importance of this bias error has to be assessed in relation to the use we make of the analytical results. If the coefficients a_i and b_i in the TF model are themselves of relevance to the problem at hand, for example, then the bias clearly introduces some considerable difficulties. If we are given *a priori* information on the covariance statistics $R_{\underline{x},\underline{x}}$, $R_{\underline{\xi},\underline{\xi}}$ and $R_{\underline{\xi},\eta}$ (or $R_{\underline{z},\underline{z}}$ and $R_{\underline{\xi},\eta}$) then it may be possible to obviate these difficulties by computing the bias from (6.59) and so correcting the estimates (e.g. Levin, 1963). Even so, we might expect such information, even if available, to be rather poor in most practical situations and the corrected estimates are likely to be extremely sensitive to such errors.

But there are situations where the biased least squares results are useful *despite* the presence of bias. Johnston (1973; p. 163), for example, shows in a

---

†Here we assume that the order of the model is identically equal to the order of the process generating the data; problems can be encountered if this is not the case (Chapter 10).

simple error-in-variables situation that the least squares prediction based on the model is unbiased, even though the model parameter estimates are biased. And more recently, Astrom and Wittenmark (1972) have shown how the recursive least squares estimates can be used directly in noisy situations to obtain "self tuning regulator" systems for dynamic processes which exhibit optimal properties in a stochastic sense. With these examples in mind, we urge the reader to note carefully that asymptotic bias is a *potential* problem when using the least squares estimator, but not to assume that this negates its utility completely. It has the advantage of great simplicity and, *when used carefully*, can still be an important tool in time-series analysis as we see in the special case discussed in the next section. But if abused it can result in confusing and misleading results.

6.4 Least Squares Estimation: A Special Case

Before considering the general estimation problem posed by equations (6.42) and (6.45), it is instructive to consider the special degenerate case of Figure 6.2(a) where the noise model is purely autoregressive and characterized by a transfer function,

$$N(z^{-1}) = \frac{1}{A(z^{-1})}$$

In other words, $C(z^{-1}) = A(z^{-1})$ and $D(z^{-1}) = 1.0$ so that the model is in a special ARMAX form with D = 1.0, i.e.,

$$y_k = \frac{B}{A}u_k + \frac{1}{A}e_k \qquad (6.60)$$

An even more limited example of this situation is the case shown in Figure 6.2(b) where the system has no 'numerator dynamics' (i.e. $B(z^{-1}) = 1.0$) and the noise enters as serially uncorrelated 'white' noise at the input to the system.

(a)

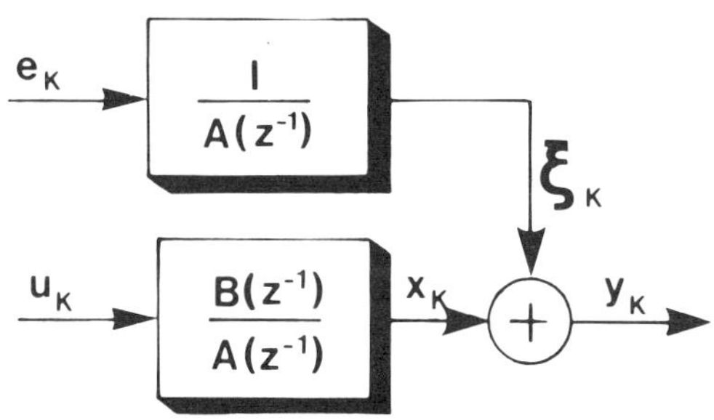

(b) 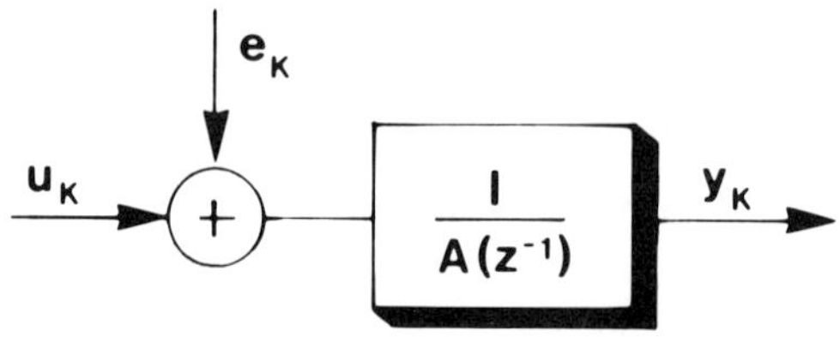

Figure 6.2 Special Time-series Models

For the model of Figure 6.2(a) equation (6.40) takes the reduced form

$$y_k = \underline{z}_k^T \underline{a} + e_k \tag{6.61}$$

where e_k is a serially uncorrelated sequence of random variables with variance σ^2 and $\underline{z}_k^T$ is given by

$$\underline{z}_k^T = [-y_{k-1} \ldots . -y_{k-n},\ u_k \ldots . u_{k-n}]$$

Since u_k is a deterministic input sequence, it is clear (Astrom, 1966) that this model is a simple extension of the simple *autoregressive time series model* with e_k, by definition, statistically independent of y_{k-1}, $i = 1,2,\ldots.,n.$ † This type of model has received considerable attention in the statistical literature (e.g. Mann and Wald, 1943; Durbin, 1960; and many others), and it is well known that an *asymptotically* unbiased and efficient estimate $\hat{\underline{a}}_k$ of the parameter vector $\underline{a}$ can be obtained by simple least squares regression analysis.

The recursive regression algorithm in this case can be obtained directly by reference to algorithm III of Chapter 4, with $\underline{x}_k$ replaced by $\underline{z}_k$.

$$\hat{\underline{a}}_k = \hat{\underline{a}}_{k-1} - \underline{k}_k^* (\underline{z}_k^T \hat{\underline{a}}_{k-1} - y_k) \qquad (1)$$ XII

---

†Although it is not, of course completely indepedent of y_k over the whole observation interval as required in true regression problems.

where

$$\underline{k}_k^* = P_{k-1}^*\underline{z}_k(\sigma^2 + \underline{z}_k^T P_{k-1}^*\underline{z}_k)^{-1}$$

and

$$P_k^* = P_{k-1}^* - P_{k-1}^*\underline{z}_k(\sigma^2 + \underline{z}_k^T P_{k-1}^*\underline{z}_k)^{-1}\underline{z}_k^T P_{k-1}^* \qquad (2) \qquad \text{XII}$$

Alternatively, (XII)(i) can be written in stochastic gradient form, i.e.

$$\hat{\underline{a}}_k = \hat{\underline{a}}_{k-1} - \frac{P_k^*}{\sigma^2}(\underline{z}_k\underline{z}_k^T\hat{\underline{a}}_{k-1} - \underline{z}_k y_k) \qquad (3) \qquad \text{XII}$$

Of course the practical relevance of algorithm XII will depend upon the application: if the limited model structure (6.60) is appropriate then the algorithm provides a quick and easy method for obtaining estimates of the parameters and their associated error covariance characteristics. But how often such a fortuitous situation will arise in practice is difficult to predict: in a considerable number of examples of time-series analysis over the past ten years, the author has found that it is the exception rather than the rule. This would suggest that the practical utility of the algorithm is rather limited in those cases where the parametric estimates are *themselves* important.

6.5 The General Case: The Structural Model

In the general TF model case, equation (6.29) can be rearranged to give

$$A^* y_k = B^* u_k + e_k$$

where $A^* \triangleq C/D$; $B^* \triangleq BC/DA$. Thus the relationship between y_k, u_k and e_k can be written

$$y_k = \frac{B^*}{A^*} u_k + \frac{1}{A^*} e_k \qquad (6.62)$$

Since equation (6.62) is structurally similar to equation (6.60) it would seem possible to use the linear regression algorithm XII to identify the parameters of the polynomials B^* and A^*. The problem is, of course, that these polynomials are nominally of *infinite* order because of the divisions inherent in their definition. Thus the model is usually inefficient in parametric terms (Box and Jenkins would say "non-parsimonious").

In addition to its parametric inefficiency, the large dimension of the model (6.62) in the general case makes it difficult to specify the polynomial orders *a priori*, and this can lead both to lengthy analysis and the possibility of poorly defined estimation results. An example of this is the "repeated least squares" analysis of Astrom and Eykhoff (1971). They consider a simulated first order ARMAX example

$$y_k - 0.5y_{k-1} = 1.0u_{k-1} + e_k + 0.1e_{k-1} \tag{6.63}$$

and then proceed to analyze the data assuming models of the form (6.62) of increasing order, each time checking the significance of the decrease in the sum of the squares with a 't' test (Kendall and Stuart, 1961). This analysis yields least squares estimates of $\hat{a}_1 = 0.643$ and $\hat{b}_1 = 1.018$ in the first order case, with $\hat{a}_1$ clearly biased. The 'final' model obtained is of 5th order with

$$A^*(z^{-1}) = 1\text{-}1.19z^{-1} + 0.81z^{-2} - 0.5z^{-3} - 0.35z^{-4} + 0.12z^{-5}$$

$$B^*(z^{-1}) = 1.08z^{-1} - 0.7z^{-2} + 0.48z^{-3} - 0.25z^{-4} + 0.12z^{-5}$$

They then point out that by dividing A(z) by B(z) one finds that

$$\frac{1.08A^*(z)}{B^*(z)} \simeq z - .495$$

In other words, the model can be written approximately as

$$y_k = \frac{1.08z^{-1}}{1\text{-}0.495z^{-1}} u_k + \frac{1}{A^*(z^{-1})} e_k$$

which can be compared with the true system (6.63). What has happened here is that the final model order is selected so that $1/A^*(z^{-1})$ approximates, in some sense, the noise TF

$$\xi_k = \frac{1 + 0.1z^{-1}}{1\text{-}0.5z^{-1}} e_k$$

$B^*(z^{-1})$ is then estimated so that the higher order factors of numerator and denominator polynomials $B^*(z^{-1})$ and $A^*(z^{-1})$ approximately cancel, to yield dynamic characteristics that resemble closely the true first order transfer function

$1.0z^{-1}/1-0.5z^{-1}$. In what sense the noise approximation occurs is not clear from the examples: presumably the model is generated so that the 'colour' is purged from the residuals leaving them approximately white in the 5th order model case. But the reader will see on division of $1 - 0.1z^{-1}$ by $1 + 0.5z^{-1}$ that $1/A^*(z^{-1})$ is not a simple approximation to the true ARMA noise model in this sense.

We see from the above example that if linear regression is applied to *low* order models of the form (6.62), then asymptotically biased estimates of the parameters in the A and B polynomials will usually result. It is necessary, therefore, to look for estimation procedures which are directly applicable to the more general TF model (6.29). The estimation problem associated with time-series models of the form (6.29) is well known and various procedures have been suggested for overcoming it. An excellent survey of these techniques is given by Astrom and Eykhoff (1971) and Eykhoff (1974). Probably the best known solution is the *Maximum Likelihood* method suggested by Box and Jenkins (e.g. 1970) and Astrom and Bohlin (1966†, 1970). Here, the problem of bias is obviated by *simultaneously* estimating the parameters of all the polynomials A, B, C and D. This is clearly a non-linear estimation problem which, in general, requires either numerical hill-climbing procedures to determine the parameters that maximize the likelihood function or some alternative iterative approach. In addition, the data are analyzed in block form so that on-line utilization and time variable parameter estimation are difficult (Goodwin and Payne, 1977).

If a recursive solution with on-line and time variable parameter estimation potential is required, then there are at present a number of alternatives. One approach is to reformulate the problem in state-space terms and consider the application of the Extended Kalman filter estimation (EKF) methods of analysis (see Jazwinski, 1970; Young, 1974 and Chapter 9). Another is to use the recursive form of Generalized Least Squares (GLS) analysis†† as suggested by Hastings-James (1969, 1970), which applies to the DA model with autoregressive residuals. Since 1970, a number of other recursive approaches have been suggested (see Eykhoff, 1974, Chapter 7) such as the Extended Matrix method (Talmon, 1971; Talmon and Van den Boom 1973; Smets, 1970) - which is related to the "Approximate Maximum Likelihood" (AML) procedure (Young, 1968b ; Panuska, 1969) and is also referred to as the "Recursive Maximum Likelihood 1" (RML1) by Soderstrom *et al.* (1974); the "Recursive Maximum Likelihood 2" (RML2) of Soderstrom *et al.* (1974), which is applied to the ARMAX model (6.18); the Prediction Error Recursion (PER) approach of Ljung (1979a), which is similar to the RML2 for the ARMAX model but can be applied to general time-series

---

† The ARMAX model was considered in this paper.

†† The iterative generalised least squares approach was discussed originally by J. Johnston(1963) and applied to time series analysis by D.W. Clarke (1967)

model forms such as the TF representation (see Chapter 9); and, finally, the various recursive algorithms suggested by Landau (e.g. 1976) which are closely related to some of the other procedures mentioned above, but are derived from an alternative theoretical standpoint based on hyperstability theory.

In the present context, it would serve little useful purpose to describe all of these alternative methods and it might well confuse the reader. It will suffice to note that all of the above procedures require *simultaneous* estimation of the system (a_i; b_i) and noise (c_i; d_i) parameters and are, therefore, relatively complex to implement. But there is one procedure which does not require this and is able to obtain consistent, asymptotically unbiased and *relatively* efficient (i.e. low but not minimum variance) estimates of the a_i and b_i parameters *without* simultaneous noise model parameter estimation. This is the instrumental variable (IV) method proposed by the author (1965b,c; 1969b,c; 1970) in connection with continuous time system estimation and by Wong and Polak (1967) and again by the author (*et al.* 1971, 1974) in relation to discrete-time systems.[†]

6.6 Summary

The purpose of this chapter has been to introduce the reader to the problem of time-series model estimation, i.e. the estimation of parameters in simple mathematical representations of linear stochastic, dynamic systems. The treatment has been limited to discrete-time (difference equation) models simply because such representations and their estimation are more straightforward than the equivalent continuous-time (differential equation) models which are mentioned in Chapter 10.

But, as we have seen, the estimation of parameters in discrete time-series models is a far from trivial problem and its solution can be approached in many different ways. In the next chapter we will consider in detail the simplest recursive approach based on the concept of instrumental variables.

[†] see also Mayne, D.Q. (1967) A method for estimating discrete-time transfer functions, *Advances in Control*, 2nd UKAC Convention, Bristol.

7 The Instrumental Variable (IV) Method of Time-Series Analysis

The recursive IV method of estimation has been discussed in Chapter 4 in relation to the estimation of parameters in structural models with errors-in-variables. In the recursive IV algorithm (IV) it is required that we generate a vector of instrumental variables $\hat{\underline{x}}_k$ at each recursive step. If this is possible, then it is easy to see why the IV modification is effective in removing the asymptotic bias from the estimates.

If the least squares equations (6.52) are modified to incorporate the IV vector, then we obtain,

$$\frac{1}{k}\sum^{k}\hat{\underline{x}}_i[\underline{x}_i+\underline{\xi}_i]^T\hat{\underline{a}} = \frac{1}{k}\sum^{k}\hat{\underline{x}}_i\{[\underline{x}_i+\underline{\xi}_i]^T\underline{a}+\eta_i\} \tag{7.1}$$

But by definition, $\hat{\underline{x}}_i$ is chosen to be independent of the noise inputs, so that

$$\lim_{k\to\infty}\frac{1}{k}\sum^{k}\hat{\underline{x}}_i\,\underline{\xi}_i^T = \underline{0}; \qquad \lim_{k\to\infty}\frac{1}{k}\sum^{k}\hat{\underline{x}}_i\eta_i = \underline{0}; \qquad \lim_{k\to\infty}\frac{1}{k}\sum^{k}\hat{\underline{x}}_i[\underline{x}_i+\underline{\xi}_i]^T = R_{\hat{\underline{x}},\underline{x}} \neq 0$$

where $R_{\hat{\underline{x}},\underline{x}}$ is the covariance matrix defined as

$$R_{\hat{\underline{x}},\underline{x}} = \lim_{k\to\infty}\frac{1}{k}\sum^{k}\hat{\underline{x}}_i\underline{x}_i^T$$

Now (7.1) can be written

$$\{\frac{1}{k}\sum^{k}\hat{\underline{x}}_i\;[\underline{x}_i+\xi_i]^T\}\;(\hat{\underline{a}}-\underline{a}) = \frac{1}{k}\sum^{k}\hat{\underline{x}}_i\eta_i$$

Thus, provided $R_{\hat{\underline{x}},\underline{x}}$ is non-singular, the limiting value, p.lim $\tilde{\underline{a}}$, to which $\tilde{\underline{a}} = \hat{\underline{a}} - \underline{a}$, tends in probability, is given by

$$\text{p.lim}\;\tilde{\underline{a}} = 0 \tag{7.2}$$

and the estimates are asymptotically unbiased, as required.

From the above simple analysis, it is clear that, by incorporating the instrumental variables in the manner suggested, it is possible to eliminate the troublesome asymptotic bias while preserving the existence of a solution. However, as Durbin has stressed (1954), the statistical efficiency of the solution is dependent upon the degree of correlation between $\hat{\underline{x}}_i$ and $\underline{x}_i$; in particular, the most efficient (i.e. the lowest variance) estimates obtained from Algorithm IV are obtained only when $\hat{\underline{x}}_i = \underline{x}_i$, i.e. when the $\hat{x}_i$ are equal to the hypothetical and nominally unavailable 'noise-free' variables x_i (although the estimates obtained in this case will not possess the minimum possible variance, as we shall see in a later chapter).

Of course the major problem with the IV method is the generation of suitable instrumental variables. In general, the difficulty in obtaining such variables has acted as a strong deterrent to the use of the IV method, as evidenced by the paucity of literature on the application of the method to the analysis of real data. Fortunately, the difficulty can be overcome fairly easily when dealing with the kind of time-series considered here, where the process is activated by a deterministic input, u_k.

7.1 The Recursive IV Algorithm for Time-Series Models

Referring to the TF model (6.29) it is seen that $\underline{x}_k$ is correlated with the input sequence u_k since x_k derives from u_k via the system transfer function, as shown in (6.30)(i) and Figure 6.1(b). As a result, the degree of correlation between u_k and $\underline{x}_k$ is a function of the form of the model and its parameter values. Also we see that the input sequence is available directly for measurement and is, by definition, uncorrelated with the noise ξ_k. Thus u_k itself satisfies the requirements of an IV and can be used to define an IV vector. Bearing in mind that the transfer function B/A will inject a dynamic lag between x_k and u_k, however, we might expect x_k to be more highly correlated with $u_{k-\tau}$ where τ is a pure time delay, and that there will be some optimum choice of τ which maximizes this correlation. This suggests that the IV vector could in general, be chosen of the form

$$\hat{\underline{x}}_k^T = [-u_{k-\tau} \ldots -u_{k-\tau-n}, u_k \ldots u_{k-n}] \tag{7.3}$$

where, ideally, τ will be chosen to maximize the elements of the covariance matrix $\hat{R}_{\hat{\underline{x}},\underline{x}}$. This approach is quite similar to that first used by Joseph *et al.* (1961).

Pursuing this line of reasoning one step further, it seems possible that instrumental variables that are even more highly correlated with x_k can be obtained by referring to equation (6.30)(i): instead of using u_k directly, we could use it to activate an 'auxiliary model' of the process (Young, 1965b) whose output $\hat{\underline{x}}_i$, $i = k-1,\ldots,k-n$, is used to define an IV vector of the form

$$\hat{\underline{x}}_k^T = [-\hat{x}_{k-1} \ldots -\hat{x}_{k-n}, u_k \ldots u_{k-n}] \tag{7.4}$$

It then makes sense to construct this auxiliary model in the same form as the model (6.30)(i), i.e.

$$\hat{A}(z^{-1})\hat{x}_k = \hat{B}(z^{-1})u_k \tag{7.5}$$

where

$$\hat{A}(z^{-1}) = 1 + \alpha_1 z^{-1} + \ldots + \alpha_n z^{-n}$$

$$\hat{B}(z^{-1}) = \beta_o + \beta_1 z^{-1} + \ldots + \beta_n z^{-n}$$

and α_i, β_i are the auxiliary model parameters chosen in some sensible manner. Clearly, the closer these parameters are to the actual, but unknown, process parameters a_i, b_i, the more highly correlated are $\hat{\underline{x}}_k$ and $\underline{x}_k$ and the lower the variance of the IV estimates; in particular, if $\alpha_i = a_i$, $\beta_i = b_i$ for all i then $\hat{\underline{x}}_k = \underline{x}_k$ and the IV estimates obtained from algorithm IV will be optimum in some sense, as discussed above (see Wong and Polak, 1967).

One off-line method of using this auxiliary model concept is easy to develop: an initial IV estimation run over a block of data is performed with $\hat{\underline{x}}_k$ defined in (7.3) and the resulting asymptotically unbiased, but probably high variance estimates are used to define the auxiliary model for a second pass through the data. Alternatively and in a rather simpler fashion, the auxiliary model could be set up on the basis of the *biased* linear regression estimates; indeed we shall see that this provides a very useful approach in practice. A second IV run is then made with an auxiliary model based on the estimates obtained in the first run being used to define improved instrumental variables. This kind of iterative procedure will yield asymptotically unbiased estimates at each iteration provided the IV's generated during the iteration possess the required properties, i.e. high correlation with $\underline{x}_k$ and statistical independence from the noise ξ_k. It can be continued until there is no significant change in the resulting estimates; at which point we should have obtained estimates which are consistent and relatively efficient (although they will not, in general, be minimum variance, as discussed later in Chapter 8). Of course this iterative procedure can use either the en-bloc solution (4.17) of the IV equations or the recursive solution given in Algorithm IV: in the latter recursive-iterative case, however, the analyst will have the added dimension of the recursive estimates to enhance his results. We will have more to say on the implementation of this iterative solution in the next section, but first let us consider an alternative *on-line* procedure.

An on-line IV procedure can be obtained by utilizing the recursive solution to the IV equations and then updating the auxiliary model continuously on the basis of these recursive estimates. In other words, the IV estimates are generated by algorithm IV, i.e.,

$$\hat{\underline{a}}_k = \hat{\underline{a}}_{k-1} - \hat{\underline{k}}_k \{\underline{z}_k^T \hat{\underline{a}}_{k-1} - y_k\} \quad (1)$$

where

$$\hat{\underline{k}}_k = \hat{P}_{k-1}\hat{\underline{x}}_k \; [1 + \underline{z}_k^T \hat{P}_{k-1}\hat{\underline{x}}_k]^{-1} \quad (2) \qquad \text{IV}$$

and

$$\hat{P}_k = \hat{P}_{k-1} - \hat{P}_{k-1}\hat{\underline{x}}_k \; [1 + \underline{z}_k^T \hat{P}_{k-1}\hat{\underline{x}}_k]^{-1} \; \underline{z}_k^T \hat{P}_{k-1} \quad (3)$$

with $\hat{\underline{x}}_k$ generated by the continuously updated auxiliary model.

In order to ensure that the auxiliary model is well behaved and that the instrumental variables satisfy the validity conditions (4.16) in this fully recursive mode of operation, it is necessary to update its parameters rather carefully, as shown in Figure 7.1. Here, the block A merely compensates for the fact that the estimate $\hat{\underline{a}}_k$ is correlated with the noise ξ_k at the same instant by passing it through a special form of time delay and a discrete low pass filter. The output of the time delay elements $\hat{\underline{a}}_k$ (t.d.), is updated only every n + q sampling instants, to the value of $\hat{\underline{a}}_{k-\tau}$ obtained $\tau = n + q$ instants previously, where q is chosen so that ξ_k can be assumed independent of ξ_{k-q} (Wong and Polak, 1967). The function of the low pass filter is to allow for any inaccuracy in the estimation of q and to smooth the delayed estimate $\hat{\underline{a}}_{k-\tau}$ so avoiding rapid changes in the auxiliary model parameters. The form of the discrete low pass filter is as follows,

$$\underline{\alpha}_k = \underline{\alpha}_{k-1} + \gamma\{\hat{\underline{a}}_{k-\tau} - \underline{\alpha}_{k-1}\}$$

or

$$\underline{\alpha}_k = (1 - \gamma)\, \underline{\alpha}_{k-1} + \gamma\hat{\underline{a}}_{k-\tau} \quad (7.6)$$

where $\gamma = 1/T$; T is the exponential smoothing time constant in sampling intervals; and $\underline{\alpha}_k = [\alpha_{1k}\ldots\alpha_{nk}, \beta_{ok}\ldots\beta_{nk}]^T$, where the subscript k denotes the values at the kth sampling instant.

The function of the second block B shown in Figure 7.1 is to prevent the auxiliary model from becoming unstable during the initial convergence of the algorithm: a check on the stability such as the Jury (1964) or Shur-Cohn (Marden, 1949) criteria is applied to the $\underline{\alpha}_k$ and the auxiliary model is only updated if stability is indicated; otherwise the model parameters are frozen at their previous stable values until a stable update is obtained.

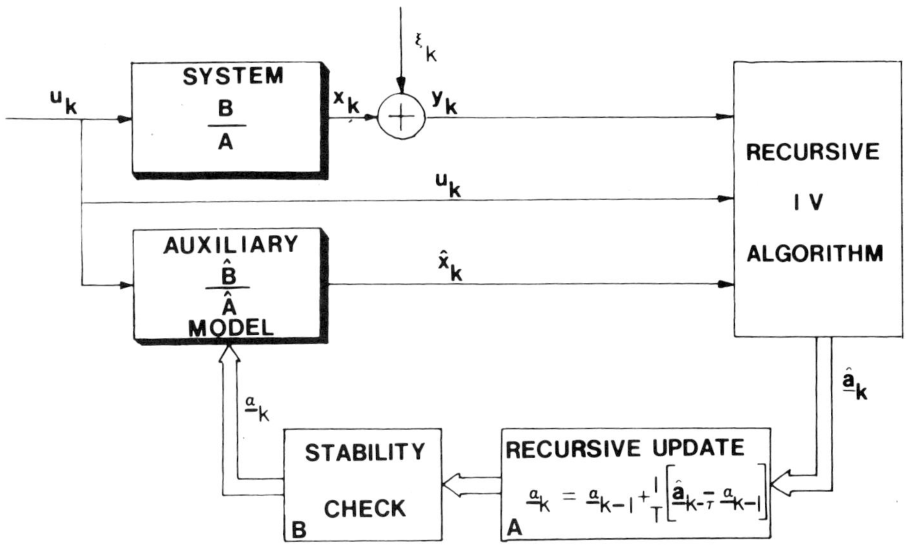

Figure 7.1 The Recursive IV Algorithm

One simple addition to the stability check is to keep the auxiliary model parameters set at their initial values $\underline{\alpha}_0$ (which can be chosen as in the off-line iterative procedure described earlier or on the basis of prior off-line analysis) until reasonable initial convergence has been achieved and the danger of obtaining an unstable set of parameters can be assumed negligible, i.e.

$$\left.\begin{aligned} \underline{\alpha}_k &= \underline{\alpha}_0; \quad k = 0, 1, \ldots, M \\ \underline{\alpha}_k &= (1 - \gamma)\,\underline{\alpha}_{k-1} + \gamma\underline{\alpha}_{k-\tau}; \quad k > M \end{aligned}\right\} \qquad (7.7)$$

where M is chosen according to the expected convergence rate and will depend upon the level of noise on the data. Naturally M is a design parameter; however, the algorithm does not appear too sensitive to its choice.

Another improvement which is possible at the cost of introducing yet other program parameters, in addition to τ and M, is to utilize an EWP version of the Algorithm IV, i.e.,

$$\left.\begin{aligned} \hat{\underline{a}}_k &= \hat{\underline{a}}_{k-1} - \hat{P}_{k-1}\,\hat{\underline{x}}_k(\delta_k + \underline{z}_k^T\hat{P}_{k-1}\hat{\underline{x}}_k)^{-1}\{\underline{z}_k^T\hat{\underline{a}}_{k-1} - y_k\} \\ \hat{P}_k &= \frac{1}{\delta_k}\{\hat{P}_{k-1} - P_{k-1}\hat{\underline{x}}_k(\delta_k + \underline{z}_k^T\hat{P}_{k-1}\hat{\underline{x}}_k)^{-1}\,\underline{z}_k^T\hat{P}_{k-1}\} \end{aligned}\right\} \qquad (7.8)$$

with the scalar δ_k defined as,

$$\delta_k = \lambda_o \delta_{k-1} + (1-\lambda_o)\delta \tag{7.9}$$

and $\delta = 1.0$. Here δ is used rather than α (as in equation (5.15)) to avoid any confusion with $\underline{\alpha}$. From equation (7.9), δ_k approaches unity as k increases so that the algorithm 'forgets' its initial (and rather poor) estimates but approaches infinite memory as time proceeds (see Chapter 5, p.63). λ_o and δ_o are the additional parameters to be chosen by the analyst: typical values are $\lambda_o = 0.99$ and $\delta_o = 0.95$.

7.2 The Recursive-Iterative IV Algorithm

Not surprisingly given its off-line nature, the recursive-iterative solution of the IV equations yields better results than the equivalent on-line version in the sense that, for a given finite set of sampled data, the recursive-iterative estimates have lower estimation error variance; they are, in other words, relatively more efficient in statistical terms. Of course, this advantage becomes less as the data base is increased in size, simply because the additional improvement obtained via iteration itself becomes less. But for normal applications we cannot assume that large sample sizes will be available and normally we must 'make do' with short runs of data obtained from limited experiments or monitoring exercises. In this situation, the recursive-iterative algorithm comes into its own and provides a natural tool for day-to-day time-series analysis where on-line processing is not demanded.

The recursive-iterative IV algorithm has been in general use for some ten years and the exact form of its implementation has changed little over this time. The procedure is broadly as we have outlined in the last section but it is worth describing the implementation in a little more detail. For a total sample size T, a typical general implementation of the algorithm is as follows:

1. The number of iterations I is specified
2. The $\hat{P}_0$ matrix is set to be diagonal in form with equal diagonal elements of 10^6 and $\hat{\underline{a}}_0$ is set to have zero elements[†]
3. The data vector $\underline{z}_k$ is initialized at k = n+1 by reading in the first n+1 samples of y_k and u_k, i.e.

$$\underline{z}_{n+1}^T = [-y_n,\ldots,-y_1, u_{n+1},\ldots,u_1] \tag{7.10}$$

[†] This assumes that, in general, little will be known of the model parameter values.

4. The recursive least squares estimates $\hat{\underline{a}}_{n+1},\ldots,\hat{\underline{a}}_T$ are obtained from the IV algorithm IV by setting $\hat{\underline{x}}_k$ equal to $\underline{z}_k$ whenever it occurs. The matrix $\hat{P}_s$ with $s = {}^T/_I$ is saved and the estimates $\hat{\underline{a}}_T$ are printed out.
5. The parameters of the auxiliary model (7.5) are set to the least squares estimates $\hat{\underline{a}}_T$ obtained at the end of the least squares pass.
6. The initial instrumental variables $\hat{x}_1, \hat{x}_2,\ldots,\hat{x}_n$ are generated by the auxiliary model with $\underline{\alpha} = \hat{\underline{\alpha}}_T$ from 4, i.e.

$$\hat{x}_k = -\alpha_1\hat{x}_{k-1} - \ldots -\alpha_n\hat{x}_{k-n} + \beta_o u_k + \ldots + \beta_n u_{k-n};\ k = 1,2,\ldots,n \qquad (7.11)$$

for $\hat{x}_o, \hat{x}_{-1},\ldots,\hat{x}_{1-n}$ assumed zero; and $\hat{\underline{x}}_{n+1}$ is then initialized with these values i.e.

$$\hat{\underline{x}}_{n+1}^T = [-\hat{x}_n,\ldots,-\hat{x}_1, u_{n+1},\ldots,u_1] \qquad (7.12)$$

while $\underline{z}_{n+1}$ is again initialised as in (7.10).

7. The recursive IV estimates $\hat{\underline{a}}_{n+1},\ldots,\hat{\underline{a}}_T$ are obtained from the IV algorithm IV with the auxiliary model supplying the instrumental variables required to define the IV vector $\hat{\underline{x}}_k$ at each recursive step $k = n+1,\ldots,T$. Here the auxiliary model parameters are maintained *constant* at the values obtained at the end of the least squares pass 4., and the $\hat{P}_o$ matrix is set to the value $\hat{P}_s$ saved in 4. Again $\hat{P}_s$ is saved and the estimates $\hat{\underline{a}}_T$ printed out.
8. The parameters of the auxiliary model are updated to the IV estimates $\hat{\underline{a}}_T$ obtained at the end of the latest IV run 7. Steps 6. and 7. are repeated but with this new auxiliary model defining the IV vector $\hat{\underline{x}}_k$.
9. Step 8. is repeated a further I-2 times.

10. A final (I+1) recursive pass through the data is completed with the auxiliary model set constant to the final estimates $\hat{\underline{a}}_T$ obtained at the end of the Ith pass, but with P_o set to be diagonal in form with diagonal elements of 10^6. Here the recursive estimates $\hat{\underline{a}}_{n+1},\ldots\hat{\underline{a}}_T$, the matrix $\hat{P}_T$, and the noise variable estimates $\hat{\xi}_k = y_k - \hat{x}_k$ $k = 1, 2,\ldots,T$ are all saved. $\hat{x}_k$, $k = 1,2,\ldots,T$ are saved.
11. $\hat{\underline{a}}_T$ saved in 9. and $\hat{P}_T$ saved in 10. are printed out. The sum of the squares of $\hat{\xi}_k$, i.e. $J_o = \sum_{i=1}^{T} \hat{\xi}_i^2$ is printed out.
12. $\hat{\underline{a}}_{n+1},\ldots,\hat{\underline{a}}_T$ saved in 9. are plotted. $\hat{x}_k$, y_k and $\hat{\xi}_k$, $k = 1,2,\ldots,T$ are plotted.

Some comments on the above implementation are necessary. First, note that *the auxiliary model parameters are always kept constant for any iteration at the values obtained at the end of the previous iteration*; the recursive estimates are *not* used to continually update the auxiliary model parameters at each recursive step as in the on-line version, so the update procedures (7.6) and (7.7) are not necessary.

Second, the reader will note that, subsequent to the first iteration and except for the final (I+1)th iteration, the $\hat{P}_0$ matrix is set to a value saved at the recursive step $s(={}^T/_I)$ of the previous iteration. This is an acknowledgement that there will be increasing confidence in the estimates as iterations proceed: bearing in mind that the equivalent P matrix in the least squares algorithm can be interpreted as a covariance matrix, this choice of $\hat{P}_0$ is a heuristic method for informing the algorithm of this increased confidence. Note also that we choose here to specify the number of iterations I, rather than examine the estimates at each iteration in order to assess whether convergence has occurred (e.g. that $\hat{\underline{a}}_T$ obtained at the Ith iteration is insignificantly different from the estimate $\hat{\underline{a}}_T$ obtained at the (I-1)th iteration). This choice is a pragmatic one: it allows for the automatic adjustment of the $\hat{P}_0$ matrix at each iteration, as described above, and has been found to work well in practice. Typically 6 iterations are sufficient, although convergence often occurs after only 2 or 3. To check that convergence has been achieved, however, the estimates should always be printed out at each iteration and examined by the analyst.

Thirdly, the (I+1)th iteration is required to generate the value of the $\hat{P}_T$ matrix (or $\hat{P}_{n+1},\ldots,\hat{P}_T$ if desired) associated with the final IV estimates. As we point out below, $\hat{P}_k$ can itself be interpreted in error covariance matrix terms and can, therefore, be used to assess the statistical properties of the $\hat{\underline{a}}$ estimates. For such purposes, it is important that it is based on a *single* run through the data from an initialization $\hat{P}_0 = 10^6$ diag; otherwise the final $\hat{P}_T$ matrix elements and, therefore, the estimated error covariance properties will be artificially depressed and can lead to overly confident assessment of the error bounds on the estimates.

Fourthly, both the estimated noise variables (or IV residuals) $\hat{\xi}_k$ and the sum of the squares of these variables are required because they provide an indication of the nature of the noise and the degree of explanation of the data. The $\hat{\xi}_i$, $i = 1,2,\ldots,T$, can be used, as we shall see in the next section, to estimate the time-series properties of the noise; whilst the sum of squares J_0 can be used to calculate the coefficient of determination R_T^2 (see e.g. Johnston 1963) where

$$R_T^2 = 1 - \frac{J_0}{\sum_{i=1}^{T} y_i^2} \qquad \text{(N.B. if necessary } y_i \text{ can be replaced by } y_i-\bar{y}\text{, where } \bar{y} \text{ is the mean value of } y_i) \qquad (7.13)$$

This is simply a normalized measure of the degree of explanation of the data: for example, if $J_0 = 0$, indicating that the data are explained perfectly without error then $R_T^2 = 1.0$; as J_0 approaches $\sum_{i=1}^{T} y_i^2$, however, R_T^2 tends to zero indicating that the model has failed completely to explain the data.

Finally, note that this is a 'typical' method for a recursive/iterative solution of the IV equations and modifications to this procedure are possible. It has been found, however, that the above implementation is very robust in day-to-day use and it is recommended on the basis of the author's experience over the past ten years.

7.3 Estimation of the Noise Model Parameters: the AML Method

The recursive IV algorithm in either its on-line or off-line form provides asymptotically unbiased and consistent estimates of the basic system model parameters. For some applications it may be that the analyst wishes to go further in his characterization of the system and obtain some evaluation of the estimated noise sequence $\hat{\xi}_k$. If forecasts of y_k are required, for instance, then the noise characteristics will be important. And again, if the model is required for stochastic control system synthesis, then a model of the noise would be an essential pre-requisite for most analyses.

Practical and simulation experience with the adaptive IV algorithm discussed in Section 7.2 indicates that the instrumental variable output of the auxiliary model quickly becomes indistinguishable from the output of a similar model constructed with the maximum likelihood estimates†; in other words, $\hat{x}_k$ converges to x_k quite rapidly. As we have pointed out, therefore, it is possible to generate a rapidly convergent estimate $\hat{\xi}_k$ of the noise ξ_k from the relationship

$$\hat{\xi}_k = y_k - \hat{x}_k \tag{7.14}$$

In the on-line IV algorithm, the time series generated from (7.14) should represent a reasonable estimate of the noise time-series ξ_k subsequent to the satisfactory convergence of $\hat{x}_k$ to x_k: while in the off-line algorithm the estimate $\hat{\xi}_k$ obtained at the Ith iteration should be a very good estimate of ξ_k, given the superior properties of the recursive-iterative algorithm. In both cases, therefore, the analyst can obtain a good idea of the nature of the noise.

In this latter connection, note that the IV approach yields good estimates of the system parameters *regardless of the noise*, provided only that it is uncorrelated with the chosen instrumental variables. As a result, reasonable results can be obtained even if ξ_k is 'peculiar' noise, e.g. constant or slowly changing bias, sinusoidal, sawtooth or even erratically behaved. If the noise ξ_k is well behaved, however, simple sample statistics will provide a reasonable description and this can be obtained by computing the sample statistics associated with its estimate $\hat{\xi}_k$ generated during the IV analysis: for example mean, variance and autocorrelation properties are easy to evaluate as follows

---

† As might be expected, convergence in this sense is nearly always faster than convergence in a purely parametric sense.

sample mean: $$\bar{\xi} = \frac{1}{T}\sum_{i=1}^{T}\hat{\xi}_i \qquad (7.15)$$

sample variance: $$\bar{\sigma}^2 = \frac{1}{T}\sum_{i=1}^{T}(\hat{\xi}_i - \bar{\xi})^2 \qquad (7.16)$$

sample autocorrelation for lag δ $$r_\delta = \frac{c_\delta}{c_o} \; ; \; c_\delta = \frac{1}{T}\sum_{i=1}^{T-\delta}(\hat{\xi}_i - \bar{\xi})(\hat{\xi}_{i+\delta} - \bar{\xi}) \qquad (7.17)$$

for $\delta = 0, 1, 2, \ldots, K$.

Alternatively or additionally, spectral properties of $\hat{\xi}_k$ could be evaluated (see e.g. Box and Jenkins, 1970, p.39).

If the noise $\hat{\xi}_k$ has Gaussian amplitude distribution and rational spectral density, then the autocorrelation or spectral properties of the estimated noise are useful in providing a description of the noise ξ_k. But they are both (nominally) infinite dimensional representations and so are neither particularly efficient in parametric terms nor convenient in analytical terms. The alternative is to attempt to model the noise ξ_k directly in ARMA terms via equation (6.30)(ii): this is clearly a parametrically efficient (parsimonious) representation and can be of immediate use in forecasting and control system design.

If it is assumed that the statistical properties of $\hat{\xi}_k$ closely resemble those of ξ_k, then the problem of estimating $\underline{c}$ in equation (6.45), i.e.

$$\xi_k = \underline{n}_k^T \underline{c} + e_k \qquad (7.18)$$

reduces to the well known problem of estimating the parameters of an ARMA model, but with the actual, though unmeasurable, noise sequence ξ_k replaced by the available estimate, $\hat{\xi}_k$. In other words, it is necessary to consider the estimation of $\underline{c}$ in the model

$$\hat{\xi}_k = \hat{\underline{n}}_k^T \underline{c} + e_k \qquad (7.19)$$

where

$$\hat{\underline{n}}_k^T = [-\hat{\xi}_{k-1} \ldots -\hat{\xi}_{k-n}, e_{k-1} \ldots e_{k-n}]$$

There are a number of methods available for estimating the parameters of a model such as (7.19): in the case where the model is purely autoregressive (i.e. $D(z^{-1}) = 1.0$), the problem is simple since e_k is independent of $\hat{\underline{n}}_k$ by definition and consistent estimates can be obtained by simple linear regression analysis in a manner similar to that discussed previously (page 123). As with the basic system discussed in the last section, however, this approach will yield biased estimates if it is applied in the more general case where there are 'numerator dynamics' in the transfer function, e.g. in the case, where there are moving average residuals of the form (6.22).

$$\eta_k = d_1 e_{k-1} + \ldots + d_n e_{k-n} + e_k$$

Once again, an IV procedure can be used for obviating this difficulty and obtaining estimates of the autoregressive parameters: here the IV vector is defined in terms of the delayed output sequence in the following manner

$$\underline{\hat{x}}_k^T = [-\hat{\xi}_{k-2n-\delta}, -\hat{\xi}_{k-2n+1-\delta}, \ldots, \hat{\xi}_{k-n-1-\delta}]$$

This is an interesting approach which the author (Young, 1972) has shown is equivalent to the solution of the well known Yule-Walker equations (Mann and Wald, 1943; Box and Jenkins, 1970) when $\delta = 0$. Unfortunately it suffers from the same disadvantages as the Yule-Walker method, namely, it only provides estimates of the AR parameters c_i, $i = 1, 2, \ldots, n$ and these estimates can possess very poor statistical efficiency.[†]

A statistically more efficient and algorithmically more attractive solution to the complete ARMA estimation problem is the recursive Approximate Maximum Likelihood[††] (AML) procedure: this was suggested first by Panuska (1968) in an ARMAX model context using a scalar gain stochastic approximation algorithm; it was then modified to the more rapidly convergent recursive least squares form by Young (1968) and later by Panuska (1979). It was subsequently discussed in more detail within a TF model context by Young et al (1971) and Young (1974).

The AML method, as its name implies, owes much in conceptual terms to the Maximum Likelihood procedure of Astrom and Bohlin (1966). In order to understand the motivation behind the method, it is necessary to consider the ARMA model (7.19). If the $e_{k-1}, \ldots, e_{k-n}$ in this model could be measured, then the problem is similar in estimation terms to the purely autoregressive model case and asymptotically efficient estimates could be obtained by linear regression analysis. Bearing this in mind, an approximate solution is possible in which the parameter vector $\underline{c}$ is estimated using the following least squares algorithm,

---

† The d_i, $i = 1, 2, \ldots, n$, could be obtained once the c_i are known (e.g. Box and Jenkins, 1970) but this is unwieldy and again statistically inefficient.

†† A term first used by Astrom and Eykhoff (1971); sometimes the method when used with an ARMAX model is termed "Extended Least Squares" or RML1 (see Chapter 6).

$$\hat{\underline{c}}_k = \hat{\underline{c}}_{k-1} - \underline{k}_k \hat{\underline{n}}_k \{\hat{\underline{n}}_k^T \hat{\underline{c}}_{k-1} - \hat{\xi}_k\} \qquad (1) \qquad \text{XIII}$$

where

$$\underline{k}_k^n = P_{k-1}^n \hat{\underline{n}}_k [1 + \hat{\underline{n}}_k^T P_{k-1}^n \hat{\underline{n}}_k]^{-1}$$

and

$$P_k^n = P_{k-1}^n - P_{k-1}^n \hat{\underline{n}}_k [1 + \hat{\underline{n}}_k^T P_{k-1}^n \hat{\underline{n}}_k]^{-1} \hat{\underline{n}}_k^T P_{k-1}^n \qquad (2) \qquad \text{XIII}$$

Here $\hat{\underline{n}}_k$ is defined as

$$\hat{\underline{n}}_k^T = [-\hat{\xi}_{k-1} \dots -\hat{\xi}_{k-n}, \hat{e}_{k-1} \dots \hat{e}_{k-n}] \qquad (7.20)$$

and $\hat{e}_k$ is an estimate of e_k based on the current estimate $\hat{\underline{c}}_k$, i.e. from (7.18)[†]

$$\hat{e}_k = \hat{\xi}_k - \hat{\underline{n}}_k^T \hat{\underline{c}}_k \qquad (3) \qquad \text{XIII}$$

At k = n+1, it is assumed that the elements $-\hat{\xi}_{k-1}, \dots, -\hat{\xi}_{k-n}$ are known, i.e. they have been read in before to the start of estimation. Since no alternative *a priori* information is available, it is assumed that $\hat{e}_n, \dots, \hat{e}_1$ are all zero; i.e. they are set to their expected values. After the first recursive step at k = n+1, however, an estimate $\hat{e}_{n+1}$ can be computed from III(3) and substituted in (7.20), i.e. XIII(3)

$$\hat{\underline{n}}_{n+2}^T = [-\hat{\xi}_{n+1} \dots -\hat{\xi}_2, \hat{e}_{n+1}, 0, \dots, 0]$$

This is then used to obtain $\hat{\underline{c}}_{n+2}$ and the whole procedure is repeated until, at the (2n + 1)th step,

$$\hat{\underline{n}}_{2n+1}^T = [-\hat{\xi}_{2n}, \dots, -\hat{\xi}_{n+1}, \hat{e}_{2n}, \dots, \hat{e}_{n+1}]$$

and the vector is full. Thereafter, the procedure continues with a new $\hat{e}_{k-1}$ being added to, and $\hat{e}_{k-n-1}$ being removed from $\hat{\underline{n}}_k$ at each step.

In the on-line use of the algorithm XIII an EWP modification can be utilized with δ_k defined as in (7.9): this will often lead to worthwhile improvement since the $\hat{\xi}_k$ and $\hat{e}_k$ estimates obtained initially can be quite poor and will need to be progressively forgotten. A recursive-iterative version of algorithm XIII is also possible although it does not yield so much improvement in estimation terms. Here, at each iteration, the P_0^n matrix is adjusted as in the IV case to the value saved at recursive step s = T/I and the initial parameter estimate $\hat{\underline{c}}_0$ is set to the value at the end of the previous iteration. Also the $\hat{\underline{n}}_k$ vector is 'primed' initially by $\hat{e}_k$

† It has been suggested (Soderstrom *et al.* 1974) that the *a priori* estimate $\hat{\underline{c}}_{k-1}$ should be used in this definition but this makes little sense since the estimate $\hat{e}_k$ is not then based on the latest information. Soderstrom *et al.* refer to the algorithm with this modification as RML1 (see Chapter 6).

values generated with these latest estimates. An EWP modification may also be worthwhile in this case, although it requires the specification of the additional program parameters δ_0 and λ_0, but the author is not aware of any studies which have suggested that it makes a significant difference in practice.

7.4 Statistical Properties of the IV and AML Estimates

By analogy with the recursive regression algorithm, we can argue that the $\hat{P}$ and P^n matrices generated by the IV and AML algorithms will be related to the covariance matrices of the estimation errors associated with $\hat{\underline{a}}$ and $\hat{\underline{c}}$ respectively. As we shall see in subsequent chapters, however, it is only in certain "refined" versions of these algorithms that there is a direct relationship which is useful in evaluating accurately the error statistics. Nevertheless we can use the $\hat{P}$ and P^n matrices to give us some idea, at least, of the *relative* error covariance properties.

IV Estimates

It is well known that the problem of assessing the statistical properties of IV estimates is rather difficult (Durbin, 1954). In the regression situation, expressions for the covariance matrix of the estimation errors can be obtained as shown, for example, by Dhrymes (1970). The problem is, however, more difficult in the time-series case but, if strong stability of the adaptive IV procedure can be assured, the problem is made somewhat more tractable because of the asymptotic nature of the recursive estimates.

The en-bloc IV solution can be written (c.f. equation (4.15))

$$\hat{\underline{a}}_k = [\hat{X}_k^T Z_k]^{-1} \hat{X}_k^T \underline{y}_k \tag{7.21}$$

if $\hat{X}_k$ and Z_k are defined as kx(2n+1) matrices with rows $\hat{\underline{x}}_i^T$ and $\underline{z}_i^T$ respectively; while $\underline{y}_k$ is a k vector with elements y_i and where i is defined accordingly. Now, since in this case,

$$\underline{y}_k - Z_k\underline{a} + \underline{\eta}_k$$

where $\underline{\eta}_k$ is a k vector with elements η_i, i = 1,2,...,k, equation (7.21) can be rewritten as

$$\hat{\underline{a}}_k = [\hat{X}_k^T Z_k]^{-1} \hat{X}_k^T [Z_k \underline{a} + \underline{\eta}_k]$$

so that the estimation error $\tilde{\underline{a}}_k$ is given by

$$\tilde{\underline{a}}_k = [\hat{X}_k^T Z_k]^{-1} \hat{X}_k^T \underline{\eta}_k$$

and the covariance matrix of $\tilde{\underline{a}}_k$ is obtained as

$$P_k^* = E\{\tilde{\underline{a}}_k \tilde{\underline{a}}_k^T\} = E\{[\hat{X}_k^T Z_k]^{-1} \hat{X}_k^T \underline{\eta}_k \underline{\eta}_k^T \hat{X}_k \{[\hat{X}_k^T Z_k]^{-1}\}^T\} \tag{7.22}$$

It is difficult to progress any further with this expression since Z_k contains components of $\underline{n}_k$ and the expected value operator cannot be taken within the expression, as in the pure regression case.

A simpler but approximate expression for P_k^* can be derived, however, by taking note that, upon convergence, the adaptive IV algorithm is performing in some average manner like the linear regression solution for the equivalent regression model† (where $\underline{x}_i$ is assumed measurable), i.e.,

$$y_i = \underline{x}_i^T \underline{a} + \xi_i$$

i.e.

$$\hat{\underline{a}}_k = [X_k^T X_k]^{-1} X_k^T \underline{y}_k$$

so that, for large sample lengths, the matrix $\frac{1}{k}\hat{P}_k$ approximates the matrix $\frac{1}{k}[X_k^T X_k]^{-1}$. Thus, for the case where the ξ_i are serially independent (C = D = 1.0) in (6.30)(ii), an estimate of P_k^* can be obtained from the well known linear regression result, i.e.

$$P_k^* \simeq \sigma^2[X_k^T X_k]^{-1} \simeq \sigma^2 \hat{P}_k \tag{7.23}$$

while in the more general case where the ξ_i are serially correlated,

$$P_k^* \simeq \hat{P}_k \hat{X}_k^T W \hat{X}_k \hat{P}_k \tag{7.24}$$

where W is the covariance matrix of $\underline{\xi}_k$ which could be estimated, for example, by reference to the estimates of the noise process characteristics using AML.

To ensure reasonable results when using either of the expressions (7.23), (7.24) it is desirable to obtain $\hat{P}_k$ (and, in the case of (7.24), $\hat{X}_k$) from the final (I+1)th run through the data, with the auxiliary model parameters set to the estimates obtained from the initial analysis. In this way, $\hat{\underline{x}}_i \simeq \underline{x}_i$ for all i = 1, 2,...,T, and the effects of convergence are removed.

AML Estimates

In practice, the AML procedure appears to converge quite rapidly and e_k in XIII(3) provides a reasonable estimate of $\hat{e}_k$ even for small k. Noting this and the previously mentioned similarities between the AML procedure and regression analysis for an autoregressive model, it is possible to argue that for $k \to \infty$, the covariance matrix of the AML estimates will be given approximately by

$$P_k^{n*} = E\{\tilde{\underline{c}}_k \tilde{\underline{c}}_k^T\} \simeq \sigma^2 [\hat{N}_k^T \hat{N}_k]^{-1}$$

or

$$P_k^{n*} \simeq \sigma^2 P_k^n \tag{7.25}$$

† Since $\hat{\underline{x}}_k \to \underline{x}_k$; Young, 1969

where $\underline{\tilde{c}} = \underline{\hat{c}} - \underline{c}$ and $\hat{N}_k$ is a kx2n matrix with rows $\underline{\hat{n}}_i^T$. Moreover, the rapid convergence of the AML method should ensure that this expression is an acceptable approximation for finite k.

7.5 Convergence of the IV and AML Estimators

So far we have assumed that the recursive and recursive-iterative IV and AML procedures described in the previous sections will converge. Although it has been known for some ten years that the IV and AML algorithms do, in fact, work well in practical terms, it is clear that the problem of proving theoretically that these recursive and recursive-iterative procedures will converge globally is not a trivial one. Recent work by Ljung (1977a, 1977b, 1979a), Hannan (1976) and Solo (1980) have helped to provide a better understanding of the recursive algorithms, however, and theoretical appraisal of the convergence properties in both cases are available from these references. We will refer to this topic in more detail when "refined" IV and AML methods are discussed in the next chapter. For the moment, it will suffice to say that the IV algorithm will almost always converge to asymptotically unbiased, consistent estimates and, while there are counter examples to the *general* convergence of the AML algorithm, it should also converge most of the time to yield reasonable estimates of the ARMA model parameters, if this provides a reasonable description of the noise (see Chapter 8).

7.6 Identifiability and the Choice of Input Signals

When considering the problem of stochastic model building, it is necessary to investigate those conditions which need to be satisfied in order to obtain sensible estimates of the model parameters. A number of authors have considered this problem of *identifiability* (e.g. Astrom, 1966; Aoki and Yue, 1970; Staley and Yue, 1970) and we will consider here the definition of stochastic identifiability similar to that used by Staley and Yue:

Definition: a model parameter vector $\underline{a}$ is said to be stochastically identifiable if every asymptotically efficient estimator $\underline{\hat{a}}$ of $\underline{a}$ converges to $\underline{a}$ in mean square. By this definition, if $\underline{a}$ is constant and stochastically identifiable, then the estimation error covariance matrix converges to zero as the number of samples approaches infinity.

It is not appropriate that the problem of identifiability should be discussed in detail in this introductory text, since the quoted references give a treatment that is adequate for most purposes. Nevertheless, it is worthwhile reviewing the problem briefly to see where the question of identifiability impinges on the *practical* aspects of stochastic model building.

A special case

Let us consider first the special model with serially uncorrelated residuals outlined in Chapter 6 (page 123). In this case, the normal equations of the least squares analysis can be written in the normal manner as

$$C_k\hat{\underline{a}} = \underline{b}_k$$

with,

$$C_k = \sum^k \underline{z}_i\underline{z}_i^T = Z_k^T Z_k; \quad \underline{b}_k = \sum^k \underline{z}_i y_i = Z_k^T \underline{y}_k$$

From algorithm XII we know that, as $k\to\infty$, the covariance matrix P_k^* of the estimation errors is given by

$$P_k^* = \sigma^2 P_k = \sigma^2 (C_k)^{-1}$$

so that the parameter vector $\underline{a}$ will be stochastically identifiable provided this matrix converges to zero for $k\to\infty$. Put simply, this implies that

$$\lim_{k\to\infty} \frac{1}{k}C_k \text{ exists and is positive definite} \tag{7.26}$$

In order to examine condition (7.26) let us consider the form of C_k. Since $\underline{z}_i$ is defined in terms of the sampled output and input data it is possible to partition Z_k in the following manner

$$Z_k = [Y_k \mid U_k]$$

where Y_k is a kxn matrix with rows $[-y_{i-1},\dots,-y_{i-n}]$ and U_k is a kx(n+1) matrix with rows $[u_i,\dots,u_{i-n}]$. As a result, C_k can be written as

$$C_k = Z_k^T Z_k = \begin{bmatrix} Y_k^T \\ \hline U_k^T \end{bmatrix} \begin{bmatrix} Y_k \mid U_k \end{bmatrix} = \left[\begin{array}{c|c} Y_k^T Y_k & Y_k^T U_k \\ \hline U_k^T Y_k & U_k^T U_k \end{array}\right]$$

In order that condition (7.26) should hold, it is necessary that for $k\to\infty$,

$$\left| \frac{1}{k} C_k \right| \neq 0 \tag{7.27}$$

where $\left| \frac{1}{k} C_k \right|$ denotes the determinant of $\frac{1}{k}C_k$. Now,

$$\frac{1}{k}|C_k| = \left|\begin{array}{c|c} \frac{1}{k}Y_k{}^TY_k & \frac{1}{k}Y_k{}^TU_k \\ \hline \frac{1}{k}U_k{}^TY_k & \frac{1}{k}U_k{}^TU_k \end{array}\right|$$

and, by using a determinantal identity for partitioned matrices (see Gantmacher, 1960; p. 46), this determinant can be expressed either as

$$|\frac{1}{k}Y_k{}^TY_k| \cdot |\frac{1}{k}U_k{}^TU_k - \frac{1}{k}U_k{}^TY_k(\frac{1}{k}Y_k{}^TY_k)^{-1}\frac{1}{k}Y_k{}^TU_k|$$

or

$$|\frac{1}{k}U_k{}^TU_k| \cdot |\frac{1}{k}Y_k{}^TY_k - \frac{1}{k}Y_k{}^TU_k(\frac{1}{k}U_k{}^TU_k)^{-1}\frac{1}{k}U_k{}^TY_k|$$

As a result, condition (7.27) can be written in either of the following alternative forms, for $k\to\infty$,

$$|\frac{1}{k}Y_k{}^TY_K| \cdot |\frac{1}{k}U_k{}^TFU_k| \neq 0$$

or

$$|\frac{1}{k}U_k{}^TU_k| \cdot |\frac{1}{k}Y_k{}^TGY_k| \neq 0$$

where $F = I-Y_k(Y_k{}^TY_k)^{-1}Y_k{}^T$ and $G = I-U_k(U_k{}^TU_K)^{-1}U_k{}^T$ are both idempotent matrices (see e.g. Johnson, 1963.)

From these expressions, we see that necessary conditions for stochastic identifiability are that

$$\left.\begin{array}{ll} \text{(A)} & \lim\limits_{k\to\infty}\frac{1}{k}U_k{}^TU_k \\ & \\ \text{(B)} & \lim\limits_{k\to\infty}\frac{1}{k}Y_k{}^TY_k \end{array}\right\} \text{ exist and are positive definite} \qquad (7.28)$$

Although in this case it is possible to show that these conditions are also sufficient for stochastic identifiability (e.g. Astrom, 1966), we will not pursue the matter any further at this time. It will suffice here to examine how the conditions (7.28) (A) and (B) impose restrictions both on the nature of the input signal u_k and the characteristics of the system being modelled.

Choice of input signals

The matrix $\frac{1}{k} U_k^T U_k$ can be written as

$$\frac{1}{k}\begin{bmatrix} \sum^k u_i^2 & \cdots\cdots\cdots & \sum^k u_i u_{i-n} \\ \sum^k u_{i-1}u_i & \sum^k u_{i-1}^2 \quad \cdots & \sum^k u_{i-1}u_{i-n} \\ \vdots & \ddots & \vdots \\ \sum^k u_{i-n}u_i & \cdots\cdots\cdots & \sum^k u_{i-n}^2 \end{bmatrix} \tag{7.29}$$

If this is to exist and be positive definite for $k \to \infty$ then, as Astrom has pointed out (Astrom, 1966), the u_i must be *persistently exciting* of order n+1 in the sense that

$$\bar{u} = \lim_{k\to\infty} \frac{1}{k} \sum^k u_i \tag{7.30}$$

and

$$r_\delta = \lim_{k\to\infty} \frac{1}{k} \sum^{k-\delta} (u_i - \bar{u})(u_{i+\delta} - \bar{u})$$

both exist and the covariance matrix

$$R = \begin{bmatrix} r_o & r_1 & \cdots\cdots & r_n \\ r_1 & r_o & \cdots\cdots & r_{n-1} \\ \vdots & & \ddots & \vdots \\ r_n & \cdots\cdots & & r_o \end{bmatrix} \tag{7.31}$$

is positive definite. In other words, u_i should remain bounded both in mean and variance, and should continuously perturb or 'excite' the system. These seem very reasonable, almost intuitive conditions and clearly provide a good criterion for judging the adequacy of input signals. For example, a pseudo-random binary (PRBS) input signal with finite variance σ^2 provides persistent excitation since conditions (7.30) and (7.31) are clearly satisfied and

$$r_\delta = \begin{cases} \sigma^2 & ; \quad \delta = 0 \\ 0 & ; \quad \delta \neq 0 \end{cases}$$

Similarly, a square wave of an appropriate frequency provides a useful perturbing signal. On the other hand, a purely sinusoidal signal will only provide persistent excitation of order two† and it is well known (e.g. Young, 1968c) that such a signal can only be used in the identification of a first order process.

It is interesting to note that a number of authors have addressed the problem of optimal choice of input signals (e.g. Aoki and Staley, 1970; Goodwin, 1969, 1971; Box and Jenkins, 1970; Goodwin and Payne, 1977). Here the book by Goodwin and Payne is probably the most useful text for general reading although it is a little esoteric at times for the introductory reader. Note one problem, however: the system characteristics need to be known to define the optimal input! This 'chicken and egg' situation is quite understandable given the nature of the problem but it is inconvenient as regards the application of the analysis: in effect some iterative solution must be defined in which the analysis converges on the optimal input in some manner (see Goodwin and Payne, 1977).

Restrictions on the system to be identified

Turning now to the matrix $\frac{1}{k} Y_k^T Y_k$, we see that this can be written as

$$
\frac{1}{k}\begin{bmatrix}
\sum^{k} y^2_{i-1} & \sum^{k} y_{i-1}y_{i-2} & \cdots\cdots & \sum^{k} y_{i-1}y_{i-n} \\
\sum^{k} y_{i-2}y_{i-1} & \sum^{k} y^2_{i-2} & \cdots\cdots & \sum^{k} y_{i-2}y_{i-n} \\
\vdots & & \ddots & \vdots \\
\sum^{k} y_{i-n}y_{i-1} & \cdots\cdots & \cdots\cdots & \sum^{k} y^2_{i-n}
\end{bmatrix}
$$

If this is to exist and be positive definite for $k\to\infty$ then, as Staley and Yue (1970) have pointed out, the system must not only be excited by a persistently exciting input signal u_k but it should also possess the following characteristics:

1. The system must be stable in the sense that the roots of $A(z^{-1})$ lie outside the unit circle. This ensures that y_k remains bounded provided the noise ξ_k remains bounded.

2. The numerator coefficients b_i, $i = 0, 1, \ldots, n$ should not all be zero. This is obviously necessary, because otherwise the input u_k would have no effect on the output.

---

† The reader can verify that the rank of R in this case is 2.

3. $A(z^{-1})$ and $B(z^{-1})$ should have no common factors. This is the *observability* condition: if common factors exist, then a mode of the system will be masked due to the cancellation of the factors and will be unidentifiable from the input/output measurements (see Ogata, 1967).

The general case

So far we have considered the problem of identifiability only in the case of the special model with serially uncorrelated residuals. In the more general case, simple least squares analysis will not yield consistent estimates and recourse must be made either to the IV-AML method discussed in this chapter or to other approaches such as maximum likelihood. When using the IV-AML procedure, it seems reasonable to conjecture that persistent excitation by the input signal will again be required. However, when considering the system requirements, we are now concerned with the matrix $\frac{1}{k}\hat{X}_k^T Z_k$ rather than $\frac{1}{k} Z_k^T Z_k$. Partitioning as before, this matrix can be written

$$\frac{1}{k}\begin{bmatrix} \hat{X}^T \\ \hline U_k^T \end{bmatrix} [Y_k \mid U_k]$$

Consequently, it is necessary to consider the conditions for existence and positive definiteness of the matrix

$$\lim_{k\to\infty} \frac{1}{k}\hat{X}_k^T Y_k$$

Now, Y_k can be written as

$$Y_k = X_k + \Xi_k$$

where X_k is a kxn matrix with rows $[x_{i-1},\ldots,x_{i-n}]$ and Ξ_k is a kxn matrix with rows $[\xi_{i-1},\ldots,\xi_{i-n}]$. Since by definition Ξ_k is independent of $\hat{X}_k$, it is clear that

$$\lim_{k\to\infty} \frac{1}{k}\hat{X}_k^T Y_k = \lim_{k\to\infty} \frac{1}{k}\hat{X}_k^T X_k$$

In the general IV case, it is difficult to progress any further because the $\hat{X}_k$ matrix of instrumental variables is too loosely defined. However, it seems reasonable to suppose that for $k\to\infty$ the matrix $\frac{1}{k}\hat{X}_k X_k$ will exist and be positive definite provided both that the conditions 1. to 3. (page 147) are satisfied and the instrumental variables are generated satisfactorily for all k. In this sense the adaptive selection of the auxiliary model, either recursively or iteratively, is once again an obvious advantage.

Noise process identifiability

When considering identifiability in the general case, it is necessary to consider the identifiability of the autoregressive moving average noise process. Mehra has addressed this problem (Mehra, 1971) and indicated that, in addition to restrictions on $C(z^{-1})$ and $D(z^{-1})$ similar to those placed on $A(z^{-1})$ and $B(z^{-1})$, it is necessary to assume that the noise model is "minimum phase" or "invertible", i.e. that the roots of $D(z^{-1})$ are outside the unit circle. The term minimum phase comes from the control systems literature (see e.g. Truxal, 1955) while "invertible" is a statistical term (Box and Jenkins, 1970). The noise is "non-invertible" if the roots of $D(z^{-1})$ are inside the unit circle because an attempt to generate e_k from ξ_k via the *inverse* of the transfer function will fail since the resulting inverse transfer function is unstable (with the roots of the denominator inside the unit circle) and the calculation would 'blow up' (see also Young and Jakeman, 1979c).

Some concluding comments on identifiability

The various identifiability conditions discussed briefly in this section should only be considered as a set of rules to be *borne in mind* when attempting to estimate the parameters of a dynamic process: it should be remembered, for instance, that the definition of identifiability on which these rules are based is somewhat restrictive, relying as it does on purely asymptotic characteristics. Before leaving the subject of identifiability, therefore, it is worth noting the following points:

1. Although certain signals may not be persistently exciting, for example a single step input, this does not necessarily mean that sensible estimates of the system parameters cannot be obtained when using such signals. It merely means that when the signal no longer continues to excite the system, no more useful information will be received and the estimates will not improve beyond that point. Nevertheless, the estimates obtained up to the point where the input signal loses its persistent excitation may well be *good enough* for the purposes of the experiment.

2. Again, while a stable system is certainly necessary to prevent the matrices becoming unbounded asymptotically, it may still be possible to obtain estimates of an unstable system provided sufficient estimation accuracy is achieved before the matrices 'blow up' (see e.g. Lee, 1964).

3. It is true that it is not possible to fully identify a non-minimum phase (n.m.p.) noise process, since the covariance generating function is the same for both the non-minimum and equivalent minimum phase systems (see e.g. Box and Jenkins, 1970). But this does not mean that the n.m.p. process cannot be identified at all; it means only that the *equivalent minimum phase* model will be estimated from the data. Whether such information is sufficient will be very problem-dependent: in general, however, it will be sufficient for most purposes.

4. Finally, it should be remembered that, as in any least squares fitting procedure, there is an implicit assumption that the variables associated with the unknown parameters are not either linearly dependent (when there is complete ambiguity and the estimates do not exist) or highly correlated. It is possible that this problem of collinearity (see Chapter 4) could arise in the sort of models considered here; and if collinearity in the form of very high correlation (but not complete linear dependence) is present, it can make the system "poorly identifiable"[†] in the sense that the estimates will exist but have extremely high variance. In the situation where the sampling interval has been chosen too small, for example, the system dynamics can be poorly identified, as we see if we consider a first order stable system of the form

$$x_k = a_1 x_{k-1} + b_o u_k \,; \quad y_k = x_k + \xi_k$$

If the sampling interval T_s is extremely small in comparison to the time constant $\tau = -T_s/\log_e a_1$, then a_1 will be close to unity and even small uncertainty in its estimation will lead to large variations in the estimate of τ (e.g. for $a_1 = 0.99$ and $T_s = 0.1$ secs then $\tau = 9.95$ secs; while for $a_1 = 0.97$ and $T_s = 0.1$ secs, $\tau = 3.28$ secs); this is because the change in x_k is very small between samples in this situation. This is generally true and it emphasizes how the model sampling interval must be chosen carefully if a well defined model is to be obtained and poor identifiability avoided (see e.g. Gustavsson, 1971). Of course, if *full* linear dependence is present (as in the case of a second order system excited by a pure sinusoid) then the estimates will not exist and the system will not be identifiable at all (unless *a priori* information on some of the parameters is available which can remove the ambiguity; see Johnston, 1963; Young, 1968c, 1969c).

7.7 Parametric Variations

Up to this point, it has been assumed that, except for the special Yaglom type of non-stationarity, the time-series data have time-invariant statistical characteristics and can be explained adequately by difference equation models with time-invariant parameters; in other words, it has been assumed implicitly that

$$\underline{a}_k = \underline{a}_{k-1} \,; \quad \underline{c}_k = \underline{c}_{k-1}$$

If this assumption is not correct, then the proposed methods of time-series analysis may fail to yield satisfactory results. There are various methods of obviating this difficulty; Box and Jenkins (1970) for instance, suggest analysis of different

---

† This has been discussed by Young (1968c, 1969c) in connection with differential equation model estimation.

portions of the data to see if there is any indication of significant inconsistency in the parameter estimates obtained from each portion. When recursive estimation is being used, however, it is possible to use a more sophisticated approach based on the techniques of time variable parameter estimation discussed in Chapter 5.

Although the 'dynamic' least squares regression algorithms VI and VII in Chapter 5 are only directly applicable to true regression problems, the similarities between the IV and AML algorithms and the least squares regression algorithm suggest that these algorithms may be modified in a similar manner in order to detect parameter variation. Although such an extension is quite heuristic, it does appear to work well. In the case of the random walk model of the parameter variations, for example, the IV algorithm can be modified to become

$$\hat{P}_{k/k-1} = \hat{P}_{k-1} + D \tag{1}$$

$$\underline{\hat{a}}_k = \underline{\hat{a}}_{k-1} - \underline{\hat{k}}_k \{\underline{z}_k^T \underline{\hat{a}}_{k-1} - y_k\} \tag{2}$$

where

$$\underline{\hat{k}}_k = \hat{P}_{k/k-1}\underline{\hat{x}}_k [1 + \underline{z}_k^T \hat{P}_{k/k-1}\underline{\hat{x}}_k]^{-1}$$

and

$$\hat{P}_k = \hat{P}_{k/k-1} - \hat{P}_{k/k-1}\underline{\hat{x}}_k [1 + \underline{z}_k^T \hat{P}_{k/k-1}\underline{\hat{x}}_k]^{-1} \underline{z}_k^T \hat{P}_{k/k-1} \tag{3}$$

XIV

Here D is a (2n+1)x(2n+1) matrix which allows for possible parameter variation: it is chosen initially in a similar manner to Q_p† but may need to be adjusted by experimentation. A similarly motivated algorithm can be generated in the AML case. We will see later that these algorithms make good sense and can be further justified in statistical terms when considered in relation to the statistically more sophisticated "refined" IV and AML algorithms discussed in Chapters 8 and 10.

7.8 A Time-Series Analysis Procedure Based on IV-AML Estimation

The instrumental variable and approximate maximum likelihood procedures combine to form the basis for a single, flexible method of statistical analysis for linear stochastic dynamic systems with either deterministic or purely stochastic inputs. This IV-AML procedure is not, in itself, a complete approach to the time-series analysis; it merely provides a central computational tool and must be supplemented by additional statistical procedures for preliminary model assessment and *a posteriori* model validation. A complete time-series analysis procedure of this type was outlined first by Young and Hastings-James (1970). It has now been implemented in a modified form as a suite of user orientated computer programs called

† see equation (5.27)

"CAPTAIN"† incorporating storage tube display, graphical and line printer output facilities (Young, Shellswell and Neethling, 1971; Young and Jakeman, 1979b). These programs have several features in common with the alternative approaches of Box-Jenkins (1970) and, to a lesser extent, Astrom and Bohlin (1966). In fact the proposed analysis procedure may be considered as an amalgam of the best aspects of both of these approaches suitably extended and made more flexible by the use of the IV-AML method of recursive estimation.

7.9 Representative Practical Results

The method of stochastic model building discussed in previous sections of this chapter has been applied to various sets of real data, and two typical examples are discussed below. Other examples are given in Young et al (1971).

The gas furnace data of Box-Jenkins

It is appropriate that the first example we should discuss here is the analysis of the gas furnace input-output data considered originally by Box and Jenkins (1970). This is interesting not only because it is the central example of transfer function estimation cited by Box and Jenkins, but also because the results of recursive IV analysis highlight certain important aspects of the data not noted by them.

The gas furnace data consists of 296 samples of the input gas rate, U_k, output CO_2 concentration, Y_k, which are plotted in Figure 7.2. The model structure ascertained by Box-Jenkins, using correlation arguments, is of the form

$$y_k = \frac{(b_0 + b_1 z^{-1} + b_2 z^{-2})}{(1 + a_1 z^{-1})} u_{k-3} + \frac{1}{(1 + c_1 z^{-1} + c_2 z^{-2})} e_k \tag{7.32}$$

where $y_k = Y_k - \bar{Y}_k$, $u_k = U_k - \bar{U}_k$ and $\bar{Y}_k$, $\bar{U}_k$ are the arithmetic means of the Y_k and U_k sequences respectively. The IV-AML estimates obtained for the A, B and C polynomials are set out below in Table 7.1 together with the corresponding results using the maximum likelihood technique of Box and Jenkins.

Comparison between the actual output y_k and the output from the two models in Table 7.1 can be accomplished by setting the auxiliary model parameters to the process model estimates and then observing the auxiliary model output $\hat{x}_k$ over one cycle of the data (Figures 7.3 and 7.4). In both cases, there is close agreement between the actual output and that of the process model, except for a short period at the very end of the data, when quite large errors are observed.

---

† *Computer Aided Program for Time-Series Analysis and Identification of Noisy Systems.*

TABLE 7.1

	IV-AML Estimation		Maximum Likelihood Estimation	
	Parameter	Approx. Standard Error	Parameter	Standard Error
a_1	-0.59	0.03	-0.57	0.21
b_0	-0.59	0.12	-0.53	0.08
b_1	-0.37	0.23	-0.37	0.15
b_2	-0.37	0.19	-0.51	0.16
c_1	-1.53	0.02	-1.53	0.05
c_2	0.65	0.02	0.63	0.05

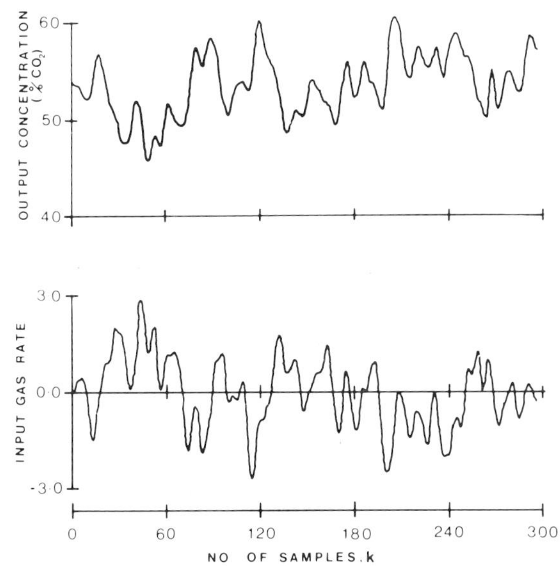

Figure 7.2 Gas Furnace Data

These results are significant for two reasons. First, since the output $\hat{x}_k$ of the auxiliary model is so similar to the actual *noisy* output y_k of the gas furnace, it is clear that there is very little noise on the data and that the stochastic aspects of the model are relatively unimportant! Secondly, those differences that do exist between $\hat{x}_k$ and y_k occur mainly at the very end of the data set.

The first point is confirmed if a simple recursive least squares analysis is applied to the data: the output of an auxiliary model in this case is compared with the actual gas furnace output in Figure 7.5 and a visual comparison shows that the output errors, $y_k - \hat{x}_k$, are about one half of those obtained with the IV-AML and Maximum Likelihood methods! Certainly the model obtained in this simple manner

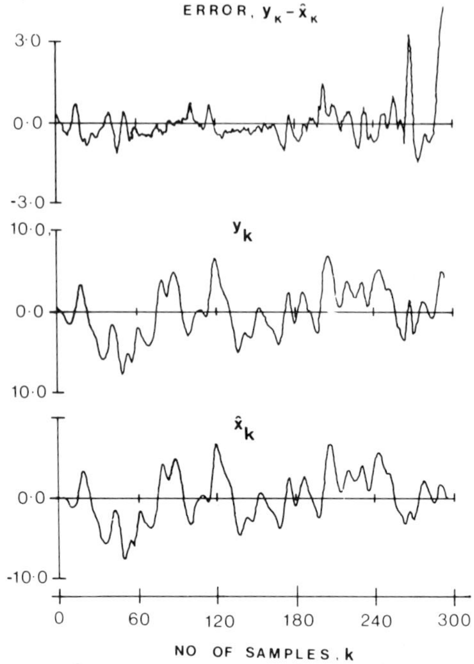

Figure 7.3 Box-Jenkins Results for Gas Furnace

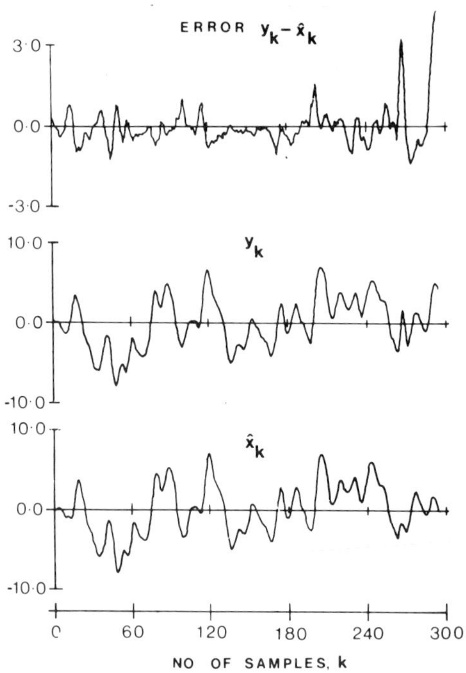

Figure 7.4 IV-AML Results for Gas Furnace

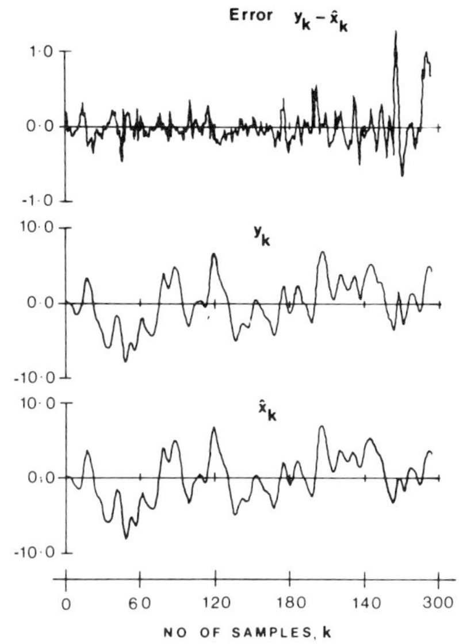

Figure 7.5 Least Squares Results for Gas Furnace

would be good enough for most applications and the use of the more sophisticated and complicated analysis would be difficult to justify in this case.

As far as the second observation is concerned, we see that while the least squares model provides a reasonable fit over the whole 296 sample set, the output error during the final portion of the data is still significantly larger than that observed during the first 270 samples. Such deviation could arise for several reasons, but the most likely explanations are either that the parameter estimates are badly defined in some way (perhaps due to poor identifiability arising from a poor choice of model structure) or that the data are non-stationary, particularly towards the end of the experiment.

There is some evidence for the former explanation, since solution of the least squares normal equations by direct matrix inversion is made difficult because the determinant of the matrix is quite small, indicating near singularity. There is also evidence of parametric non-stationarity: the dynamic least squares regression algorithm VII with $\phi = \Gamma = I$ was applied to the data with Q_p defined as

$$Q_p = 0.705 \begin{bmatrix} .003 & & & \\ & .09 & & \\ & & .32 & \\ & & & .16 \end{bmatrix} \qquad (7.33)$$

The diagonal elements of Q_p were chosen to be equal to the diagonal elements of the error covariance matrix P^*_{296} obtained at the end of the normal least squares analysis. (It was not possible to choose Q_p on the basis of the expected rates of change of the parameters since such information is not available in this case).

The recursive estimates with Q_p set to (7.33) are given in Figure 7.6. There is some tendency for the estimates a_1 and b_2 to 'drift' consistently in one direction during the first 270 samples (the physical outcome of the ill-conditioning discussed above). During the last 30 samples, however, there are significant changes in all of the estimates as the algorithm attempts to correct for the change in response characteristics noted in Figures 7.3 to 7.5.

It would be wrong to draw any firm conclusions from the results obtained here because we have no information on the detailed nature of the experiment. A number of possibilities exist: for instance it could be that something went wrong at the end of the experiment (could this have led to the termination after 296 samples?); alternatively, there could have been actual changes in the dynamic characteristics of the gas furnace arising from some associated changes in the physical characteristics of the process itself, although this seems unlikely. But these are hypotheses; the only real conclusion we can make is that the data are somewhat suspect and that utilization of the estimated model for such applications as automatic control system synthesis should be preceded by further confirmatory experiments. If this is not possible, then it makes sense to utilize estimates obtained by reference to the first 200 samples, where the parameters seem reasonably invariant.

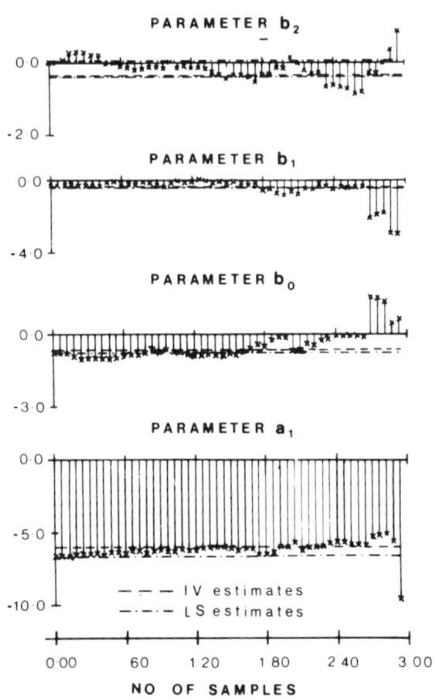

Figure 7.6 Recursive Estimates of Gas Furnace Parameters

The story of this gas furnace data is, however, not yet complete: as we have implied, the recursive estimates in Figure 7.6 do not seem too well defined and tend to 'drift', even during the first 270 samples. Could this mean that identifiability problems are present? On the surface of it, the inputs appear to be 'persistently exciting' and there do not appear to be any problems of this type. But is the model structure (7.32) selected by Box and Jenkins satisfactory and, if not, could this be causing problems? The results of Young *et al.* (1980) show that there is evidence that the structure is indeed over-parameterized and that the estimates are rather poorly defined for this reason.

Rainfall-runoff modeling the Bedford Ouse River Basin

This analysis was required as part of a comprehensive study of water quality in the Bedford-Ouse River System of Central Eastern England and full details of the hydrological modeling are given in Whitehead, Young and Hornberger (1979). The rainfall-runoff analysis has been discussed in a general recursive time-series context by Young (1975) and is concerned with estimating a TF time-series model of the rainfall and streamflow data shown in Figure 7.7. Here the rainfall is averaged from three rain gauges in the Bedford-Ouse area over 1972, while the streamflow is an estimate of the runoff flow due to rainfall over the same period. This estimate was obtained by subtracting the output of a deterministic model for streamflow at a gauging station on the main Bedford-Ouse River system from the measured streamflow at this point: since the deterministic model provided a 'prediction' of the downstream flow based *only* on upstream measured flow, it was assumed that the difference could be accounted for by rainfall-runoff processes between the upstream and downstream flow measurement stations.

It is apparent from Figure 7.7 that the relationship between y_k and u_k is nonlinear with minimum flows occurring at periods of highest rainfall and vice-versa. We might suspect, therefore, that a linear TF model with constant coefficients would be an inappropriate description of the data. But a similar model with *time-variable coefficients* should provide a suitable candidate model structure, and can be evaluated by application of the 'dynamic' IV algorithm XIV. Initial identification studies based on the Box-Jenkins methods of correlation analysis assuming constancy of parameters (Box and Jenkins, 1970) provide a rough[†] indication of the impulse response[*] or the unit hydrograph characteristics between rainfall and runoff flow. On the basis of this initial analysis, a candidate model for rainfall-flow is suggested of the form

---

† Clearly this identification of model structure is not necessarily very satisfactory because the linear constant coefficient model is not strictly applicable. But it has been found that this approach often yields reasonable results in practice.

* see Appendix 1

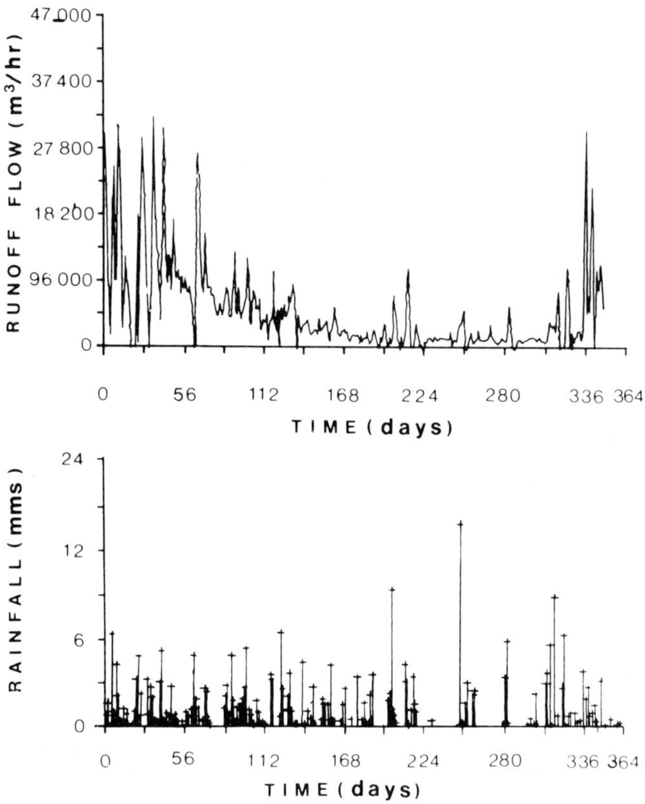

Figure 7.7 Rainfall and Estimated Runoff Data for Bedford-Ouse River, U.K., 1972

$$y_k = \frac{b_1 z^{-1} + b_2 z^{-2}}{1 + a_1 z^{-1} + a_2 z^{-2}} u_k + \xi_k \tag{7.34}$$

The recursive estimates of the coefficients in this model obtained using the recursive IV-AML algorithm display clear signs of temporal variability: in particular, the estimates of b_1 and b_2 appear to vary considerably over the observation interval, even when the algorithm is informed that they are constant. To investigate this further, the b parameters are estimated using the IV-AML algorithm XIV using the assumption that the 'b' coefficients may vary over the observation interval, while the 'a' coefficients are constant.

The estimation results obtained in this manner are shown in Figure 7.8(a). It is clear that both 'b' parameters show distinct variations over the year: in addition to a long-term variation which seems indicative of seasonal influences, there are also short-term effects which appear to occur predominantly when storms follow periods of dry weather. From the hydrological point of view these changes are most probably due to 'evapotranspiration' effects in the long term and 'soil moisture' in the short term: if rain occurs when the temperature is hot, it is quite likely that some evaporation will occur so reducing the rainfall which is 'effective' in inducing river flow; similarly, if it rains when the earth is very dry, we would expect the soil to absorb moisture, whereas if the earth is wet because of previous rainfall, then the rainfall is more likely to be converted into flow effects. The reader will notice that the soil moisture effect is particularly apparent on day 250, for example, where b_1 falls from 1000 to 120 following the end of a dry period.

The concept of an "effective rainfall" measure has been developed by hydrologists in order to allow for factors such as evapotranspiration and soil moisture. Penman (1950) and Grindley (1967) have proposed methods for estimating runoff for rural areas where the evaporation and transpiration effects are calculated as a direct loss depending upon temperature, relative humidity and vegetation types; and where the soil moisture effect is calculated as the cumulative balance of inflow and outflow from the soil storage zones.

In the Bedford-Ouse study, two simple relationships are utilized to allow for evapotranspiration and soil moisture effects. In order to allow for the predominantly temperature dependent evapotranspiration effects, a modified rainfall measure r_k^* is formed by modulating the basic rainfall r_k by a temperature dependent factor proportional to the difference between the prevailing mean monthly air temperature T_i and the overall maximum temperature T_m, i.e.

$$r_k^* = \gamma(T_m - T_i) r_k; \; \gamma = \text{prop.constant} \tag{7.35}$$

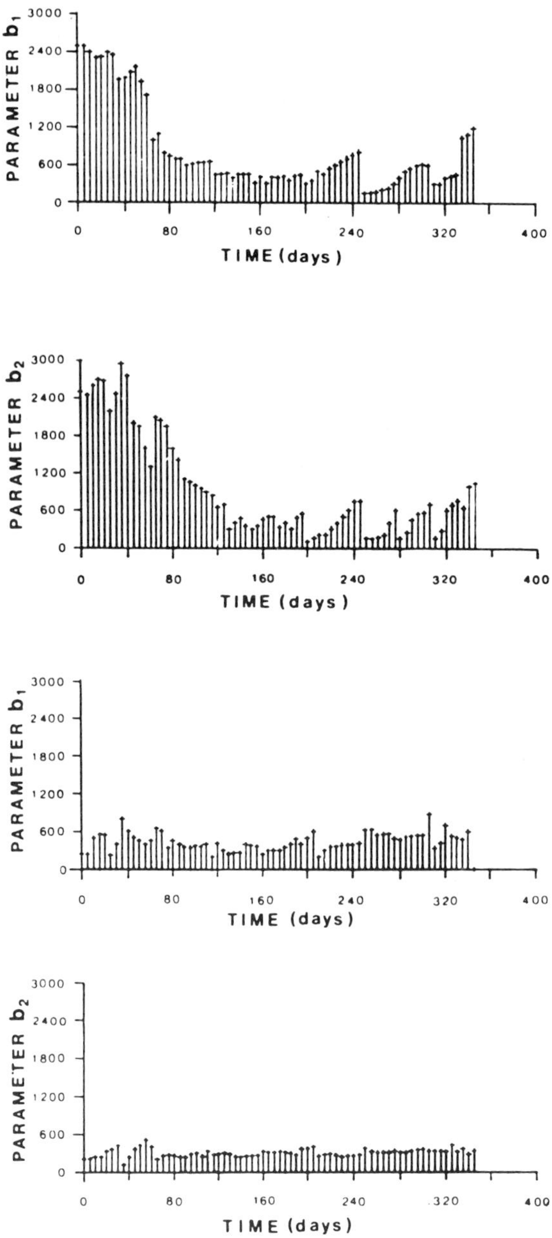

a)

b)

Figure 7.8 Recursive Estimates of Parameters in Rainfall-Runoff Model
(a) using original rainfall series
(b) using effective (filtered) rainfall series

The soil moisture compensation is motivated by a simple analysis which has much in common with the so-called Antecedent Precipitation Index (API) approach[†], but which uses a much simpler and parametrically more efficient approach to exponential weighting into the past. A measure of soil moisture content is obtained by filtering r_k^* by means of a discrete first order filter of the form

$$S_k = S_{k-1} + \frac{1}{T_s}(r_k^* - S_{k-1}) \tag{7.36}$$

where S_k is the output of the filter and T_s is a time constant to be chosen empirically. The final 'effective' rainfall measure u_k is then obtained by modulating the modified rainfall r_k^* by S_k, i.e.

$$u_k = S_k r_k^* \tag{7.37}$$

In this manner, a prolonged period of dry weather will lead to a low value of S_k which, because of the exponential weighting into the past imposed by the filter mechanism (7.36), will not change dramatically after a single storm and will achieve higher sustained values only after continued rainfall. Thus u_k should provide an indication of the effective rainfall, i.e., the rainfall which results in runoff. In effect (7.36) is a crude model of the wetting and drying 'lag' characteristics of the soil; clearly more sophisticated models might be possible (but would they be justified by the data, given that they may require more parameters? - we leave the reader to ponder on this point).

The filter time constant T_s is selected by a trial and error process with γ chosen as a scaling factor. u_k generated from equation (7.37) utilizing these values of γ and T_s, is then used, together with the run-off flow estimate y_k, as the basis for the recursive estimation of the time-series model parameters. This procedure is repeated with different values of T_s *until relatively time-invariant estimates of the* b_1 *and* b_2 *parameters are obtained*, as shown in Figure 7.8(b). Here T_s is set at 25 days which is, presumably, indicative of the slow dynamics associated with soil wetting and drying processes in the Bedford-Ouse catchment.

At this point, the time-series model between the 'effective' rainfall u_k and runoff flow y_k is characterized by coefficients which appear predominantly constant over the observational interval. Thus we have 'identified' a candidate time-invariant parameter model structure which can now form the basis of more statistically efficient parameter estimation *under the assumption of parametric invariance*. Such estimation was carried out using the recursive-iterative version of the IV-AML algorithm. The final estimates obtained in this manner, together with their estimated approximate standard errors (in parentheses) are as follows:

$$a_1 = -0.09 \quad (0.017); \quad b_1 = 275 \quad (1.01);$$
$$a_2 = -0.36 \quad (0.02)\ ; \quad b_2 = 383 \quad (4.40).$$

---

† See e.g. Weyman (1975).

In other words, the transfer function model between y_k and u_k is:

$$y_k = \frac{275z^{-1} + 383z^{-2}}{1 - 0.09z^{-1} - 0.36z^{-2}} u_k + \xi_k \tag{7.38}$$

Finally correlation analysis applied to the estimated noise term ξ_k in this equation indicates that it is a serially uncorrelated series of random variables, thus eliminating the need for a more complicated ARMA noise model in this case.

The nature of the impulse response for this model is shown in Figure 7.9. In hydrological terms, this impulse response would be referred to as a unit hydrograph (or unit graph). Note the parametric efficiency of the model: a normal unit hydrograph model of this type, which is commonly used in hydrology, would probably have required at least 8 - 10 coefficients for adequate characterization, rather than the 4 required in (7.38). Note also the slight oscilliatory nature of the response: this is probably due to the existence of different lagged pathways between rainfall and its effect on flow. In other words, the unit hydrograph obtained in Figure 7.9 is probably an approximation to a 'true' hydrograph which is the sum of a number of non-oscillatory unit hydrographs with different lag effects (see inset to Figure 7.9). This demonstrates how the procedure proposed here can be a useful prelude to more detailed hydrological analysis, if this is demanded by the objectives of the study.

Finally note that, for simplicity of analysis, T_s was "estimated" here by a trial and error procedure. Clearly it might be analytically and statistically superior to estimate T_s and the TF model parameters *together* in a single algorithm (see later).

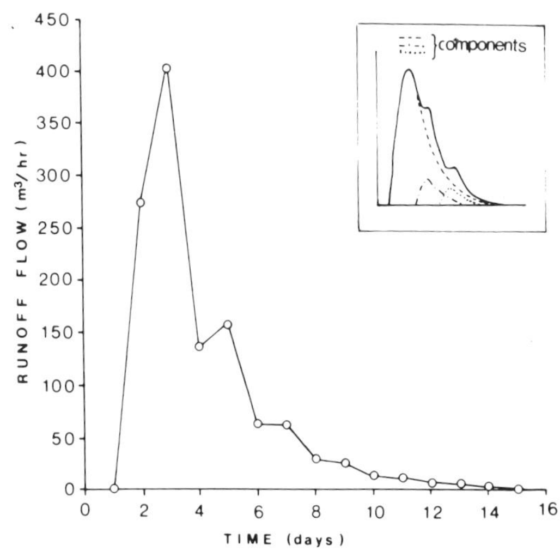

Figure 7.9 Impulse Response of Rainfall Runoff Model

Having identified and estimated the rainfall-runoff model characteristics, it is now possible to combine this model with the basic stream flow model to produce the complete characterization of the hydrological system. Figure 7.10 compares the river flow estimated from this overall model with the observed flow used in the estimation of the model. It is clear from the residual error plot that the estimation is fairly good; the mean percentage error is 8.6% which is well within the accuracy of the flow gauging station, which is estimated at 10% by the river authority. In addition, the model explains 99% of the variance of the original flow series and the errors are within 10% of the observed flows for 70% of the time.

The final and continuing stage in model building is validation: here the model's forecasting ability is evaluated on data *other than that used in the identification and estimation studies*. If the model continues to forecast well over this test data interval, it is assumed that it is *conditionally* acceptable, in the sense that *as far as it is possible to test*, the model seems satisfactory. Figure 7.11 compares the observed stream flow measured at Harrold during 1973 with the flow generated by the model estimated previously over 1972. As might be expected, the fit is not as good as in 1972 but, except for the large storm on day 175, the forecasted flow behaviour is satisfactory with good general agreement between observed and simulated flows.

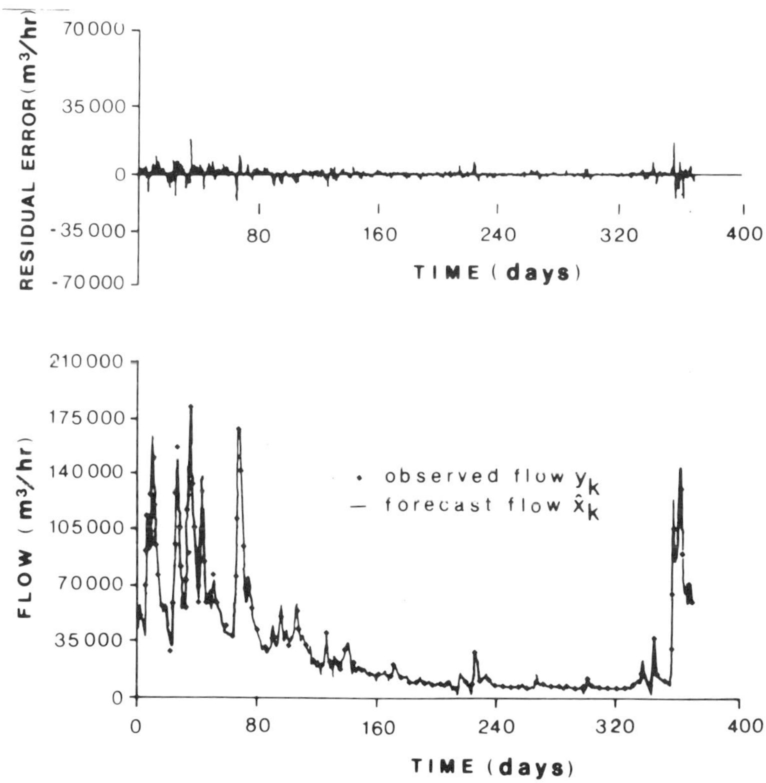

Figure 7.10 Final Model Predicted Flow in 1973

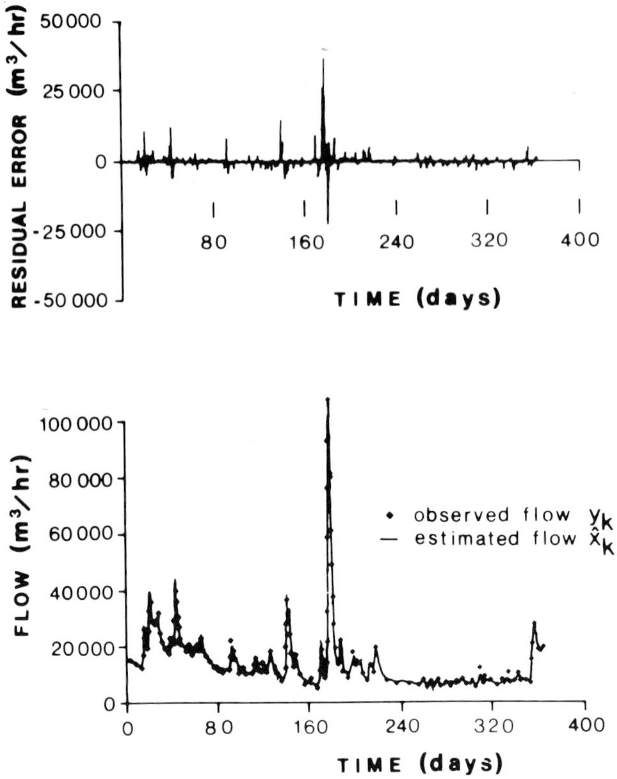

Figure 7.11 Model Predicted Flow in 1973 Using 1972 Model

Flow data were also available for 1973 at Bedford and provided an additional check on the model validity. The *same* rainfall-runoff model for the river section between Harrold and Bedford was applied between Tickford Abbey and Harrold and provides a satisfactory flow forecast at Bedford, as shown in Figure 7.12. This kind of validation exercise is, of course, a continuing procedure since the model would need to be reassessed in the light of future developments and additional data.

This hydrological example demonstrates how a simple but inherently stochastic model for a rather badly defined and nonlinear system can be obtained by application of recursive time-series analysis. It is clear that ordinary en-bloc analysis, such as that described in most texts on time-series analysis, would be poorly suited to the analysis of these data. At the very least, the analyst would take a lot longer to carry out the analysis if he did not have access to the 'added dimension' of recursive estimation and its ability to indicate both the existence and nature of parametric variability or nonlinearity.

The reader should note the following points about the analysis:

1. The nonlinearity was of a special form in which only the b_i parameters of the linear time-variable parameter model changed significantly: in other words, the nonlinear model could be decomposed into a dynamic nonlinearity (described by equations (7.35) - (7.37)) *in series* with a *linear* system, so simplifying the analysis.

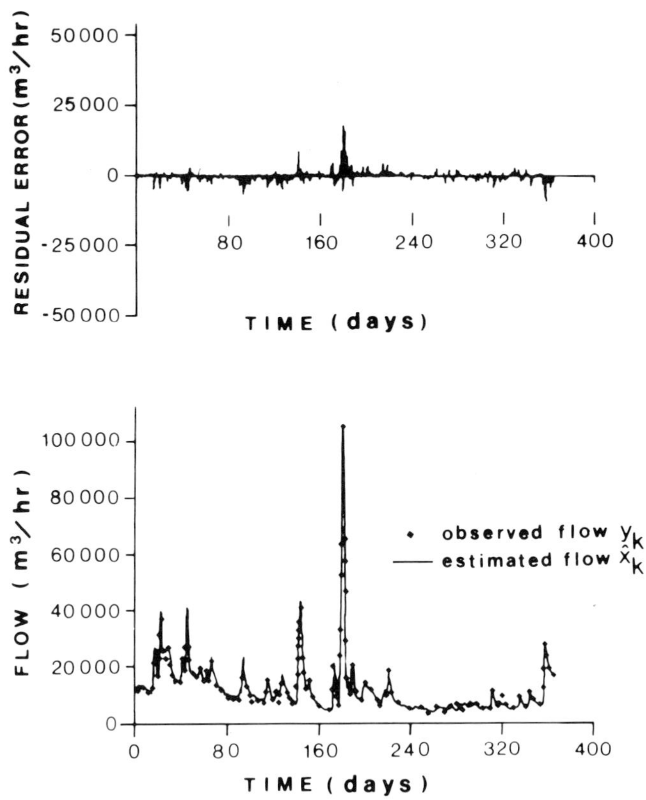

Figure 7.12 Model Predicted Flow at Bedford in 1973 Using 1972 Model for Flow at Harrold

2. This decomposition is quite pleasing in physical terms. It means that there is a constant parameter linear system which characterizes what we might consider as the 'invariant' aspects of the catchment behaviour. These can be described either by the low order TF model itself or, alternatively, by the infinite dimensional impulse response (unit hydrograph) defined completely by this model. But these invariant characteristics are 'hidden' behind the observed overall behaviour of the catchment which is a function of the series nonlinearity. And this nonlinearity is, in part, a function of the antecedent rainfall behaviour, nominally into the infinite past. In effect, therefore, the nonlinear unit hydrograph generated by the entire model is a whole (infinite dimensional) ensemble of impulse responses: at any time k, the basic form of the instantaneous impulse response will be defined by the linear TF model characteristics, but the exact magnitude of its ordinates will be multiplied by a gain which is dependent upon the nonlinearity and, therefore, on the history of the catchment up to time k. And all of this is neatly encapsulated in the five parameters (a_1, a_2, b_1, b_2 and T_s) of the nonlinear model; a quite efficient parameterization for the description of a catchment of this size. The reader will note how neatly the recursive estimation discloses the presence of the hidden invariant catchment characteristics and so enables the analyst to achieve this efficient parameterization.

3. The "tuning" of the soil moisture lag time constant T_s in (7.36) was justified because only a single unknown parameter was involved; it is, however, another unknown parameter and, strictly, it should be estimated simultaneously with the a_i and b_i TF model parameters. This would involve considerable additional complication in parameter estimation terms (particularly if recursive estimation was demanded) since it would pose a problem of nonlinear estimation. But this would undoubtedly be a more correct approach and might result in improved parameter estimates. Whether this possibility would justify the increased complexity is, however, difficult to assess in practical terms.

4. The approach to modelling nonlinear systems exemplified by this hydrological example has much wider implications, as discussed by Young (1978) and Priestley (1980 Almost any nonlinear system can be considered as a time-variable parameter, linear system with parameters either the function of other "external" variables (as in the case of temperature in the hydrological model) or other state variables, i.e. variables that directly describe the behaviour of the system (as in the case of soil moisture S_k in the hydrological model). Priestley, in an elegant description of such models (1980) describes the first kind of characterisation as a "dynamic linear model' (DLM) and reserves the description "state dependent model"(SDM) for the second type. Young (1978) prefers to consider all such models simply as time-variable parameter (TVP) models and assumes that the recursive analysis will reveal, during a "structure identification" phase, whether the parametric variations are related to external (i.e. exogenous) variables or whether they result from interrelationships *within* the system and are of the SDM type.† Basically, however, both approaches are philosophically and conceptually identical: the major differences are methodological with Priestley suggesting the use of recursive "Extended Kalman Filter" type estimators (see Chapter 9); whilst Young proposes the IVAML type procedures described in this Chapter and Chapter 8, which appear more strongly consistent and efficient, particularly in small-medium sample situations (see e.g. Young and Whitehead, 1977).

7.10 Summary

In this chapter, we have seen how general algorithms for recursive time-series analysis can be developed which are only a little more complicated than the recursive least squares algorithm. The recursive-iterative IV procedure provides a robust method for generating asymptotically unbiased and consistent estimates of the

---

† Of course, if we are prepared to extend the system description widely enough, it is possible to consider all variables as "state variables"; but it is usual in systems analysis to define as state variables only those which are essential to the definition of the "dominant dynamic modes" of the system behaviour (Young, 1978) and to consider other variables which affect the variations of the state variables as exogenous variables which are not themselves influenced by the state variables (e.g. by an important feedback mechanism).

basic system parameters in the time-series transfer function (or Box Jenkins) model, even in situations where the noise may not conform to any simple statistical description. When the noise is better behaved and has rational spectral density, however, the recursive AML algorithm can provide a useful adjunct to the IV estimation and yield estimates of the parameters in an ARMA model for the noise. The overall system-noise model in this form provides a parametrically efficient (parsimonious) representation of a linear, stochastic dynamic system which can be useful in many applications such as forecasting, control, management and planning.

But the major advantage of the recursive IV-AML method over alternative procedures, such as en-bloc maximum likelihood, is its extra dimension; its ability to handle the possibility of parametric variability in the time-series model. And we have seen in the hydrological example that this makes it a valuable analytical tool in the modelling of "badly defined" and possibly nonlinear dynamic systems (Young, 1978; Priestley, 1980), as well as a potentially valuable tool in on-line forecasting and control. In the next chapter, we will see how more sophisticated but complex IV procedures can be developed within a maximum likelihood context. Here the AML noise estimates are used to "refine" the IV estimates on a recursive or iterative basis in order to induce asymptotic efficiency, i.e. minimum variance estimation.

8 Optimum Instrumental Variable Methods of Time-Series Model Estimation

Occupying as it does such a curious position at the crossroads between theory and practice, the IV technique described in previous Chapters has tended to be frowned upon by certain statisticians and control theorists because it lacks the statistical niceties of its more sophisticated relations such as maximum likelihood. At the same time it is often misunderstood by many practising engineers and systems analysts, who sometimes tend not to appreciate fully the relevance of the structural model estimation problem and, at the same time, are apt to view the recursive IV algorithm as an over complicated approach to dynamic system estimation. This latter point is accentuated by the fact that many of the publications on the IV method in the control literature are couched in rather abstract mathematical terms which few practitioners find very digestible. And this is a pity since papers of this type can have practical significance, albeit obscured behind a formidable front of mathematical symbolism (see e.g. Finnegan and Rowe, 1974).

In this chapter, we will attempt to correct this situation by showing in simple terms how the IV and AML procedures described in the previous chapter can be considered within the context of maximum likelihood (ML) estimation and so can be *refined* to yield superior performance in statistical terms. In particular, we formulate the estimation problem for the TF time-series model in ML terms and then decompose it into two coordinated sub-problems, the recursive and recursive-iterative solutions of which resemble closely the equivalent IV and AML algorithms discussed in Chapter 7. In this manner, both the virtues and limitations of the *basic* algorithms become more apparent and the reader should be better able to judge when these simple estimation procedures are sufficient for his purposes or, conversely, when the more sophisticated but more complex refined algorithms may prove necessary in order to obtain greater statistical efficiency (Young, 1976a).

8.1 The Maximum-Likelihood Method

Once again let us consider the TF time-series model (6.30) which can be written as

$$y_k = \frac{B(z^{-1})}{A(z^{-1})} u_k + \frac{D(z^{-1})}{C(z^{-1})} e_k \tag{8.1}$$

For simplicity of exposition, let us assume initially that the C and D polynomials are known *a priori*. Suppose that we then further assume that the sequence e_k, $k = 1,2,\ldots,T$, has a Gaussian amplitude distribution, i.e.

$$\underline{e} \sim N(\underline{0}, \sigma^2\underline{I}); \quad \underline{e} = [e_1, e_2, \ldots, e_T]^T \tag{8.2}$$

The Maximum Likelihood (ML) approach is well known in the literature (see e.g. Johnston (1963) and Young (1968a) for a simple exposition, and Dhrymes (1970) for a more detailed discussion). Following these authors, the ML approach requires the formulation of a log-likelihood function for the observations given in (8.1) of the form (see Appendix 1),

$$L(\underline{a}, \underline{b}, \sigma^2, \underline{y}, \underline{u}) = -\frac{T}{2}\log_e(2\pi) - \frac{T}{2}\log_e \sigma^2 - \frac{1}{2\sigma^2}\left[\frac{C}{D}\underline{y} - \frac{BC}{AD}\underline{u}\right]^T \left[\frac{C}{D}\underline{y} - \frac{BC}{AD}\underline{u}\right] \qquad (8.3)$$

Here $\underline{a} = [a_1, a_2, \ldots, a_n]^T$; $\underline{b} = [b_0, b_1, \ldots, {}_n]^T$; $\underline{y} = [y_1, y_2, \ldots, y_T]^T$; $\underline{u} = [u_1, u_2, \ldots, u_T]^T$ and, once again, the argument z^{-1} is omitted from the polynomials A, B, C and D for convenience. Maximization of this log likelihood function clearly requires the minimization of the final term on the RHS of (8.3), which the reader will recognize as simply the sum of the squares of the residuals e_i. Thus we need to find those estimates $\hat{A}$ and $\hat{B}$ of the polynomials A and B which minimize the sum of the squares of the residuals - i.e. a classical least squares problem. The conditions for this are obtained in the usual manner by partially differentiating with respect to $\hat{a}_i$, $i = 1,2,\ldots,n$; $\hat{b}_i$, $i = 0,1,2,\ldots,n$ and $\hat{\sigma}^2$ respectively and setting these derivatives to zero. This yields the following three equations (see Appendix 1).

$$\left.\begin{aligned}
\frac{\partial L}{\partial \hat{a}_i} &= \frac{1}{\sigma^2}\sum_{k=2n+1}^{T}\left[\frac{C}{D}y_k - \frac{\hat{B}C}{\hat{A}D}u_k\right]\frac{\hat{B}C}{\hat{A}^2D}z^{-i}u_k = 0; \quad \text{for } i = 1,2,\ldots,n && (1)\\
\frac{\partial L}{\partial \hat{b}_i} &= \frac{1}{\sigma^2}\sum_{k=2n+1}^{T}\left[\frac{C}{D}y_k - \frac{\hat{B}C}{\hat{A}D}u_k\right]\frac{C}{\hat{A}D}z^{-i}u_k = 0; \quad \text{for } i = 0,1,2,\ldots,n && (2)\\
\frac{\partial L}{\partial \hat{\sigma}^2} &= -\frac{T}{\sigma^2} + \frac{1}{\sigma^4}\sum_{k=2n+1}^{T}\left[\frac{C}{D}y_k - \frac{\hat{B}C}{\hat{A}D}u_k\right]^2 = 0 && (3)
\end{aligned}\right\} \quad (8.4)$$

These equations, for given C and D (assumed known) polynomials, are highly nonlinear in the a_i and b_i parameters but are *linear* in the b_i for a given a_i; and they are also linear in the σ^2. And, moreover, if estimates of the a_i and b_i are available, an estimate of σ^2 can be obtained easily from condition (3) of (8.4).

One method of obtaining estimates of the a_i and b_i that satisfy conditions (8.4) is to search the parameter space in some manner; for example, using a hill-climbing scheme such as the Newton-Raphson procedure employed by Astrom and Bohlin (1966) or a "relaxation" technique such as that described by Goodwin and Payne (1977).

An alternative procedure becomes apparent if we define the following discrete "pre-filtered" variables as shown in Figure 8.1 (Young, 1976a).

$$y_k^* = \frac{C}{AD}y_k; \quad u_k^* = \frac{C}{AD}u_k; \quad \hat{x}_k^* = \frac{B}{A}u_k^* \tag{8.5}$$

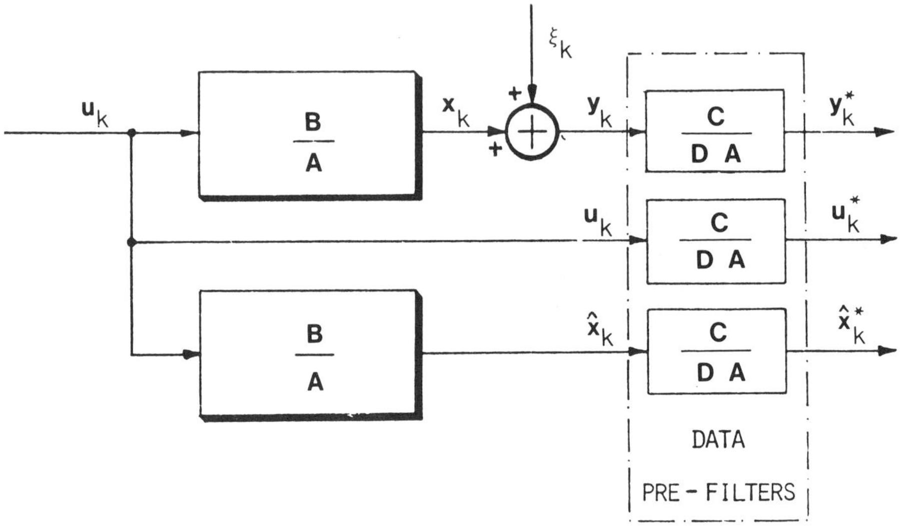

Figure 8.1 Data Pre-filters

and note then that equations (1) and (2) of (8.4) can be rewritten in terms of these prefiltered variables (by replacing A and B in (8.5) by $\hat{A}$ and $\hat{B}$) as shown below

$$\left.\begin{aligned} &\Sigma[\hat{A}y_k^* - \hat{B}u_k^*]\hat{x}_{k-i}^* = 0 \text{ for } i = 1,2,\ldots,n && (1) \\ &\Sigma[\hat{A}y_k^* - \hat{B}u_k^*]u_{k-i}^* = 0 \text{ for } i = 0,1,2,\ldots,n && (2) \end{aligned}\right\} \quad (8.6)$$

which are *linear* in the $\hat{a}_i$ and $\hat{b}_i$ parameters, provided we assume knowledge of the pre-filtered or transformed variables y_k^*, u_k^* and $\hat{x}_k^*$.

8.2 IV Within the Context of Maximum Likelihood

Perhaps the most interesting aspect of the equations (8.6) is their remarkable similarity to the equivalent IV equations for the same problem. This becomes clear if we examine the IV normal equations (4.13) i.e.

$$[\sum_{k=1}^{N} \hat{\underline{x}}_k \underline{z}_k^T]\hat{\underline{a}} - \sum_{k=1}^{N} \hat{\underline{x}}_k \underline{y}_k = 0$$

where $\hat{\underline{x}}_k$ and $\underline{z}_k$ are defined as in the previous chapter

$$\hat{\underline{x}}_k = [-\hat{x}_{k-1},\ldots, -\hat{x}_{k-n}, u_k, u_{k-1},\ldots,u_{k-n}]^T$$

$$\underline{z}_k = [-y_{k-1},\ldots, -y_{k-n}, u_k, u_{k-1},\ldots,u_{k-n}]^T$$

We have seen that this vector-matrix equation represents a set of 2n+1 linear simultaneous equations. Upon expansion the first n equations can be written in the form

$$\Sigma[y_k + \hat{a}_1 y_{k-1} + \ldots + \hat{a}_n y_{k-n} - \hat{b}_o u_k - \ldots - \hat{b}_n u_{k-n}]\hat{x}_{k-i} = 0,\ i=1,2,..,n$$

while the subsequent n+1 equations become

$$\Sigma[y_k + \hat{a}_1 y_{k-1} + \ldots + \hat{a}_n y_{k-n} - \hat{b}_o u_k - \ldots - \hat{b}_n u_{k-n}]u_{k-i} = 0;\ i=0,1,\ldots,n$$

which, on introduction of the operator notation and the A and B polynomials, becomes

$$\left.\begin{aligned} &\Sigma[\hat{A}y_k - \hat{B}u_k]\hat{x}_{k-i} = 0; \quad i = 1,2,\ldots,n \\ &\Sigma[\hat{A}y_k - \hat{B}u_k]u_{k-i} = 0; \quad i = 0,1,\ldots,n \end{aligned}\right\} \quad (8.7)$$

Comparison of equations (8.6) and (8.7) shows immediately that equation (8.6) can be interpreted as the IV normal equations for the system with the input u_k, the output y_k and the auxiliary model output $\hat{x}_k$ replaced by their pre-filtered equivalents u_k^*, y_k^* and $\hat{x}_k^*$ respectively. As a result, it is clear that the iterative and recursive methods of solution used in the basic IV case can also be utilized in the present context: it is necessary merely to introduce the additional pre-filtering operations and to update the parameters of these prefilters either iteratively or recursively using similar adaptive mechanisms to those used to update the auxiliary model parameters in the basic IV case.

In the recursive-iterative solution shown diagrammatically in Figure 8.2 , the filters are denoted by $C/D\hat{A}_{j-1}$ where $\hat{A}_{j-1}$ denotes the estimate of the A polynomial based on the parameter estimates $(\hat{a}_i)_{j-1}$ obtained at the previous (j-1)th iteration. As in the basic IV analysis, the instrumental variables $\hat{x}_k$ are generated from an 'auxiliary model" of the system, that is

$$\hat{x}_k = \frac{\hat{B}_{j-1}}{\hat{A}_{j-1}} u_k$$

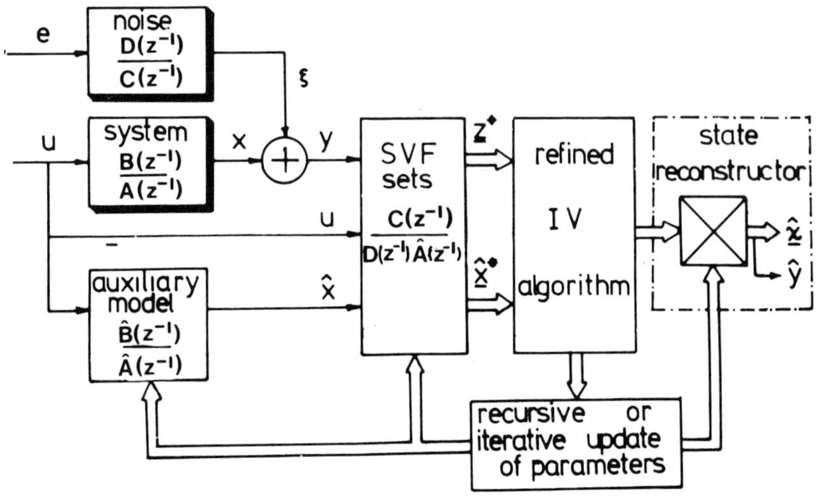

Figure 8.2 Recursive-Iterative and Recursive Implementations of the Refined IV Algorithm

(a) Recursive-Iterative: auxiliary model and prefilter parameters updated *iteratively* with $\hat{A}$ and $\hat{B}$ fixed at the jth iteration to estimates $\hat{A}_{j-1}$ and $\hat{B}_{j-1}$ obtained at previous iteration, i.e. auxiliary model $\hat{B}_{j-1}/\hat{A}_{j-1}$; prefilters $C/D\hat{A}_{j-1}$

(b) Recursive: auxiliary model and prefilter parameters updated *recursively* using latest recursive estimates suitably filtered and checked to ensure stability of auxiliary model

Note: the 'State Reconstructor' block is not important in this Chapter but is discussed briefly in Chapter 10, page 235.

where $\hat{B}_{j-1}$ is defined in a similar manner to $\hat{A}_{j-1}$ and is based on the estimates $(\hat{b}_i)_{j-1}$. The fully recursive solution follows directly by reference to the basic IV recursive algorithm, except that it is now necessary to update recursively not only the auxiliary model but also the pre-filters as shown in Figure 8.2(b). As in the basic IV algorithm, the recursive estimates would themselves need to be pre-filtered prior to their use in updating the pre-filter and auxiliary model parameters in order to remove any correlation between the estimates and the noise ξ_k, as required in the IV formulation. In line with previous analysis, such filtering would include a pure time delay, low-pass filtration and a stability check (Chapter 7).

The block diagram interpretation shown in Figure 8.2 is also interesting because the estimation procedure considered in this manner shows some similarity with a previous estimation scheme designed on a more heuristic basis: in this case, an IV approach suggested previously for the estimation of the coefficients in a continuous-time differential equation model of a dynamic system (Young, 1965a,b;1970). Although in this former case the choice of the equivalent continuous-time filters was more arbitrary and they were not normally updated in an iterative manner, they were chosen to have somewhat similar frequency response characteristics to those suggested in the present scheme. In particular, and on a completely heuristic basis, the band-pass of the filters was chosen to match closely as possible the band-pass of the system under study; in this manner, signals important to the estimation problem were 'passed' by the filters but high frequency noise outside the system band pass was attenuated by the operation of the filters. It is clear that this is entirely consistent with the present filtering operation, where the signals are passed through a discrete-time filter C/AD which will, in simplistic terms, tend to pass signals with components within the frequency band defined by the transfer function 1/A while, at the same time, 'pre-whitening' the noise components ξ_k by the introduction of the 'inverse' noise filter C/D (a procedure which, as we have pointed out previously, tends to simplify the estimation problem because it yields serially uncorrelated residuals).

This similarity between techniques based on intuitive reasoning and those based on analytical investigations, although not altogether surprising, is nonetheless rather comforting. And at the same time, the analytical approach has provided us with additional insight into the estimation problem: for example, the above analysis suggests that if the noise ξ_k is serially correlated, then knowledge of its statistical properties, in the form of an auto-regressive moving average (ARMA) model, may be used to improve the statistical efficiency of the IV estimates.

8.3 The AML Method Within the Context of Maximum Likelihood

Up to this point, we have assumed prior knowledge of the noise model in the sense that we have available the coefficients characterizing the C and D polynomials. Previously, the approximate maximum likelihood (AML) method of estimating these coefficients has been used in a "boot-strap" fashion[†] subsequent to the IV estimation of the A and B polynomial coefficients. Recalling the analysis in the previous section, it would seem advantageous to examine the AML method further to see if it can also be considered within the present context of maximum likelihood estimation. Of course, since the method was originally motivated by the maximum likelihood method of Astrom and Bohlin, we might expect that this will be possible.

Bearing in mind the boot-strap technique used in the basic IV-AML approach to time-series analysis, let us now follow the procedure used in the previous section and formulate the maximum likelihood solution to the problem of estimating the c_i and d_i coefficients in the noise model, this time assuming that it is the A and B polynomials that are known *a priori*.

In this new situation, the conditions equivalent to (1) and (2) of (8.4) take the form

$$\left.\begin{aligned}
\frac{\partial L}{\partial \hat{c}_i} &= \frac{1}{\sigma^2}\sum_{k=2n}^{T}\left[\frac{\hat{C}}{\hat{D}}y_k - \frac{B\hat{C}}{A\hat{D}}u_k\right]\left[\frac{1}{\hat{D}}z^{-i}y_k - \frac{B}{A\hat{D}}z^{-i}u_k\right] = 0 \quad \text{for } i=1,2,\dots,n \qquad (1)\\
\frac{\partial L}{\partial \hat{d}_i} &= \frac{1}{\sigma^2}\sum_{k=2n}^{T}\left[\frac{\hat{C}}{\hat{D}}y_k - \frac{B\hat{C}}{A\hat{D}}u_k\right]\left[-\frac{\hat{C}}{\hat{D}^2}z^{-1}y_k + \frac{B\hat{C}}{A\hat{D}^2}z^{-i}u_k\right] = 0 \quad \text{for } i=1,2,\dots,n \qquad (2)
\end{aligned}\right\}$$

But note that ξ_k and e_k are defined as

$$\xi_k = y_k - \frac{B}{A}u_k; \quad e_k = \frac{C}{D}\xi_k \tag{8.9}$$

Furthermore $[\frac{C}{D}y_k - \frac{BC}{AD}u_k]$ can, with these same definitions, be written as

$$\frac{C}{D}y_k - \frac{BC}{AD}u_k = \frac{C}{D}\xi_k = e_k \tag{8.10}$$

And, if we note the definitions of C and D, then

$$e_k = \xi_k + c_1\xi_{k-1} + \dots + c_n\xi_{k-n} - d_1e_{k-1} - \dots - d_ne_{k-n} \tag{8.11}$$

which can be written in operator notation as

$$e_k = C\xi_k - D^*e_k \tag{8.12}$$

where

$$D^* = D^*(z^{-1}) = d_1z^{-1} + d_2z^{-2} + \dots + d_nz^{-n} \tag{8.13}$$

---

† A term first used by Rowe (1970).

Thus by noting (8.9) and (8.12) we see that the conditions (8.8) can be interpreted in the following 'simpler' form,

$$\left.\begin{aligned}\Sigma[\hat{C}\xi_k - \hat{D}^*e_k]\xi^*_{k-i} &= 0; \quad \text{for } i = 1,2,\ldots,n \\ \Sigma[\hat{C}\xi_k - \hat{D}^*e_k]e^*_{k-i} &= 0; \quad \text{for } i = 1,2,\ldots,n\end{aligned}\right\} \tag{8.14}$$

where the prefiltered variables $\xi_k{}^*$ and $e_k{}^*$ are obtained from ξ_k and e_k in the following manner:

$$\xi_k{}^* = \frac{1}{D}\xi_k \quad \text{and} \quad e_k{}^* = \frac{1}{D}e_k \tag{8.15}$$

Thus, once again, we have been able to obtain a set of linear equations in the unknown parameters by defining certain pre-filtered variables.

The situation here, however, is not quite so straightforward as in the IV context, since it is necessary to generate estimates of both ξ_k and e_k in some manner, presumably by reference to equations (8.9) and (8.12). But these equations are precisely the equations on which the AML procedure is based: in particular, with the AML algorithm an estimate $\hat{\xi}_k$ of ξ_k is first obtained by subtracting the auxiliary model output $\hat{x}_k$ from the measured output y_k; while $\hat{e}_k$ is obtained from the relationship (cf. equation (8.12)).

$$\hat{e}_k = \hat{C}\xi_k - \hat{D}^*e_k; \quad \text{for } k = 1,2,\ldots,T \tag{8.16}$$

where $\hat{C}$ and $\hat{D}^*$ are, respectively, estimates of the C and D* polynomials, the parameters of which are themselves updated recursively on the basis of the current recursive estimates of the c_i and d_i parameters. With this in mind, we see that one possible approach to the solution of equation (8.14) is a recursive algorithm similar to the basic AML algorithm but incorporating the pre-filtered variables $\hat{\xi}_k{}^*$ and $\hat{e}_k{}^*$, as shown in Figure 8.3. Here, as in the IV case, both the pre-filters and the recursive estimator for e_k (which could be considered as an 'inverse' auxiliary noise model by analogy with the IV algorithm) would need to be updated recursively on the basis of the recursive parameter estimates. The main difference between this new AML algorithm and the basic AML algorithm is, however, simply the introduction of the pre-filtered variables, $\hat{\xi}_k{}^*$ and $\hat{e}_k{}^*$, via the pre-filters $1/\hat{D}$.

8.4 A Refined Recursive IV-AML Approach to Time-Series Analysis

In previous sections we have seen how, by alternately making assumptions about prior knowledge of the system and noise model parameters, it is possible to decompose the maximum likelihood solution of the time-series problem into simpler sub-problems, as shown diagrammatically in Figure 8.4. And we have also seen how recursive solutions to both of these sub-problems have been possible; solutions

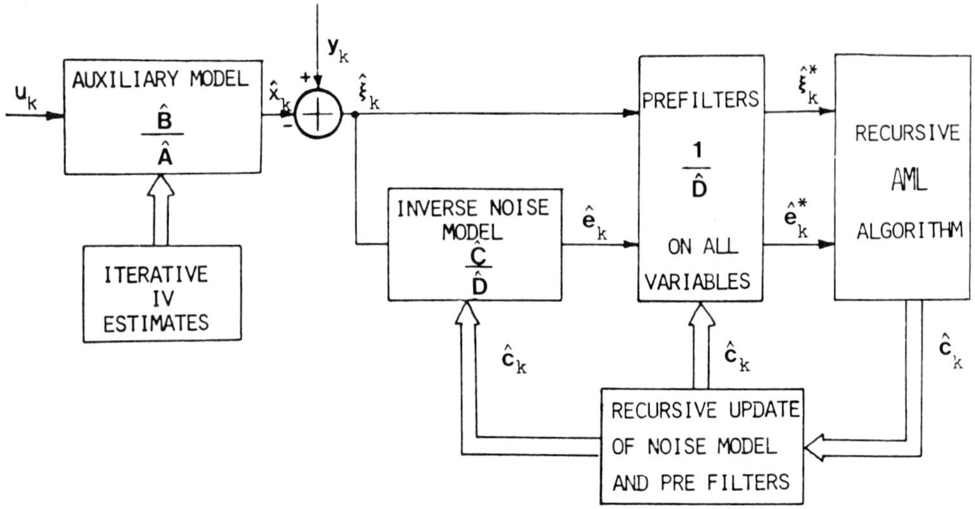

Figure 8.3 Refined AML Algorithm

which are quite similar to the IV and AML solutions discussed in previous chapters, except that they include additional data pre-filtering operations. This similarity between the presently proposed algorithms and the basic IV and AML algorithms suggests that simultaneous recursive estimation of the system and noise model parameters can be achieved by allowing direct communication (that is coordination) between each algorithm as the solution proceeds. In other words, initial consistent estimates of the system and noise model parameters can be used to define the initial auxiliary model and pre-filter parameters and, as the subsequent recursive estimates are obtained, they can be used to update or adapt these parameters as shown in Figure 8.4. Such a procedure is directly analogous to the basic recursive IV-AML algorithm, differing only by the presence *and added complication* of the adaptive pre-filters.

The implementation of this 'refined' IV-AML algorithm in iterative or recursive form is fairly obvious but, for clarity, we will consider it in some detail. In order to simplify the exposition, however, we will consider in detail only the iterative version of the algorithm. Here, at the kth sampling instant on some arbitrary jth iteration, the refined IV sub-algorithm generates an estimate $\underline{\hat{a}}_k = [\hat{a}_1,\ldots,\hat{a}_n, \hat{b}_0,\ldots,\hat{b}_n]_k^T$ of the system model parameter vector $\underline{a} = [a_1,\ldots,a_n, b_0,\ldots b_n]^T$ from the following recursive algorithm (cf. algorithm IV, Chapter 7, noting algorithm XII, Chapter 6).

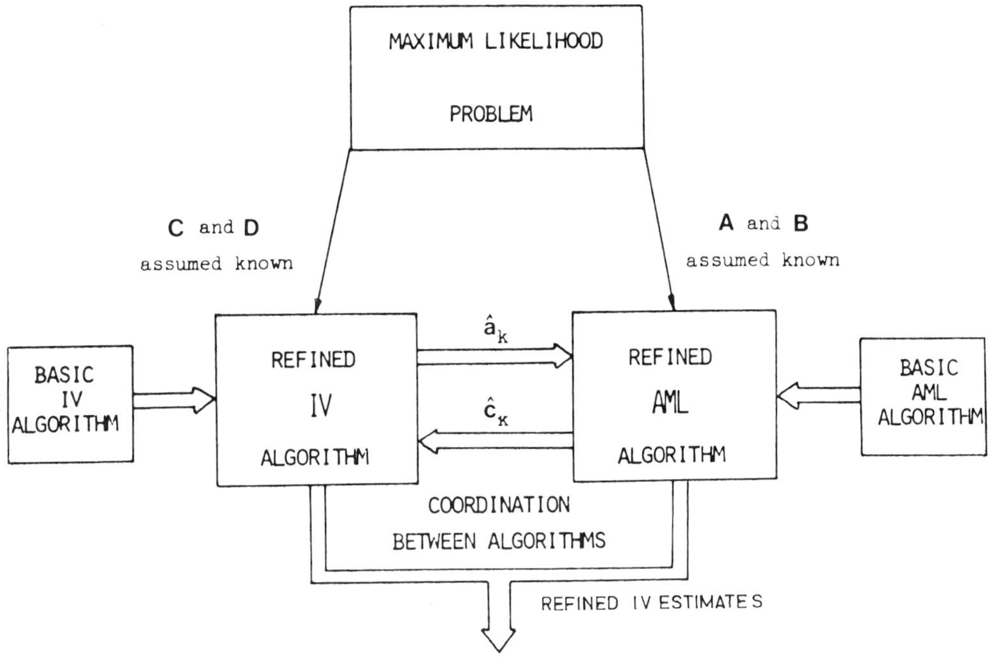

Figure 8.4 Decomposition of the Maximum Likelihood Problem

$$\hat{\underline{a}}_k = \hat{\underline{a}}_{k-1} - \hat{P}_{k-1}\hat{\underline{x}}_k{}^*[\hat{\sigma}^2 + \underline{z}_k{}^{*T}\hat{P}_{k-1}\hat{\underline{x}}_k{}^*]^{-1}\ (\underline{z}_k{}^{*T}\ \hat{\underline{a}}_{k-1} - y_k{}^*) \quad (1) \qquad \text{XV}$$

where

$$\hat{P}_k = \hat{P}_{k-1} - \hat{P}_{k-1}\hat{\underline{x}}_k{}^*[\hat{\sigma}^2 + \underline{z}_k{}^{*T}\ \hat{P}_{k-1}\hat{\underline{x}}_k{}^*]^{-1}\underline{z}_k{}^{*T}\hat{P}_{k-1} \quad (2) \qquad \text{XV}$$

Here $\hat{\sigma}^2$ is an estimate of σ^2 while the vectors $\underline{z}_k{}^*$ and $\hat{\underline{x}}_k$ are defined as

$$\begin{aligned} \underline{z}_k{}^* &= [-y_{k-1}{}^*, \ldots., -y_{k-n}{}^*, u_k{}^*, \ldots., u_{k-n}{}^*]^T \\ \hat{\underline{x}}_k{}^* &= [-\hat{x}_{k-1}{}^*, \ldots., -\hat{x}_{k-n}{}^*, u_k{}^*, \ldots., u_{k-n}{}^*]^T \end{aligned} \quad (3) \qquad \text{XV}$$

where $\hat{x}_k$ is the instrumental variable generated by the following auxiliary model:

$$\hat{x}_k = \frac{\hat{B}_{j-1}}{\hat{A}_{j-1}}u_k \quad (4) \qquad \text{XV}$$

while the star superscript indicates that the variables are prefiltered as shown in Figure 8.1 by a filter F_j of the form

$$F_j = \frac{\hat{C}_{j-1}}{\hat{D}_{j-1}\hat{A}_{j-1}} \qquad (5) \qquad XV$$

In (4) and (5) of XV $\hat{A}_{j-1}$, $\hat{B}_{j-1}$, $\hat{C}_{j-1}$ and $\hat{D}_{j-1}$ are, respectively, estimates of the polynomials $A(z^{-1})$, $B(z^{-1})$, $C(z^{-1})$ and $D(z^{-1})$ obtained at the end of the previous (j-1)th iteration and are denoted by

$$\begin{aligned}\hat{A}_{j-1} &= 1 + \alpha_1 z^{-1} + \dots + \alpha_n z^{-n}\\ \hat{B}_{j-1} &= \beta_o + \beta_1 z^{-1} + \dots + \beta_n z^{-n}\\ \hat{C}_{j-1} &= 1 + \gamma_1 z^{-1} + \dots + \gamma_n z^{-n}\\ \hat{D}_{j-1} &= 1 + \delta_1 z^{-1} + \dots + \delta_n z^{-n}\end{aligned} \qquad (6) \qquad XV$$

In other words, α_i, β_i, γ_i and δ_i are set equal to the estimates of the equivalent a_i, b_i, c_i and d_i parameters obtained at the *previous* (j-1)th iteration of the complete algorithm.

The estimates of the noise model parameters required for the adjustment of the prefilters F_j are obtained from the refined AML sub-algorithm. Here, at the kth sampling instant on the jth iteration, the estimate $\hat{\underline{c}}_k = [\hat{c}_1, \dots, \hat{c}_n, \hat{d}_1, \dots, \hat{d}_n]^T$ of the noise model parameter vector $\underline{c} = [c_1, \dots, c_n, d_1, \dots, d_n]^T$ is given by

$$\hat{\underline{c}}_k = \hat{\underline{c}}_{k-1} - \hat{P}^n_{k-1}\hat{\underline{m}}_k[\hat{\sigma}^2 + \hat{\underline{n}}_k^T\hat{P}^n_{k-1}\hat{\underline{m}}_k]^{-1}(\hat{\underline{n}}_k^T\hat{\underline{c}}_{k-1} - \hat{\xi}_k) \qquad (1)$$

where

$$\hat{P}_k^{\,n} = \hat{P}^n_{k-1} - \hat{P}^n_{k-1}\hat{\underline{m}}_k[\hat{\sigma}^2 + \hat{\underline{n}}_k^T\hat{P}^n_{k-1}\hat{\underline{m}}_k]^{-1}\hat{\underline{n}}_k^T\hat{P}^{\,n}_{k-1} \qquad (2)$$

XVI

and the vectors $\hat{\underline{n}}_k$ and $\hat{\underline{m}}_k$ are defined as follows:

$$\begin{aligned}\hat{\underline{n}}_k &= [-\hat{\xi}_{k-1}, \dots, -\hat{\xi}_{k-n}, \hat{e}_{k-1}, \dots, \hat{e}_{k-n}]^T\\ \hat{\underline{m}}_k &= [-\hat{\xi}_{k-1}^*, \dots, -\hat{\xi}_{k-n}^*, \hat{e}_{k-1}^*, \dots, \hat{e}_{k-n}^*]^T\end{aligned} \qquad (3) \qquad XVI$$

Here, $\hat{\xi}_k$ is an estimate of the noise variable ξ_k generated by subtracting the auxiliary model output $\hat{x}_k$ from the observed output y_k, i.e.

$$\hat{\xi}_k = y_k - \hat{x}_k \qquad (4) \qquad XVI$$

and $\hat{e}_k$ is a recursive estimate of the "white noise" input e_k obtained from the equation

$$\hat{e}_k = \hat{\xi}_k - \hat{\underline{n}}_k^T \hat{\underline{c}}_k \qquad (5) \qquad \text{XVI}$$

which can be considered as an 'inverse noise model'. Once again, the star superscript indicates that the variables generated in (4) and (5) are prefiltered, but here the filter is simply F_j^n, where

$$F_j^n = \frac{1}{\hat{D}_{j-1}} \qquad (6) \qquad \text{XVI}$$

Note also that, while the auxiliary model and prefilter parameters are updated iteratively and kept constant throughout each iteration, the estimate $\hat{e}_k$ in (5) is generated with the help of the *current recursive* estimate of $\underline{c}$.

In the iterative solution suggested here, the sub-algorithms XV and XVI are utilized alternately, starting with the refined IV estimator XV. During the first iteration through the complete data set (say T samples), it is clear that this estimator must be initialized in some manner. As indicated in Figures 8.2 and 8.3, we believe that this is best achieved by reference to the basic IVAML results obtained in prior analysis. Here the noise variance estimate $\hat{\sigma}^2$, the elements of $\hat{\underline{a}}_0$ and the parameters of both the auxiliary model XV (4) and the prefilters XV (5) are all set to the appropriate prior IVAML estimates, while the matrix $\hat{P}_0$ is assumed to be diagonal with all elements set to a high level, say 10^6, as in the ordinary IVAML case.

We have seen that this latter assumption is equivalent to informing the algorithm that the analyst has little confidence in the accuracy of the initial estimates $\hat{\underline{a}}_0$ and $\hat{\sigma}^2$. While this is not the case if the estimates have been obtained by prior IVAML analysis, it does prevent over-confidence in the estimates, which could in turn lead to poor convergence. It is, in other words, a conservative safety measure which seems worthwhile if the algorithm is to be used in routine day to day application. As in the basic IVAML mechanization, however, this setting for $\hat{P}_0$ is not used on iterations subsequent to the first and it is also not advisable in the purely recursive form of this refined algorithm (see Chapter 7).

In addition to the initialization of those aspects of the algorithm linked with the system and noise model parameters, it is also necessary to initialize the elements of the vectors $\underline{z}_k^*$ and $\hat{\underline{x}}_k^*$ before estimation can proceed. This is achieved by solving the filtering equations for y_k^*, u_k^* and $\hat{x}_k^*$ recursively until all required elements of $\underline{z}_k^*$ and $\hat{\underline{x}}_k^*$ are fully specified. In discrete-time terms, these filtering operations can be written:

$$\left.\begin{aligned} y_k^* &= \underline{y}_k^T \underline{p} + y_k \\ u_k^* &= \underline{u}_k^T \underline{p} + u_k \\ \hat{x}_k^* &= \underline{\hat{x}}_k^{*T} \underline{\alpha} \end{aligned}\right\} \quad (7) \qquad \text{XV}$$

where

$$\underline{y}_k = [-y_{k-1}^*, \ldots, -y_{k-2n}^*, y_{k-1}, \ldots, y_{k-n}]^T$$

$$\underline{u}_k = [-u_{k-1}^*, \ldots, -u_{k-2n}^*, u_{k-1}, \ldots, u_{k-n}]^T$$

$$\underline{\alpha} = [\alpha_1, \ldots, \alpha_n, \beta_o, \ldots, \beta_n]$$

$$\underline{p} = [\alpha_1+\delta_1, \alpha_2+\alpha_1\delta_1 + \delta_2, \alpha_3+\alpha_1\delta_2+\alpha_2\delta_1+\delta_3, \ldots, \alpha_n+\delta_n, \gamma_1, \ldots, \gamma_n]^T$$

Here the definition of the 3n dimensional vector $\underline{p}$ arises because the denominator of the prefilter F_j in (5) of XV is defined as the product of the two polynomials $\hat{D}_{j-1}$ and $\hat{A}_{j-1}$.

The recursive algorithm XV is not called until the initialization of the vectors $\underline{z}_k^*$ and $\underline{\hat{x}}_k^*$ is complete. As a result, $\underline{\hat{a}}_1$ to $\underline{\hat{a}}_n$ are all set to $\underline{\hat{a}}_o$ and only at sample n+1 is this initial estimate updated. Application of the algorithm XV then continues until all the available data have been processed: at this point $\underline{\hat{a}}_T$ provides the estimate of $\underline{a}$ at the end of the first refined IV iteration and allows for the updating of the polynomials $\hat{A}_o$ and $\hat{B}_o$ to $\hat{A}_1$ and $\hat{B}_1$, respectively. The updated auxiliary model based on A_1 and B_1, i.e.

$$\hat{x}_k = \frac{\hat{B}_1}{\hat{A}_1} u_k$$

can then be used to generate $\hat{x}_k$, k = 1, 2, ..., T, which, in turn, can be incorporated in equation XVI (4) to provide the estimated noise sequence $\hat{\xi}_k$, k = 1, 2, ..., T.

It is this estimated noise sequence $\hat{\xi}_k$ which provides the coordination between the IV and AML sub-algorithms. $\hat{\xi}_k$, so obtained, provides the input to the filtering operations for $\hat{\xi}_k^*$ and $\hat{e}_k^*$. As in the IV case, these filtering operations can be written as follows in discrete-time terms:

$$\left.\begin{aligned} \hat{\xi}_k^* &= \underline{\hat{\xi}}_k^T \underline{\gamma} + \hat{\xi}_k \\ \hat{e}_k^* &= \underline{\hat{e}}_k^T \underline{\gamma} + \hat{e}_k \end{aligned}\right\} \quad (8) \qquad \text{XVI}$$

where

$$\hat{\underline{\xi}}_k = [0,0,\ldots,0, -\hat{\xi}_{k-1}^*,\ldots, -\hat{\xi}_{k-n}^*]^T$$

$$\hat{\underline{e}}_k = [0,0,\ldots,0, -\hat{e}_{k-1}^*,\ldots, -\hat{e}_{k-n}^*]^T$$

$$\underline{\gamma} = [\gamma_1,\gamma_2,\ldots,\gamma_n,\delta_1,\ldots,\delta_n]^T$$

Here, the zero elements in the vectors $\hat{\xi}_k$ and $\hat{e}_k$ arise because the filter F_j^n in (6) of XVI has a numerator of unity, so that no moving average operations are specified.

The filtering operation for $\hat{\xi}_k^*$ is carried out over the first n+1 samples in a similar manner to the equivalent operation in the IV sub-algorithm and this yields the $\hat{\xi}_k^*$ elements required for the initialization of the $\hat{\underline{m}}_k$ vector in (3) of XVI. This is not possible in the case of the $\hat{e}_k^*$ in XVI (7), however, since $\hat{e}_k$, k=1,2,...,n are not known. As in the basic AML algorithm, therefore, these initial elements in the $\hat{e}_k$ sequence are set to their expectation of zero and this in turn means that $\hat{e}_k^*$, k=1,2,...,n needs also to be set to zero.[†]

Following the initialization of $\hat{\underline{n}}_k$ and $\hat{\underline{m}}_k$, $\hat{\underline{c}}_0$ is defined by reference to the prior IVAML results and recursive estimation proceeds from sample n+1 using algorithm XVI. The $\hat{e}_k$ and $\hat{e}_k^*$ variables required at each recursive step in these equations are obtained by the use of equations XVI (5) and XVI (8) respectively, and recursion continues until $\hat{\underline{c}}_T$ is obtained. This marks the end of the first complete refined IVAML iteration. At this point, the $\hat{\underline{c}}_T$ vector provides the required coordination between the AML and IV sub-algorithms: the noise model parameters defined by $\hat{\underline{c}}_T$ are communicated to the IV sub-algorithm where they define the filter parameters γ_i and δ_i, i = 1,2,...,n, needed for the second IV iteration.

The second and subsequent iterations through the complete refined IVAML procedure are similar to the first, with the initial estimates and prefilter parameters being provided by the estimates obtained at the previous iteration. It has been found by experience, however, that advantage is gained after the first iteration by initializing the $\hat{P}_0$ and $\hat{P}_0^n$ matrices to the values $\hat{P}_k$ and $\hat{P}_k^n$ saved at a specified recursive step in the previous iteration. As in Chapter 7, it has been found that, as a rule of thumb, an appropriate recursive step for saving the matrices is given by T/I, where I is the number of iterations specified by the programme user, which will normally be less than six. Indeed, we have found in practical applications that a need for greater than six iterations is normally associated with poor prior model structure identification.

[†] This is in contrast to an "exact maximum likelihood" approach where these initial values are also estimated (see, e.g. Pagan and Nicholls, 1976).

The iterative version of refined IVAML described above includes recursive parameter estimation at each iteration. But this recursive solution does *not* involve recursive updating of the prefilter and the auxiliary model parameters. A fully recursive solution which does include such operations is clearly feasible, however, and it has maximum potential utility in on-line applications, such as forecasting and adaptive control (see Chapter 10). In such a recursive algorithm, prefilter and auxiliary model parameters are updated in the same manner as in the iterative version but updating takes place after each *recursive* step.

Unfortunately, the flexibility of an on-line recursive algorithm of this type is obtained at some cost. The rapid adaptation of the auxiliary model and the prefilters coupled with the complex coordination between the IV and AML sub-algorithms will undoubtedly tend to make the recursive algorithm less robust than the iterative algorithm when in general use. It must, for example, be tuned carefully in any particular situation in order to obtain good results and may need modifications, such as the introduction of an EWP memory factor, as discussed previously in connection with the basic IV-AML algorithms. Also, since the fully recursive algorithm involves only a single passage through the data, it will not be as statistically efficient over a *finite* data set as the iterative algorithm. It seems a reasonable conjecture, however, that the statistical efficiency will approach that of the iterative algorithm for large data sets. This has been confirmed in the Monte-Carlo stochastic simulation results (Young and Jakeman, 1979a).

It is worth noting again that the recursive estimates generated throughout each iteration in the iterative solution can be useful in their own right. Indeed here, as in the basic IVAML case, it is advisable to make an additional iteration through the data in which the auxiliary model and prefilter parameters are maintained constant at their converged values and the recursive estimation is initiated from initial $\hat{P}_0$ and $\hat{P}_0^n$ matrices set to the values used in the first iteration (i.e. diagonal with elements set to 10^6). The recursive estimates obtained in this manner can then be examined to see if there is any evidence of parametric non-stationarity or other statistical problems such as collinearity. Also, as we shall see in Section 8.5, the final $\hat{P}_T$ and $\hat{P}_T^n$ matrices obtained in this additional iteration are useful in assessing the error-covariance matrices associated with the parameter estimates.

Finally on the basis of the Monte-Carlo simulation results (Young and Jakeman, 1979a; Jakeman and Young, 1979a) and theoretical analyses of the refined IVAML algorithm based on the approach of Ljung (see Chapter 9 and Holst 1977) and Solo (1980), it would appear that certain improvements to the basic estimation procedure described above are possible. These improvements can result in superior performance of the *refined AML sub-algorithm* in certain applications and avoid possible convergence problems that can affect the refined AML algorithm in certain rare circumstances. The modifications, which also are suggested by the theoretical results of Pierce (1972) in connection with the non-recursive Maximum Likelihood method of Box and Jenkins (1970), involve converting the $\hat{P}_k$ and $\hat{P}_k^n$ matrices into *symmetric* form.

The use of symmetric matrix gains in recursive IV algorithms has been discussed previously by Young (1969c; see also 1965c) who suggested a symmetric matrix version of the basic IV algorithm. An alternative symmetric matrix version suggested by the results of Pierce (1972) involves replacing the refined IV algorithm XV by the following equations:

$$\left.\begin{aligned} \hat{\underline{a}}_k &= \hat{\underline{a}}_{k-1} - \hat{\underline{P}}_{k-1}\hat{\underline{x}}_k^*[\hat{\sigma}^2 + \hat{\underline{x}}_k^{*T}\hat{\underline{P}}_{k-1}\hat{\underline{x}}_k^*]^{-1}(\underline{z}_k^{*T}\hat{\underline{a}}_{k-1} - y_k^*) \quad &(1)\\ \text{where}\quad \hat{\underline{P}}_k &= \hat{\underline{P}}_{k-1} - \hat{\underline{P}}_{k-1}\hat{\underline{x}}_k^*[\hat{\sigma}^2 + \hat{\underline{x}}_k^{*T}\hat{\underline{P}}_{k-1}\hat{\underline{x}}_k^*]^{-1}\hat{\underline{x}}_k^{*T}\hat{\underline{P}}_{k-1} \quad &(2)\end{aligned}\right\}\ \text{XVII}$$

In other words, all $\underline{z}_k^*$ vectors, other than that appearing in the error term $(\underline{z}_k^{*T}\hat{\underline{a}}_{k-1} - y_k^*)$ in XVII(1), are replaced by $\hat{\underline{x}}_k^*$. The advantages of this algorithm over the non-symmetric matrix equivalent will become clear when the statistical properties of the estimates $\hat{\underline{a}}_k$ are discussed in the next section. The algorithm seems to perform well in general applications where there is a reasonable sample size, *but does not appear so reliable as the non-symmetric gain algorithm for lower sample sizes* (Young and Jakeman, 1979a). It cannot, therefore, be recommended for general use.

Theoretical analysis by Holst (1977) based on the approach by Ljung (1977b) indicates that a symmetric matrix version of the refined AML algorithm can have definite theoretical advantages in certain circumstances. While the non-symmetric matrix refined IVAML procedure should normally have good convergence characteristics when applied to ARMAX time-series models (i.e. model (8.1) with $C(z^{-1})$ assumed equal to $A(z^{-1})$), the *refined AML part of the algorithm* may, in certain rare circumstances, experience some minor convergence problems when applied to the errors-in-variables transfer-function model preferred here. For this reason, it may be better, in general applications, to convert the refined AML algorithms XVI to the following symmetric matrix form, which is similar to the Recursive Maximum Likelihood (RML) algorithm originally suggested by Soderstrom (1973) as an approximation to the non-recursive ML method of Astrom and Bohlin (1966).†

$$\left.\begin{aligned} \hat{\underline{c}}_k &= \hat{\underline{c}}_{k-1} - \hat{P}^n_{k-1}\hat{\underline{m}}_k[\hat{\sigma}^2 + \hat{\underline{m}}_k^T\hat{P}^n_{k-1}\hat{\underline{m}}_k]^{-1}(\hat{\underline{n}}_k^T\hat{\underline{c}}_{k-1} - \hat{\xi}_k) \quad &(1)\\ \text{where}\quad \hat{P}_k^n &= \hat{P}^n_{k-1} - \hat{P}_{k-1}^n\hat{\underline{m}}_k[\hat{\sigma}^2 + \hat{\underline{m}}_k^T\hat{P}^n_{k-1}\hat{\underline{m}}_k]^{-1}\hat{\underline{m}}_k^T\hat{\underline{P}}^n_{k-1} \quad &(2)\end{aligned}\right\}\ \text{XVIII}$$

Although this algorithm is undoubtedly superior on theoretical grounds avoiding, as it does, the possible convergence problems of the alternative refined AML algorithm and yielding superior statistical efficiency to the basic AML algorithm, it has been found rather unreliable both in Monte Carlo Simulation analysis and when used with real data (see Young and Jakeman, 1979a; Norton 1977). For these reasons, we recommend that the reader wishing to implement algorithms of this more sophisticated kind should carefully evaluate them in this regard and, if he is going to use

† Here, however, refined AML is applied to the purely stochastic ARMA model estimation whereas RML is applied to the ARMAX model (see Jakeman and Young, 1981a,b).

the algorithms for day-to-day applications work, he may well feel that the basic AML has certain computational and reliability advantages which outweigh the theoretical superiority of the RML alternative. Certainly if it is the statistical efficiency of the A and B polynomial coefficients that is of prime concern, then the results of Young and Jakeman (1979a) suggest that the use of the symmetric gain refined AML algorithm XVIII is *not essential*: minimum variance estimates of the A and B parameters are obtained from refined IV even if rather poorer estimates of the C and D polynomial coefficients are obtained, for example from basic AML or the non-symmetric gain refined AML algorithms.

8.5 The Statistical Properties of the Refined IVAML Estimates

Pierce (1972) has shown that the large sample properties of the maximum likelihood (ML) estimates of the parameters in the time-series model (8.1) can be evaluated in a manner which is pertinent to our present discussion. In the nomenclature of the present paper, and using systems terminology, his results are as follows.

Theorem

If in (8.1):

(i) the e_k are independent and identically distributed with zero mean, variance σ^2 and skewness and kurtosis κ_1 and κ_2;

(ii) the parameter values are admissible (i.e. the model is stable and identifiable - see Chapter 7), and

(iii) the u_k are persistently exciting (see Chapter 7), then the ML estimates $\hat{\underline{a}}_T$, $\hat{\underline{c}}_T$ and $\hat{\sigma}^2$ obtained from a data set of T samples, possess a limiting normal distribution, such that the following results hold:

(a) the asymptotic convariance matrix of the estimation errors associated with the estimate $\hat{\underline{a}}_T$ is of the form

$$P^* = \frac{\sigma^2}{T}\,[\text{plim}\,\frac{1}{T}\,\Sigma\,\hat{\underline{x}}_k^*\,\hat{\underline{x}}_k^{*T}]^{-1} \tag{8.17}$$

(b) the estimate $\underline{c}_T$ is *asymptotically independent* of $\underline{a}_T$ and has an error covariance matrix of the form

$$P^{n*} = \frac{\sigma^2}{T}\,[E\{\hat{\underline{m}}_k\hat{\underline{m}}_k^T\}]^{-1} \tag{8.18}$$

and

(c) the estimate $\hat{\sigma}^2$ has asymptotic variance $(2\sigma^4/T)(1 + \tfrac{1}{2}\kappa_2)$ and, if $\kappa_1 = 0$, is independent of the above estimates.

In (8.17) and (8.18), $\hat{\underline{x}}_k^*$ and $\hat{\underline{m}}_k$ are defined as the values of these variables which would result if the auxiliary model, inverse noise model and prefilter parameters were all set at values based on the *true* model parameters $\underline{a}$ and $\underline{c}$. The relevance of these results is obvious: if it is assumed that in all identifiable situations the refined IVAML algorithms given by the interactive use of equations XVII and XVIII converge in the sense that $\hat{\underline{a}}_k \to \underline{a}$ and $\hat{\underline{c}}_k \to \underline{c}$ for increasing k, then the matrices $\hat{P}_k$ and $\hat{P}_k^n$ will provide good empirical estimates of P^* and P^{n*}. And, noting the Monte Carlo results of Pierce (1972), we might also assume that $\hat{P}_T$ and $\hat{P}_T^n$ would provide good indication of the error covariance properties of $\hat{\underline{a}}_T$ and $\hat{\underline{c}}_T$ for even small sample size T. Monte Carlo simulation experiments (next section and Young and Jakeman, 1979a) confirm that this is indeed the case for $\hat{\underline{a}}_T$ but the results are not so satisfactory as regards $\hat{\underline{c}}_T$, although they appear acceptable for large sample sizes. In the recursive-iterative case, an alternative to the use of algorithms XVII and XVIII for the generation of P^* and P^{n*} estimates is the use of the equivalent non-symmetric algorithms XV and XVI to achieve convergence and then to run an *additional* iteration after convergence with the algorithm modified to incorporate the symmetric matrices with the auxiliary model, inverse noise model and pre-filter parameters fixed at their converged values. This has the advantage of greater robustness: as we have pointed out, it has been found that the non-symmetric IV algorithm is more reliable in practice than the equivalent symmetric algorithm, which can fail to converge for numerical reasons associated with the complicated adaptive prefiltering operations. When used in the additional iteration, where auxiliary model and prefilters are set up on the basis of the *converged* parameter values, however, the symmetric version does not exhibit problems of this type and yields better estimates of the P^* and P^{n*} covariance matrices.

8.6 Performance of the Refined IVAML Algorithms

Theoretical analysis of recursive algorithms is possible and is mentioned briefly in Chapter 9. Up to the present, however, such analyses have only addressed problems such as large sample convergence and asymptotic statistical properties: Solo (1978, 1980), for example, has shown that the recursive, refined IV algorithm has good convergence characteristics and optimal statistical properties. Also, as we have pointed out in the previous section, analysis by Ljung (1977b) and Holst (1977) has helped in evaluating the convergence properties of the refined AML algorithm. But what of practical performance in small to medium sample-size situations? We might assume that an algorithm which has good asymptotic performance would have reasonable finite sample-size properties. On the other hand, a simpler algorithm with poorer asymptotic properties might have certain computational advantages which make it attractive in practice *provided* its finite sample performance is acceptable in relation to the finite sample performance of the 'optimum' algorithms.

As theoretical analysis is largely asymptotic in character, it is clearly necessary to be able to evaluate small to medium sample performance in some other, reasonably objective, manner. And since purely analytic approaches are not available except in simple regression situations, the only reasonable alternative is to use some form of simulation: in other words, to assess the performance of the estimator when applied to data generated from the theoretical stochastic model; in the present context, the TF model (8.1) i.e.

$$y_k = \frac{B(z^{-1})}{A(z^{-1})} u_k + \frac{D(z^{-1})}{C(z^{-1})} e_k \tag{8.1}$$

where e_k is, by definition, a discrete, zero mean, white noise sequence with variance σ^2.

Because e_k is a stochastic variable, deterministic simulation of this equation is not meaningful: we could use a computer to generate a *deterministic* zero mean sequence e_k^1 with say T elements, which had the requisite autocorrelation properties of white noise (e.g. from a random number generator); and then use this, together with a chosen deterministic sequence u_k, to generate y_k. But the input-output data (u_k, y_k) so constructed would then represent only one 'realization' of the stochastic system (8.1) and the single set of estimates of the (a_i, b_i, c_i, d_i) parameters would give no information on the overall statistical properties of the estimates $(\hat{a}_i, \hat{b}_i, \hat{c}_i, \hat{d}_i;$ i.e. $\hat{\underline{a}}$ and $\hat{\underline{c}})$.

For these reasons, in order to obtain some idea of the overall statistical properties of the estimates using simulation techniques, it is necessary to conduct N *multiple* simulations of (8.1) with the e_k^i sequence, i = 1,2,...,N chosen for each simulation as an *independent* random realization of the white noise sequence e_k, i.e. each of the N sequences e_k^i should represent T sequential samples from a zero mean, white noise source and it should be chosen to be statistically independent of all other sequences e_k^j, $j \neq i$, used in the simulation experiments.

For each set of data $(u_i, y_i, i = 1,2,...,T)$ so obtained, it is possible to obtain a set of estimates for the unknown parameters. If the N sets of parameter estimates obtained in this manner are denoted by the vectors $\hat{\underline{a}}^i$ and $\hat{\underline{c}}^i$ where

$$\hat{\underline{a}}^i = [\hat{a}_1^i, \hat{a}_2^i, \ldots, \hat{a}_n^i, \hat{b}_0^i, \ldots \hat{b}_n^i]^T$$

$$\hat{\underline{c}}^i = [\hat{c}_1^i, \hat{c}_2^i, \ldots, \hat{c}_n^i, \hat{d}_1^i, \ldots \hat{d}_n^i]^T$$

and i = 1,2,....,N, then the sample-average statistical properties of the estimates can be computed quite straight forwardly. For example, in the case of the first two statistical moments, namely the mean and covariance matrix, the sample average mean and covariance matrix are defined for $\hat{\underline{a}}^i$ (similarly for $\hat{\underline{c}}^i$) as,

$$\left.\begin{aligned} &\text{mean: } \bar{\hat{\underline{a}}} = \frac{1}{N}\sum_{i=1}^{N}\hat{\underline{a}}^i && (1) \\ &\text{covariance matrix: } \operatorname{cov}(\hat{\underline{a}}-\underline{a}) = R_{\tilde{\underline{a}},\tilde{\underline{a}}} = \frac{1}{N}\sum_{i=1}^{N}(\tilde{\underline{a}}^i)(\tilde{\underline{a}}^i)^T && (2) \end{aligned}\right\} \quad (8.19)$$

where the estimation error, $\tilde{\underline{a}}^i = \hat{\underline{a}}^i - \underline{a}$. The covariance matrix $\operatorname{cov}(\tilde{\underline{a}})$ is, from (2) in (8.19), a (2n+1) x (2n+1) symmetric matrix of the form

$$R_{\tilde{\underline{a}},\tilde{\underline{a}}} = \begin{bmatrix} C_{11} & C_{12} \\ C_{12} & C_{22} \end{bmatrix} \quad (8.20)$$

where the blocks C_{11}, C_{22} and C_{12} have elements of the form,

$$C_{11}: \quad c_{jk} = \frac{1}{N}\sum_{i=1}^{N}(\tilde{a}_j^i)(\tilde{a}_k^i) \quad \text{for } \left.\begin{matrix} j \\ k\end{matrix}\right\} = 1,2,\ldots,n$$

$$C_{22}: \quad c_{jk} = \frac{1}{N}\sum_{i=1}^{N}(\tilde{b}_j^i)(\tilde{b}_k^i) \quad \text{for } \left.\begin{matrix} j \\ k\end{matrix}\right\} = 0,1,\ldots,n$$

$$C_{12}: \quad c_{jk} = \frac{1}{N}\sum_{i=1}^{N}(\tilde{a}_j^i)(\tilde{b}_k^i) \quad \text{for } \left.\begin{matrix} j \\ k\end{matrix}\right\} = \begin{matrix} 1,2,\ldots,n \\ 0,1,\ldots,n\end{matrix}$$

Similarly, the sample-average covariance matrix related to the mean of the parameter estimates (rather than the true value) can be calculated using analogous expressions with $\tilde{\underline{a}}^i$ defined by $\tilde{\underline{a}}^i = \hat{\underline{a}}^i - \bar{\hat{\underline{a}}}$. The standard error of the estimates about either the true value or the mean can then be obtained as the square root of the diagonal elements of the appropriate covariance matrices. Comparison of these standard errors provides a good indication of whether the estimates may be biased significantly from the true value (see e.g. Dhrymes, 1970).

Using the above Monte Carlo (MC) approach (Hammersley and Handscombe, 1964), we are effectively treating each set of estimates obtained from each random experiment as an 'observation' from a population whose probability distribution is the distribution of the given estimator for a sample of size N. Of course, since the statistical properties obtained in this way are based on the sample-average properties of the estimates for a *finite* sample size, they are themselves only estimates of the asymptotic statistical properties obtained when $N\to\infty$. Nevertheless, they provide a reasonably objective evaluation of the estimation performance and should demonstrate whether there is any strong indication of problems such as bias, high estimation error variance, or lack of convergence of the recursive algorithm in small to medium sample situations.

The nature and limitations of Monte Carlo (MC) analysis are outside the scope of the present text and for discussion on this matter the reader is directed to Dhrymes (1970). It is sufficient here to emphasize that the MC approach does not entirely solve the small-medium sample distribution problem. Rather it provides a means of assessing the estimation performance of an algorithm for a *given* set of parameter values ($\underline{a},\underline{c}$). It does *not* follow that the conclusions derived from the analysis are *necessarily* valid for all possible values of the parameters and for all model forms (i.e. over the entire space of model forms). We can assess this validity to some extent, however, by conducting the experiments using different simulation model forms and different sets of parameters which, in some sense, span some admissable range.

Here "admissable" is usually interpretable in relation to the kind of problems met in practice: for example, if the dynamic systems being studied are guaranteed to be non-oscillatory and stable, then we need only conduct MC experiments on a set of representative simulation models for this class of system. Even then, we could not be completely sure that the performance would be satisfactory for *all* models. An example is the system proposed by Ljung *et al.* (1975) as a counter example to the generally good convergence characteristics of the AML algorithm; although MC analysis and extensive practical application had indicated the generally good performance of the AML algorithms, these authors were able to show how, in certain (fortuitously) rare circumstances, convergence of the algorithm was not fully satisfactory. We will, however, refer to this example again later.

Notwithstanding these limitations, MC simulation analysis is a powerful tool in the evaluation of statistical algorithms simply because, in this age of the electronic computer, it is so straight-forward to use and allows us to address problems for which no adequate theory is available. It can be computationally inefficient, of course, but improvements in this regard can sometimes be made by resorting to "variance reduction" techniques (Hammersley and Handscombe, 1964) such as the "control variate" and "antithetic variate" techniques (see e.g. Moore and Clarke (1978) and Beaumont (1980) for a discussion within an hydrological context).

MC simulation analysis is, first and foremost, a computational tool in 'experimental statistics' providing a quick and useful method of assessing statistical methodology when analytical results are either not available or difficult to obtain. It has been used in this spirit to obtain some idea of the performance of the various IV estimators discussed in this and previous chapters. The results of these comprehensive simulations are discussed in detail in Young and Jakeman (1979a, 1980), Jakeman and Young (1979a, 1981a, b), and Jakeman, Steele and Young (1980). Here, we will consider in detail only one set of results for a given stochastic model of the form (8.1): this will serve to exemplify the general results.

The specific model was first used by Isermann *et al.* (1973), and is of the form,

$$y_k = \frac{-0.102z^{-1} + 0.173z^{-2}}{1 - 1.425z^{-1} + 0.496z^{-2}}u_k + \frac{1 - 0.775z^{-1}}{1 - 1.027z^{-1} + 0.264\,z^{-2}}e_k$$

The input u_k was selected as a random binary signal with levels ±1.0 and e_k was generated as a sequence of normally distributed random variables with zero mean and variance σ^2, which was adjusted to give signal/noise ratios S of either 10 (low noise) or 1.0 (high noise), where S is defined as

$$S \triangleq \frac{\sum^{T} x_k^2}{\sum^{T} \xi_k^2} \tag{8.21}$$

The results of the MC analysis are shown in Table 8.1 where the mean and ± standard error about the true value (obtained as the square root of the diagonal elements of the C_{11} and C_{22} matrices for both $\underline{a}$ and $\underline{c}$) of the parameter estimates for sample sizes of 75, 300 and 1000, respectively, are given in the first and second unlabelled columns. Also shown in the third unlabelled column are the standard errors computed by reference to the theoretical results of Pierce (equations (8.17) and (8.18)), with the expected value in (8.18) replaced by its sample-average in order to allow for numerical computation. Table 8.1A and 8.1C supply the results obtained from the refined IVAML algorithms XV and XVI, whilst 8.1B and 8.1D supply the analogous results for the ordinary or basic IVAML algorithms discussed in Chapter 7.

It is clear from these results that refined IVAML yields extremely good estimates of the system model parameters (i.e. a_i and b_i) at all sample sizes and noise levels, with zero estimation bias and close to minimum variance all the time. In other words, the algorithm appears to converge on the maximum likelihood estimates as conjectured by Young (1976a). The ordinary IVAML algorithm also performs very well with zero bias and relatively good statistical efficiency (standard errors up to approximately twice those of refined IV). These results are typical and suggest that the ordinary IVAML provides a simple general approach to the problem of estimating the system model parameters and that refined IVAML provides a method of improving the ordinary IVAML estimates of a_i and b_i when greater statistical efficiency is demanded.

The estimates of the noise model parameters obtained from either refined or ordinary IVAML are not nearly so good as those obtained for a_i and b_i, as might be expected,† although they improve with increased sample size. Indeed other simulation

---

† Note that the estimation algorithm is attempting to detect *statistical patterns* of behaviour on the basis of only the 'noise output' estimate $\hat{\xi}_k$; patterns which may not become clearly apparent until a relatively large time-series sequence of $\hat{\xi}_k$ is available. The situation is much better in the case of the system model because we can observe *both* the input and output variables and so more readily estimate the relationship between them.

Table 8.1(A) and (B) Monte Carlo Simulation Results for Model 1, S = 10

Table 8.1A Refined IVAML, $S=10$.

Parameter	True value	Number of samples		
		75	300	1000
a_1	−1·425	−1·4240 ± 0·0488 ± 0·0448	−1·4195 ± 0·0217 ± 0·0224	−1·4239 ± 0·0124 ± 0·0126
a_2	0·496	0·4952 ± 0·0459 ± 0·0420	0·4913 ± 0·0202 ± 0·0210	0·4949 ± 0·0117 ± 0·0118
b_1	−0·102	−0·1019 ± 0·0128 ± 0·0127	−0·1011 ± 0·0064 ± 0·0062	−0·1022 ± 0·0031 ± 0·0034
b_2	0·173	0·1731 ± 0·0154 ± 0·0149	0·1729 ± 0·0071 ± 0·0073	0·1733 ± 0·0036 ± 0·0039
c_1	−1·027	−0·3174 ± 0·8496 ± 0·4311	−0·5289 ± 0·5609 ± 0·1979	−0·6242 ± 0·4975 ± 0·1057
c_2	0·264	0·1356 ± 0·2379 ± 0·1259	0·1566 ± 0·1398 ± 0·0599	0·1760 ± 0·1217 ± 0·0324
d_1	−0·775	−0·0884 ± 0·8082 ± 0·4375	−0·2771 ± 0·5614 ± 0·1995	−0·3641 ± 0·5037 ± 0·1064

Table 8.1B IVAML, $S=10$.

Parameter	True value	Number of samples		
		75	300	1000
a_1	−1·425	−1·4104 ± 0·0844	−1·4214 ± 0·0381	−1·4271 ± 0·0238
a_2	0·496	0·4808 ± 0·0889	0·4943 ± 0·0413	0·4989 ± 0·0262
b_1	−0·102	−0·1036 ± 0·0193	−0·1009 ± 0·0095	−0·1022 ± 0·0053
b_2	0·173	0·1732 ± 0·0147	0·1725 ± 0·0071	0·1731 ± 0·0036
c_1	−1·027	−0·3327 ± 0·7936	−0·2934 ± 0·7764	−0·4725 ± 0·6420
c_2	0·264	−0·0327 ± 0·3664	0·0330 ± 0·2528	0·1122 ± 0·1860
d_1	−0·775	0·0541 ± 0·9175	0·0108 ± 0·8272	−0·1923 ± 0·6678

Table 8.1(C) and (D) Monte Carlo Simulation Results
for Model 1, S = 1

Table 8.1C Refined IVAML, $S=1$.

Parameter	True value	Number of samples		
		75	300	1000
a_1	-1.425	$-1.4167 \pm 0.1809 \pm 0.1418$	$-1.4090 \pm 0.0715 \pm 0.0709$	$-1.4206 \pm 0.0406 \pm 0.0399$
a_2	0.496	$0.4904 \pm 0.1702 \pm 0.1329$	$0.4830 \pm 0.0667 \pm 0.0666$	$0.4916 \pm 0.0383 \pm 0.0374$
b_1	-0.102	$-0.0988 \pm 0.0423 \pm 0.0400$	$-0.0989 \pm 0.0204 \pm 0.0197$	$-0.1026 \pm 0.0098 \pm 0.0107$
b_2	0.173	$0.1721 \pm 0.0510 \pm 0.0470$	$0.1724 \pm 0.0227 \pm 0.0230$	$0.1740 \pm 0.0113 \pm 0.0124$
c_1	-1.027	$-0.3603 \pm 0.7986 \pm 0.4312$	$-0.4730 \pm 0.6511 \pm 0.1979$	$-0.5978 \pm 0.5461 \pm 0.1057$
c_2	0.264	$0.1602 \pm 0.2025 \pm 0.1259$	$0.1450 \pm 0.1648 \pm 0.0599$	$0.1686 \pm 0.1366 \pm 0.0324$
d_1	-0.775	$-0.1407 \pm 0.7611 \pm 0.4376$	$-0.2234 \pm 0.6470 \pm 0.1995$	$-0.3384 \pm 0.5516 \pm 0.1064$

Table 8.1D IVAML, $S=1$.

Parameter	True value	Number of samples		
		75	300	1000
a_1	-1.425	-1.2301 ± 0.3876	-1.3486 ± 0.1348	-1.3753 ± 0.0822
a_2	0.496	0.3650 ± 0.3448	0.4482 ± 0.1446	0.4628 ± 0.0879
b_1	-0.102	-0.1008 ± 0.0591	-0.0988 ± 0.0285	-0.1023 ± 0.0164
b_2	0.173	0.1604 ± 0.0614	0.1648 ± 0.0253	0.1678 ± 0.0134
c_1	-1.027	-0.3595 ± 0.7601	-0.3146 ± 0.7553	-0.4307 ± 0.6729
c_2	0.264	-0.0248 ± 0.3490	0.0139 ± 0.2735	0.0723 ± 0.2233
d_1	-0.775	0.0443 ± 0.8868	0.0097 ± 0.8252	-0.1238 ± 0.7250

results for larger sample size (>1000) have indicated that the estimates are unbiased and tend to have minimum variance properties in the asymptotic situation (>2000 samples is a typical case, although this will vary with the nature of the model). For example, Table 8.2 shows the refined IVAML results for this model

$$y_k = \frac{1.0z^{-1} + 0.5z^{-2}}{1 - 1.5z^{-1} + 0.7z^{-2}} u_k + \frac{1 - 1.0z^{-1} + 0.2z^{-2}}{1 - 1.5z^{-1} + 0.7z^{-2}} e_k$$

and the refined AML algorithm is clearly performing in a close to minimum variance sense for the 2000 sample size case.

In relation to the above remarks on the noise model parameter estimates, it is important to note that the rather poor estimation performance does not seem to affect the statistical efficiency of the a_i and b_i parameter estimates. This is surprising at first sight since the noise model parameters are utilized as an aid in achieving the statistical efficiency of the (a_i, b_i) estimates via their use in the adaptive pre-filters. It can only be assumed that, while the *parametric* estimates associated with the noise model are not particularly good for low or moderate sample size, the spectral representation resulting from these estimates is sufficiently accurate to ensure good filtering characteristics. In a sense, therefore, one might conjecture that the estimated noise model, although *parametrically* ill-defined, is defined sufficiently well in *frequency domain terms* for most filtering (and, therefore, control) applications.

The good estimation of the $(a_i\ b_i)$ parameters, despite the rather poor estimation of the (c_i, d_i) noise parameters, is very important in most engineering-type applications since it is the dynamic characteristics of the basic system that are often, although not always, of most importance: in a sense the 'noise' is a nuisance tending to prevent good estimation of the basic system characteristics. Only in certain automatic control and forecasting applications are the noise model properties themselves of major importance. And even then, if our conjecture on the adequacy of the spectral characterization proves correct, it could well be that the estimated noise model will often prove satisfactory, albeit non-optimal. Once again, however, the efficacy of the estimated model in this particular application will tend to depend upon the specific nature of that application.

An interesting example of the recursive estimates obtained during the MC simulation experiments for refined IVAML is given in Figure 8.6, which shows the results obtained for the ARMAX model:

$$y_k = \frac{1.0z^{-1}}{1 - 0.9z^{-1} + 0.95z^{-2}} u_k + \frac{1 + 1.5z^{-1} + 0.75z^{-2}}{1 + 0.9z^{-1} + 0.95z^{-2}} e_k \qquad (8.22)$$

with S = 10. As might be expected, the recursive estimates of the a_1, a_2 and b_1 parameters converge much faster than those of the noise parameters c_1 and c_2 (note the model

Table 8.2 Morte Carlo Simulation Results for Model 2, S = 1

(A) Refined IV-AML Results, (B) Equivalent Results obtained by Ljung et al. (1975) using 'RML2' Algorithm.

Table 8.2A Refined IVAML, $S=1$.

Parameter	True value	Number of samples		
		100	500	2000
a_1	$-1\cdot5$	$-1\cdot5060\pm0\cdot0842\pm0\cdot1145$	$-1\cdot4913\pm0\cdot0581\pm0\cdot0506$	$-1\cdot5067\pm0\cdot0182\pm0\cdot0249$
a_2	$0\cdot7$	$0\cdot7209\pm0\cdot0813\pm0\cdot0936$	$0\cdot6972\pm0\cdot0445\pm0\cdot0416$	$0\cdot7081\pm0\cdot0183\pm0\cdot0208$
b_1	$1\cdot0$	$1\cdot0170\pm0\cdot3698\pm0\cdot3600$	$1\cdot0204\pm0\cdot1096\pm0\cdot1590$	$0\cdot9872\pm0\cdot0829\pm0\cdot0785$
b_2	$0\cdot5$	$0\cdot5472\pm0\cdot3645\pm0\cdot4938$	$0\cdot5267\pm0\cdot2454\pm0\cdot2134$	$0\cdot5190\pm0\cdot0662\pm0\cdot1048$
c_1	$-1\cdot0$	$-0\cdot8972\pm0\cdot3145\pm0\cdot1011$	$-0\cdot9762\pm0\cdot0782\pm0\cdot0442$	$-0\cdot9993\pm0\cdot0205\pm0\cdot0220$
c_2	$0\cdot2$	$0\cdot1599\pm0\cdot3933\pm0\cdot1014$	$0\cdot1926\pm0\cdot0222\pm0\cdot0442$	$0\cdot2035\pm0\cdot0078\pm0\cdot0220$

Table 8.2B RML2, $S=1$†.

Parameter	True value	Number of samples		
		100	500	2000
a_1	$-1\cdot5$	$-0\cdot888\pm0\cdot445$ (0·083)	$-1\cdot460\pm0\cdot060$ (0·037)	$-1\cdot497\pm0\cdot023$ (0·019)
a_2	$0\cdot7$	$0\cdot299\pm0\cdot287$ (0·071)	$0\cdot678\pm0\cdot054$ (0·032)	$0\cdot698\pm0\cdot021$ (0·016)
b_1	$1\cdot0$	$1\cdot135\pm0\cdot532$ (0·348)	$0\cdot992\pm0\cdot231$ (0·156)	$0\cdot970\pm0\cdot077$ (0·078)
b_2	$0\cdot5$	$0\cdot790\pm0\cdot576$ (0·425)	$0\cdot661\pm0\cdot259$ (0·190)	$0\cdot549\pm0\cdot110$ (0·095)
c_1	$-1\cdot0$	$-0\cdot204\pm0\cdot389$ (0·134)	$-0\cdot888\pm0\cdot093$ (0·060)	$-0\cdot980\pm0\cdot038$ (0·030)
c_2	$0\cdot2$	$0\cdot092\pm0\cdot219$ (0·122)	$0\cdot149\pm0\cdot035$ (0·054)	$0\cdot182\pm0\cdot039$ (0·027)

† Quantities in brackets denote relevant Cramer–Rao lower bounds on the standard deviation of the parameter estimates, as quoted by Ljung *et al.* (1975).

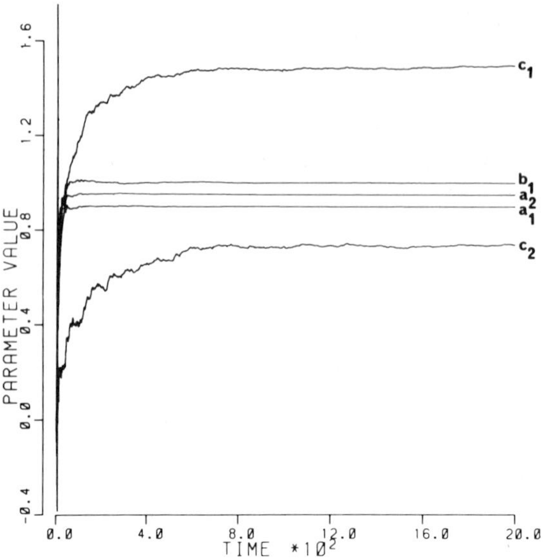

Figure 8.6 Recursive Estimates for Parameters in Model 3, Equation (8.22)

is of the ARMAX form with $C(z^{-1}) = A(z^{-1})$ so that there are no d_i parameters). The *noise* model in (8.22) is, in fact, the counter example to the general convergence of the AML algorithm suggested by Ljung *et al.* (1975). The reason that there are no convergence problems in this case, even though the refined AML algorithm has not been modified to the symmetric matrix form, lies in the ARMAX type of model. Since the refined IV algorithm, which has no convergence problems, supplies an estimate of the $A(z^{-1}) = C(z^{-1})$ polynomial coefficients, the refined AML algorithm is effectively applied to only a *pure, moving average* process (provided $C(z^{-1})$ is constrained to be equal to the $A(z^{-1})$ estimate obtained from the refined IV algorithm as estimation proceeds). And the theoretical analysis indicates that either AML or the non-symmetric gain refined AML converge satisfactorily in the case of pure moving-average models.[†]

This latter example indicates, once again, the difficulty of completely evaluating the performance of any estimation algorithm: limited theoretical results, when available, can help to provide a framework for assessment of the (usually asymptotic) behaviour, provided they are applied carefully; and MC analysis can give an indication of small-moderate sample size characteristics, but without any guarantee that the simulated behaviour is globally representative. However, if *both* types of

---

† This secondary result receives only brief mention in the report by Soderstrom *et al.* (1974) although, as we see, it is an important point.

analysis produce encouraging results, as in the case of IVAML and refined IVAML, then the user can normally utilize the estimation procedure with some confidence in practical applications.

Nevertheless the analyst should always, as a matter of course, evaluate any results obtained from the estimation procedures to ensure that they 'make sense': it is unusual indeed for the analyst to be in a situation where the system under study is a complete 'black-box'; and if *a priori* and other information are available about the system then it should *always* be used to help confirm the efficacy of the results obtained from the estimator.

8.7 A Practical Example: Analysis of Time-Series Tracer Data in Translocation Studies

One practical example which benefits by the use of the refined IVAML algorithm is the analysis of tracer data in translocation studies. Such studies, which are common in many areas of science from hydrology to medicine (see e.g. Jakeman and Young, 1980a) involve the measurement of the changes in concentration of some tracer material (e.g. dye in river systems, radioactive tracers in the circulatory systems of animals and plants) at different locations in a fluid pathway. In a typical experiment, tracer material is injected as an impulse input at one location and its concentration variations are monitored at some 'downstream' location. The problem is then to infer from this experimental data the transfer characteristics between either the injection point and a downstream location, or between two of the locations downstream.

In physical terms, it can be shown that an appropriate model for concentration variations is the "aggregated dead-zone" model (Beer and Young, 1981) which, in chemical engineering terms, is simply a combination of "plug" flow (representing the basic transportation mechanism) and a series of "continuous stirred tank reactors" (CSTR), which provide for the dispersive mechanism. In discrete-time terms, this model (with one CSTR section) can be written as

$$c_k = -a_1 c_{k-1} + b_0 u_{k-\tau} \qquad (8.23)$$

where u_k is the upstream concentration at time k, c_k is the 'downstream' concentration at time k, a_1 and b_0 are unknown coefficients, and τ is a pure time delay, usually in an integral number of sampling intervals, which accounts for the purely translational plug flow. Clearly this model is of the general time-series model form considered in this book and recursive methods of time-series analysis can be applied to the problem of identifying τ and estimating the a_1 and b_0 coefficients.

The advantage of refined IVAML in this context lies in its automatic use of prefilters on the input-output data. Since u_k may be defined simply as an impulsive input, the ordinary IVAML estimates may be rather poor and the pre-filtering improves the situation quite a lot. The ordinary and refined IVAML results for a simulation example which demonstrate this are given in Figure 8.7: here the model was chosen

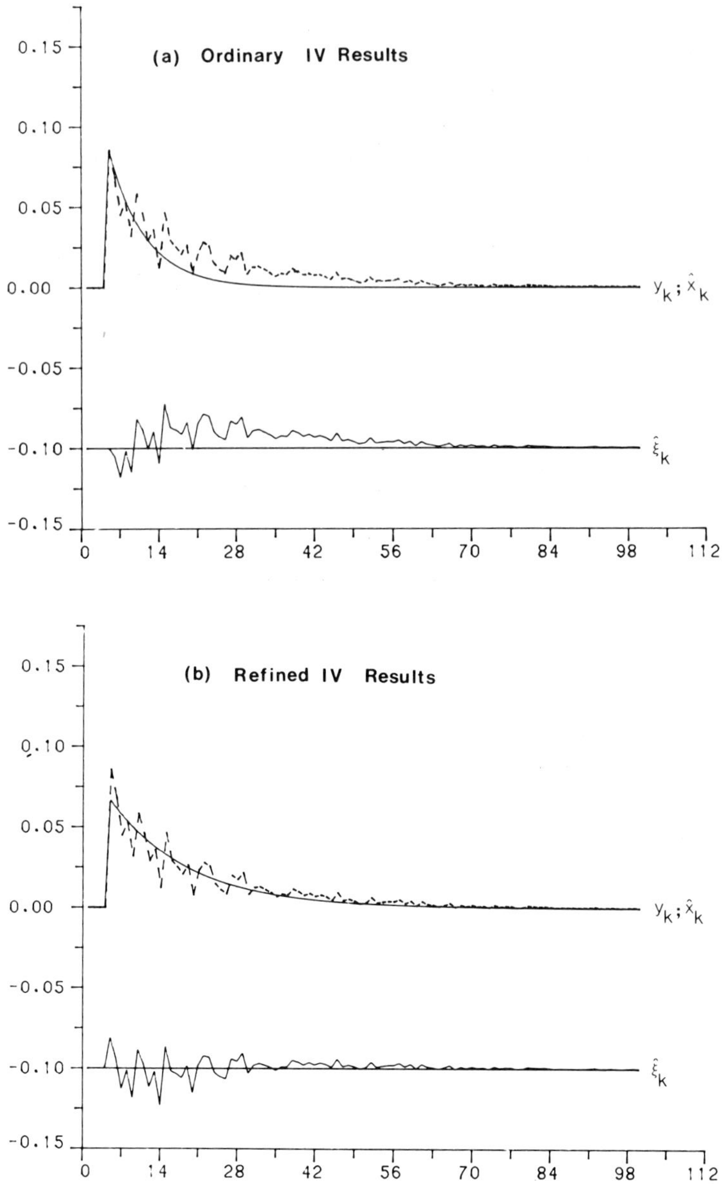

Figure 8.7 Ordinary and Refined IV Estimation Results for Simulated Translocation Expt. Data

with $a_1 = -0.95$, $b_0 = 0.05$ and $\tau = 1$ and the noise in the observations was white noise, with the variance made a function of c_k (i.e. the noise level is higher the larger the output signal). The degree of explanation of the data is better for refined ($R_T^2 = 0.89$) than ordinary IVAML ($R_T^2 = 0.74$) and both the estimates and their standard errors show a considerable improvement in the refined case as shown in Table 8.3. Note here that the standard errors in the refined case are overly optimistic; this is the consequence of the poor input signal in this example which is not persistently exciting (see Chapter 7). Note also the considerable improvement in the estimate of the steady state gain in the refined IV case; this is a consequence of the considerably better fit to the data obtained with the refined IV model as shown in Figure 8.7

TABLE 8.3

Estimation Algorithm	Parameter	True Value	Estimate	Estimated Standard Error	R_T^2	Estimated Steady State Gain
Ordinary IV	a_1	-0.95	-0.859	± 0.04	0.74	0.61
	b_0	0.05	0.086	± 0.007		
Refined IVAML	a_1	-0.95	-0.932	± 0.004	0.89	0.985
	b_0	0.05	0.067	± 0.003		

In this example $C(z^{-1}) - D(z^{-1}) = 1.0$ so that the pre-filter is simply low pass of the form $1/A(z^{-1})$ with $A(z^{-1}) = 1 - 0.95z^{-1}$. It is interesting to note that the physical effect of this pre-filter on the input time series u_k is to convert it from a single unit impulse at $k = 1$ to a decaying exponential sequence $u_k^* = 1.0, 0.95, 0.9025, 0.8574, 0.8145, \ldots$, as generated from the relationship

$$u_k^* = 0.95u_{k-1}^* + u_k$$

In other words, the information in the signal is effectively 'spread out' by the pre-filter and this helps to increase the statistical efficiency of the estimates, with a clear improvement in the estimated values. The reader will obtain some insight into the reason for this improvement in *algorithmic terms* if he examines the behaviour of the $\hat{P}_k$ matrix where, by definition, $\hat{P}_k = \hat{C}_k^{-1}$ with, for refined IVAML,

$$\hat{C}_k = \begin{bmatrix} \sum_{i=1}^{k} \hat{x}_i^* y_i^* & \sum_{i=1}^{k} \hat{x}_i^* u_i^* \\ \sum_{i=1}^{k} u_i^* y_i^* & \sum_{i=1}^{k} u_i^{*2} \end{bmatrix}$$

For ordinary IVAML, $\hat{C}_k$ is given by a similar expression without star superscripts. Note particularly that the lower diagonal term will have a value of unity for *all k* in the ordinary IVAML case, while in the refined case (with perfect pre-filtering), it will increase with increasing k as 1.0, 1.9025, 2.717, 3.452, etc., with the corresponding effect of continually reducing the associated $\hat{P}_k$ element to some extent. Thus for ordinary IVAML, the b_0 estimate is defined *totally* by the first sample value of the input time-series, whilst with the presence of filters on the data, this initial estimate is continually refined as the time-series samples are processed recursively.

A typical example of the analytical results obtained with a real set of data is shown in Figure 8.8, where the tracer data were obtained from a dye experiment in the Tuggeranong Creek, a rural/urban drain in the Australian Capital Territory. Here an impulse or "gulp" input of the red dye tracer Rhodamine WT was used in the experiment and it is clear that the data between two downstream sampling locations are explained well by a model of the form (8.23).

One additional advantage of refined IVAML in this kind of planned experimental application is that its improved statistical basis allows the analyst to consider the question of experimental planning in greater depth. For example, in the tracer experiment case, it helps him to evaluate how factors such as the nature of the input perturbation and the location of sampling points can effect the quality of the parameter estimates (e.g. Jakeman and Young, 1980a).

8.8 The Optimum Generalized Equation Error (OGEE) Approach to Time-Series Analysis

So far in our discussion of time-series analysis we have concentrated largely on a particular time-series model form, namely the TF model (8.1) i.e.

$$y_k = \frac{B(z^{-1})}{A(z^{-1})} u_k + \frac{D(z^{-1})}{C(z^{-1})} e_k \tag{8.1}$$

Other time-series model forms have been suggested, however, of which the most popular are the following

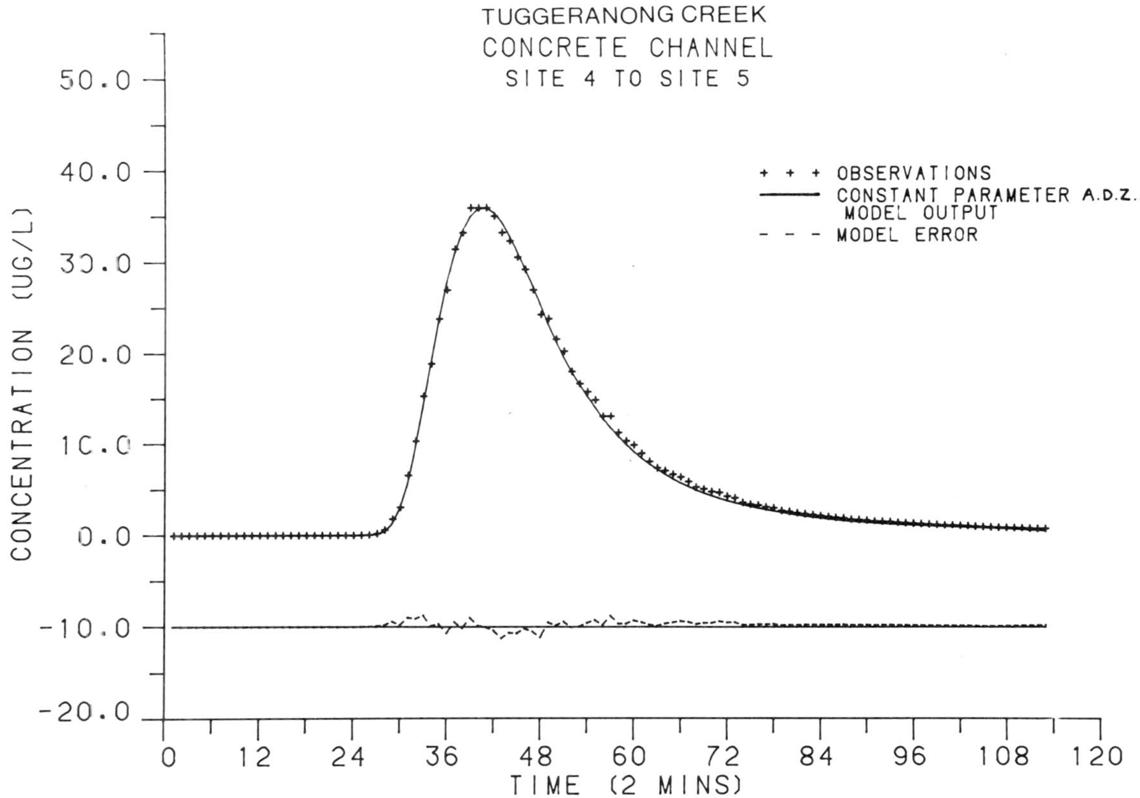

Figure 8.8 ADZ Model Results Obtained from IV Estimation of Dye Tracer Data from Tuggeranong Creek, ACT, Australia

. Autoregressive Moving Average Exogenous Variables (ARMAX):

$$A(z^{-1})y_k = B(z^{-1})u_k + D(z^{-1})e_k \tag{8.24}$$

. Dynamic Adjustment (DA) with ARMA noise:

$$A(z^{-1})y_k = B(z^{-1})u_k + \frac{D(z^{-1})}{C(z^{-1})} e_k \tag{8.25}$$

. Dynamic Adjustment (DA) with AR noise:

$$A(z^{-1})y_k = B(z^{-1})u_k + \frac{1}{C(z^{-1})} e_k \tag{8.26}$$

where the variables and polynomials are all as previously defined in relation to the TF model.

Of these models, the ARMAX model (8.24), which we have discussed briefly in Chapter 6, is probably the most popular in both the time-series and systems literature (Hannan and Tanaka, 1976; Astrom and Bohlin, 1966), although the TF model (8.1) has risen in popularity since the publication of Box and Jenkin's book on time-series analysis, forecasting and control in 1970. The DA model with AR noise is sometimes preferred, however, due to the simplicity of its noise structure (e.g. Johnston, 1963; Clarke, 1967). The DA model with ARMA noise is the least popular since it appears to offer few advantages in analytical terms over the other models.

Another model which has received attention recently (Young and Jakeman, 1980; Jakeman *et al.*, 1980) is the multiple input-single output (MISO) version of the TF model i.e.

$$y_k = \frac{B_1(z^{-1})}{A_1(z^{-1})}u_{1k} + \ldots . + \frac{B_m(z^{-1})}{A_m(z^{-1})}u_{mk} + \frac{D(z^{-1})}{C(z^{-1})}e_k \tag{8.27}$$

where u_{ik}, $i = 1, 2, \ldots., m$ are m deterministic inputs, all of which affect the output y_k, and (A_i, B_i) are polynomials in z^{-1} which define the transfer characteristics for each input. A simpler version of this model is the multiple input ARMAX model where $A_i = C = A$ for $i = 1, 2, \ldots., m$ i.e.

$$A(z^{-1})y_k = B_1(z^{-1})u_{1k} + \ldots . + B_m(z^{-1})u_{mk} + D(z^{-1})e_k \tag{8.28}$$

Alternatively, in the multi-input case, it is possible to consider full multivariable (multiple input-multiple output) versions of all the models discussed above (Jakeman and Young, 1979a).

One advantage of the refined IV type of algorithm is that it suggests a general *approach* to stochastic estimation that can be applied to all of these and other linear time-series models and which we will term 'optimal generalized equation error'

(OGEE) minimization (Young and Jakeman, 1980). This approach is related to the optimal prediction error (PE) minimization method discussed in the next chapter.

The OGEE approach to time-series analysis consists of the following three steps.

(1) Formulate the stochastic, dynamic model so that the stochastic characteristics are defined in relation to a 'source' term consisting of a white noise innovations process e_k (or $\underline{e}_k$ in the multivariable case), and then obtain an expression for e_k in terms of all the other model variables.

(2) By defining appropriate prefilters, manipulate the expression for e_k until it is a linear relationship (or set of relationships) in the unknown parameters of *the basic, deterministic part of the model* (i.e. the A and B polynomial coefficients in all examples considered so far). Because of their similarity to the equation-error relationships used in ordinary IV analysis, these linear expressions can be considered as 'generalized equation-error' (GEE)[†] functions.

(3) Apply the recursive or recursive-iterative IV algorithms to the estimation of the parameters in the GEE model(s) obtained in step (2), with the IV's generated in the usual manner as functions of the output of an adaptive auxiliary model (in the form of the deterministic part of the system model). If prefilters are required in the definition of GEE, then they will also need to be made adaptive, if necessary by reference to a noise model parameter estimation algorithm (e.g. the refined AML algorithm) utilized in parallel and co-ordinated with the IV algorithm.

This decomposition of the estimation problem into parallel but co-ordinated system and noise model estimation, as outlined in step (3), is central to the concept of GEE minimization: and it contributes to the robustness of the resultant algorithms in comparison with equivalent prediction error (PE) minimization algorithms as discussed in the next chapter (which, in general, will be nominally more statistically efficient).

The robustness is enhanced further by the IV mechanization, which ensures that the algorithm is not susceptible to contravention of theoretical assumptions about the nature of the noise process. In particular, the supposition that the noise arises from a white noise source, usually via some dynamic model (e.g. a rational transfer function, as in the ARMA model) is not restrictive: provided that the system input signals (u_k or $\underline{u}_k$) are independent of the noise, then the refined IV algorithms will yield estimates which are asymptotically unbiased and consistent *even if the noise assumptions are incorrect*. This remains true even if the noise is highly structured, e.g. a periodic signal or d.c. bias. On the other hand, if the assumptions *are* valid, then the resulting estimates will, as we have seen in this chapter, have the additional desirable property of improved statistical efficiency.

---

† The term GEE has been used previously in the literature to denote any EE function defined in terms of prefiltered variables: here we use it more specifically to mean an optimal GEE function with prefilters chosen to induce asymptotic statistical efficiency. (See Eykhoff, 1974; Young, 1968a).

To exemplify the OGEE minimization approach, let us consider the MISO transfer function model (8.27). Most time-series research in the MISO case has been directed towards the ARMAX representation (8.28), where the transfer functions relating each separate input to the output have *common* denominator polynomials. The model (8.27) is potentially more useful, however, and can be considered as the 'dynamic' equivalent of regression analysis, with the regression coefficients replaced by transfer functions.

Considering the two input case for convenience, we note from (8.27) that the white noise source e_k is defined as (removing the z^{-1} argument for convenience),

$$e_k = \frac{C}{D} y_k - \frac{CB_1}{DA_1} u_{1k} - \frac{CB_2}{DA_2} u_{2k} \tag{8.29}$$

It is now straightforward to show that (8.29) can be written in two GEE forms. First, if a single star superscript is utilized to denote prefiltering by C/DA_1, then

$$e_k = A_1 \xi_{1k}^{*} - B_1 u_{1k}^{*} \tag{8.30}$$

Here ξ_{1k} is defined as

$$\xi_{1k} = y_k - \hat{x}_{2k} \tag{8.31}$$

where $\hat{x}_{2k}$ is the output of the auxiliary model between the second input u_{2k} and the output, i.e. it is that part of the output 'explained' by the second input *alone*. Similarly, e_k can be defined in terms of ξ_{2k} where

$$\xi_{2k} = y_k - \hat{x}_{1k} \tag{8.32}$$

In this case,

$$e_k = A_2 \xi_{2k}^{**} - B_2 u_{2k}^{**} \tag{8.33}$$

where the double star superscript denotes prefiltering by C/DA_2.

By decomposing the problem into the two expressions (8.30) and (8.33), we have been able to define two separate GEE's which are linear in the unknown model parameters for each transfer function in turn. Now let us define,

$$\underline{a}_1 = [a_{11}, \ldots, a_{1n}, b_{10}, \ldots, b_n]^T; \ \underline{a}_2 = [a_{21}, \ldots, a_{2n}, b_{20}, \ldots, b_{2n}]$$

where a_{ij}, $j = 1, 2, \ldots, n$; and b_{ij}, $j = 0, 1, \ldots, n$ are the jth coefficients of A_i and B_i respectively. It is now possible to obtain estimates $\hat{\underline{a}}_1$ and $\hat{\underline{a}}_2$ of $\underline{a}_1$ and $\underline{a}_2$ from the following refined IV algorithm

$$\left.\begin{aligned}
\underline{a}_{ik} &= \hat{\underline{a}}_{i,k-1} - \hat{P}_{i,k-1}\hat{\underline{x}}_{ik}^{*}[\hat{\sigma}^2 + \underline{\xi}_{ik}^{*T}\hat{P}_{i,k-1}\hat{\underline{x}}_{ik}^{*}]^{-1}\{\underline{\xi}_{ik}^{*T}\hat{\underline{a}}_{i,k-1} - \xi_{ik}^{*}\} \quad (1)\\
\hat{P}_{ik} &= \hat{P}_{i,k-1} - \hat{P}_{i,k-1}\hat{\underline{x}}_{ik}^{*}[\hat{\sigma}^2 + \underline{\xi}_{ik}^{*T}\hat{P}_{i,k-1}\hat{\underline{x}}_{ik}^{*}]^{-1}\underline{\xi}_{ik}^{*T}\hat{P}_{i,k-1} \quad (2)
\end{aligned}\right\} \text{XIX}$$

where $\hat{\underline{x}}_{ik}$ and ξ_{ik}^{*} are defined as follows

$$\hat{\underline{x}}_{ik}^{*} = [\hat{x}_{i,k-1}^{*}, \ldots, \hat{x}_{i,k-n}^{*}, u_{ik}^{*}, \ldots, u_{i,k-n}^{*}]^T$$

$$\underline{\xi}_{ik}^{*} = [\xi_{i,k-1}^{*}, \ldots, \xi_{i,k-n}^{*}, u_{ik}^{*}, \ldots, u_{i,k-n}^{*}]^T$$

Algorithm XIX is used twice for i = 1, 2; but when i = 2, the single star superscripts are replaced by double star superscripts. The adaptive prefiltering is then executed in the same manner as for the single input case, with the refined AML algorithm providing estimates of the C and D polynomial coefficients. The extension to the general case of m inputs is obvious. There is also a symmetric gain version of XIX with $\underline{\xi}_{ik}^{*T}$ being replaced by $\hat{\underline{x}}_{ik}^{*}$ everywhere except within the braces (Jakeman *et al.*, 1980).

Jakeman *et al.* (1980) have evaluated the MISO transfer function model estimation procedure using both simulated and real data. Figure 8.9 compares the deterministic output $\hat{x}_k$ of a MISO air pollution model, obtained in this manner, with the measured data. Here the data are in the form of atmospheric ozone measurements at a 'downstream' location in the San Joaquin Valley of California. These are modelled in terms of two 'input' ozone measurements at 'upstream' (in relation to the prevailing wind) locations. This analysis proved particularly useful for interpreting across gaps in downstream data. In general the MISO algorithm works well but tends to be less robust than in the single input case; in particular, it is more important to have reasonable initial estimates of the parameters. Also collinearity problems can occur, as might be expected, if the inputs u_{ik} are highly correlated.

8.9 Summary

This chapter has shown how, by considering the estimation problem for the transfer function time-series model within the context of maximum likelihood, it is possible to derive instrumental variable-like algorithms which have improved statistical efficiency in relation to their ordinary IVAML relatives. However, the development of such algorithms does not negate the usefulness of the ordinary IVAML procedures. On the contrary, it gives the analyst a choice: ordinary IVAML is computationally simple and robust, with *relatively* good statistical efficiency; refined IVAML is more complex and somewhat less robust, but it provides a method of improving the estimates of the a_i and b_i parameters in situations where greater statistical efficiency is demanded.

Refined IVAML also puts the IV estimation procedure on a sound theoretical footing so allowing for good estimation of the statistical properties of the parameter estimates and, as we shall mention in chapter 9, theoretical evaluation of the convergence properties of the recursive algorithms. One additional advantage of this improved theoretical basis is that it permits easier consideration of factors such as the design of experiments for dynamic systems, as in the case of selecting input perturbations and sampling locations in translocation experiments using tracer materials.

We have also seen in this chapter how the refined IVAML algorithm suggests a general approach to algorithmic construction for time-series models which we have termed optimal generalized equation error (OGEE) minimization. This approach is generally applicable but has been exemplified by its application to multiple-input-single-output model estimation.

Finally, it should be pointed out that a recent paper by Soderstrom and Stoica (1980) has discussed the question of optimum instrumental variable estimation in a somewhat different manner to that used here although some of their results, as we might expect, are related to those discussed in this chapter.[†]

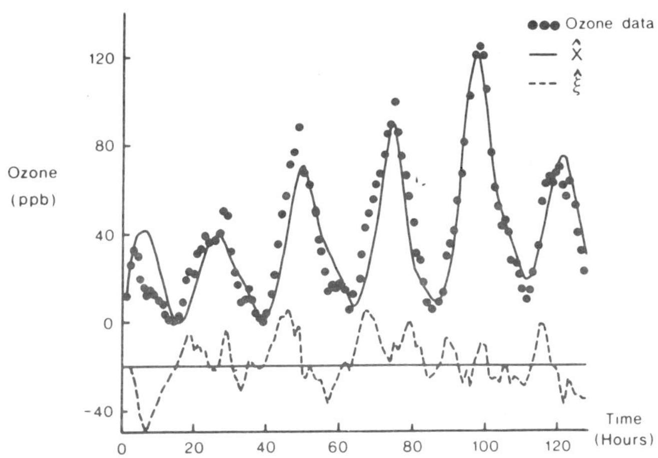

Figure 8.9 MISO Model Results Obtained from IV Estimation of Ozone Data from San Joaquin Valley, California

[†] Soderstrom and Stoica have now produced a book (1983) which comprehensively discusses IV methods from a systems theoretic standpoint (see Epilogue, page 241).

9 Alternative Recursive Approaches to Time-Series Analysis

In an attempt to aid the introductory reader in understanding a difficult subject, the previous chapters have followed sequentially from simple beginnings in deterministic, linear least squares methods; through the more complicated but still straightforward procedures of statistical regression analysis; to the quite complex techniques for solving the nonlinear problems of time-series analysis. And in this latter time-series case, we have been careful to look in detail only at those approaches which are related most closely to the simpler linear regression methods and can, to some extent, be regarded as logical developments of these methods in the context of stochastic, dynamic system estimation.

But there are other important recursive methods of time-series model estimation which, while closely related to the procedures described in the previous chapters, represent somewhat different approaches to the problem. We have referred to these procedures briefly in previous chapters and, in the present chapter, we will discuss them in sufficient detail for the reader to become acquainted with their basic form and computational nature. In this introductory text, however, no attempt will be made to evaluate them fully and the potential user will need to follow up some of the references in order to obtain the requisite understanding demanded for practical application of the various techniques. A more complete treatment will appear in Young (1984).

9.1 Prediction Error (PE) Methods

Once the reader has come to terms with the statistical basis of time-series models, the prediction-error approach to recursive estimation becomes rather obvious: in simple terms, and based on our discussion in Chapter 3, it implies the choice of a cost function such that the resultant recursive algorithm has a particular 'least squares' like form and the innovations process (or recursive residuals) in this recursive algorithm has the properties of white noise. In this way, the estimation problem is simplified in statistical terms while the resultant algorithm, if chosen correctly, is inherently convergent and has certain desirable statistical properties such as consistency and asymptotic statistical efficiency.

Perhaps the best way of introducing this PE concept in a simple manner is to consider the estimation problem in block diagram transfer function terms. It is then possible to visualize two basic approaches to the problem of estimating the TF time-series model (6.30): the first shown in Figure 9.1 has been termed the response or "output error" (OE) method; the second, illustrated in Figure 9.2 is the regression or "equation-error" (EE) method.

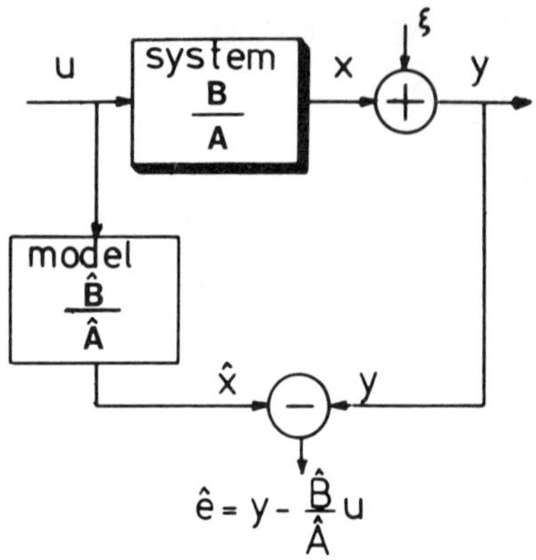

Figure 9.1 Output or Response Error

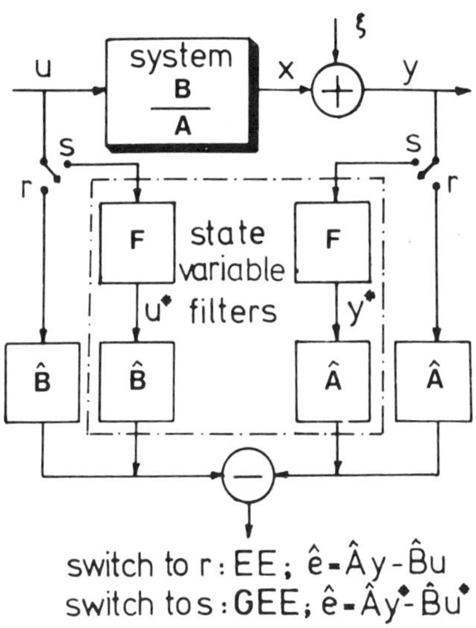

Figure 9.2 Equation or Generalised Equation Error

As its name implies, the OE method is concerned with the minimization of some cost function (normally quadratic in the present context) in the output error $\hat{e}_0$, which is defined simply as the error between the measured output y_k of the system and the output $\hat{x}_k$ of the TF model, i.e.

$$\hat{e}_{0k} = y_k - \frac{\hat{B}}{\hat{A}} u_k = y_k - \hat{y}_k \tag{9.1}$$

where $\hat{y}_k = \frac{\hat{B}}{\hat{A}} u_k$ is the model output and can be considered as an estimate of y_k. In the equation error case, on the other hand, the error $\hat{e}_E$ is defined directly in terms of the model equation, hence its name 'equation-error', i.e.

$$\hat{e}_{Ek} = \hat{A}y_k - \hat{B}u_k \tag{9.2}$$

The reader will recognize that, in the previous chapters, we have been predominantly concerned with the equation error approach; this becomes clear when (9.2) is written in its alternative vector form i.e.

$$\hat{e}_{Ek} = y_k - \underline{z}_k^T \hat{\underline{a}} \tag{9.3}$$

with the usual definitions, as given in Chapter 6. Note again that $\hat{y}_k = \underline{z}_k^T \hat{\underline{a}}$ can be considered as an estimate of y_k, although it is not the same estimate as in the OE case, except in the special situation where there is no noise on the data ($\xi_k \equiv 0$).

Both OE and EE approaches have been used in the development of estimation procedures for dynamic systems (see e.g. Young 1979a); the main difference being that e_E is linear in the unknown parameters, while e_0 is nonlinear. If there is no noise on the data ($\xi_k \equiv 0$) then there is little to choose between the two error functions in conceptual terms. On the other hand, the analysis is made much simpler in the EE case because of the linearity in the parameters; indeed, this is the major reason for choosing the EE function as the basis for our treatment of the problem in this book.

When the data are noisy ($\xi_k \neq 0$) the situation is not nearly so clear: while the EE approach still looks simpler, the noise induced asymptotic estimation bias discussed in Chapter 6 presents some problems and, as we have seen, it is necessary to introduce additional complexity into the solution to obviate this difficulty. The output error formulation does not suffer from this difficulty and the minimization of a quadratic function in the OE, although inherently a problem of nonlinear optimization with its unavoidable complexity, does yield asymptotically unbiased, consistent estimates under fairly mild, non-restrictive conditions on the nature of system and the observed data.

If there is noise on the data then, as we have seen in Chapter 8, we may be able to investigate the problem by making certain statistical assumptions about the stochastic nature of the noise. The most obvious assumptions are those used in

Chapter 8, namely that ξ_k has rational spectral density and so is derived from a white noise source e_k via an ARMA process. As we have pointed out in Chapter 3, stochastic estimation problems are made more straightforward and analytically more tractable if the error function on which minimization is based assumes the properties of zero mean, serially uncorrelated, white noise, denoted by e_k, when the estimated parameter vector, say $\hat{\underline{\theta}}$, in the definition of the error function is replaced by the true vector $\underline{\theta}$. Here $\underline{\theta}$ will, in general, be a vector of *all* the unknown parameters in the system and noise models. In such a situation, it is possible to obtain estimates with desirable statistical properties such as consistency, lack of asymptotic bias and asymptotic statistical efficiency (i.e. minimum variance estimates of $\underline{\theta}$ for large sample size).

In simple terms, this is the basic stimulus of the PE approach to estimation. As in the OE case, the error function is usually defined as $\hat{e}_k = \underline{y}_k - \hat{\underline{y}}_k$, but now $\hat{\underline{y}}_k \equiv \hat{\underline{y}}(k|\hat{\theta})$ is defined as some 'best prediction' of y_k given the current estimates of the parameter vector $\underline{\theta}$, which characterize the system *and* the noise models. In other words, $\hat{\underline{y}}(k|\hat{\underline{\theta}})$ is the conditional mean of $\underline{y}_k$ given *all* current and past information on the system; and $\hat{e}_k$ is an 'innovations' process with serially uncorrelated, white noise characteristics which converge in probability on e_k; i.e. $\underset{k\to\infty}{\text{p.lim.}}\,\hat{e}_k = e_k$. In the SISO case of equation (6.30) the most obvious PE method involves the minimization of a quadratic norm in $\hat{e}_k$, where $\hat{e}_k$ is defined as

$$\hat{e}_k = \frac{\hat{C}}{\hat{D}}[y_k - \frac{\hat{B}}{\hat{A}} u_k] \tag{9.4}$$

as illustrated in Figure 9.3. It is easy to see from the TF model equation for the system (8.1) i.e.

$$y_k = \frac{B}{A}u_k + \frac{D}{C}e_k$$

that, on convergence of $\hat{A}$, $\hat{B}$, $\hat{C}$ and $\hat{D}$ to A, B, C and D respectively, $\hat{e}_k$ assumes the same white noise properties of e_k and so satisfies the requirements of the PE method. But the class of PE methods is much larger than either the OE or EE methods; indeed, since it depends only on the definition of an error function with certain desirable statistical properties, it can be considered to contain the other two methods as special cases. We see this very easily in the OE case where, if the noise $\xi(t)$ is considered white (i.e. $C(s) = D(s) = 1.0$), then the PE and OE functions are one and the same. The link in the EE case is not quite so obvious but it becomes apparent if the noise model assumes the special form with $D(s) = 1.0$ and $C(s) = A(s)$. In the scalar case, the PE function $\hat{e}_k$ is then defined by the equation

$$\hat{e}_k = \hat{A}[y_k - \frac{\hat{B}}{\hat{A}} u_k] = \hat{A}y_k - \hat{B}u_k \tag{9.5}$$

which is simply the EE function (9.2).

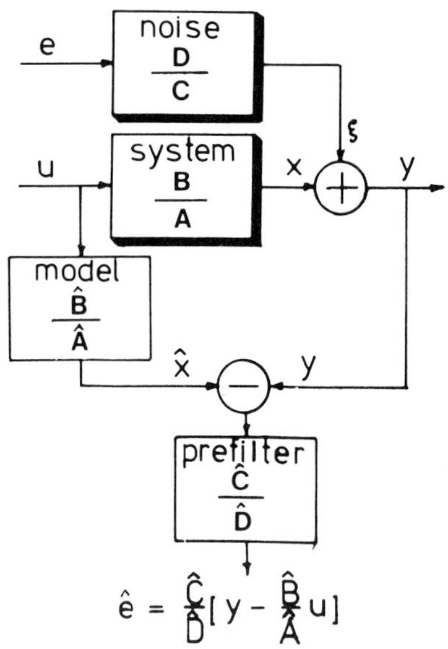

Figure 9.3 Prediction Error in OE Context

Of course, other arrangements are possible: for example, the more general PE approach within an EE context is obtained by defining $\hat{e}_k$ directly in a recognizable EE-like form. In the SISO case, for instance, $\hat{e}_k$ would be formulated as

$$\hat{e}_k = \frac{\hat{C}}{\hat{D}\hat{A}}[\hat{A}y_k - \hat{B}u_k] \tag{9.6}$$

as shown in Figure 9.4. The significance of this formulation is obvious in relation to the refined IVAML situation, since $\hat{e}_k$ in (9.6) can be rewritten as

$$\hat{e}_k = \hat{A}y_k^* - \hat{B}u_k^* = y_k^* - \underline{z}_k^{*T}\hat{\underline{a}} \tag{9.7}$$

where y_k, u_k and $\underline{z}_k$ are defined as in Chapter 8. In other words, $\hat{e}_k$ in (9.6) and (9.7) is the negative of the innovations process generated by the refined IVAML algorithm and clearly, therefore, the refined IVAML algorithm can be considered in PE terms.

In general, the solution of the PE minimization problem is more complex than the OE and EE equivalents since the formulation of the PE norm implies the concurrent estimation of the noise model parameters. As a result, the problem is inherently one of quite difficult, nonlinear estimation with its attendant practical problems. Almost all research on recursive PE methods has been in relation to purely discrete-time problems (e.g. Ljung, 1976; Caines and Ljung, 1976; Ljung, 1977a, b; and Ljung, 1978, 1979a),

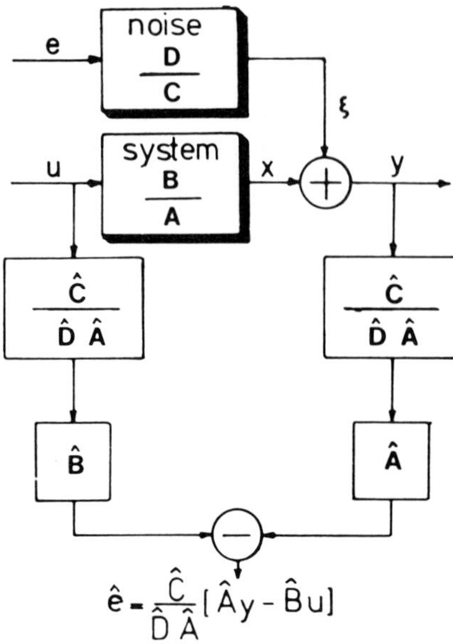

Figure 9.4 Prediction Error in EE Context

although Solo (1980) has considered both continuous and discrete-time implications. The introductory reader to whom the present book is addressed may find these papers and their theoretical treatment of PE methods a little difficult to comprehend. Nevertheless, they represent important contributions to the subject of recursive estimation since they deal with the important topics of algorithmic convergence and the asymptotic statistical properties of the estimates. Because of the difficult nature of the problem, however, we have purposefully omitted detailed discussion of these theoretical questions in this introductory text; rather we have let the reader assume that the algorithms described have generally good convergence properties unless otherwise stated (as in the AML algorithm).

This semi-theoretical approach does not imply that the convergence analysis is not important; merely it is intended to shield the reader new to the subject from rather esoteric theory before it is absolutely necessary. It also serves to emphasize that alternative *experimental* methods of evaluating convergence, such as the Monte-Carlo procedures discussed in the last chapter, are also important in assessing algorithmic performance. All purely theoretical analysis to date is based largely on asymptotic behaviour; small sample performance must, therefore, be considered by carefully conducted stochastic simulation experiments of the Monte Carlo type.

In relation to the theoretical analysis, it will suffice here to say that the theory can be based either on: (i) relating the behaviour of recursion to the solution of an ordinary, continuous-time differential equation (Ljung, 1977a,b); (ii) utilizing probability theory to reduce the analysis to the consideration of a deterministic Lyapunov function (Hannan, 1976); or (iii) analysis of the behaviour as a stochastic difference equation (Solo, 1978, 1980). This latter approach allows for second order analysis and can indicate the asymptotic covariance behaviour of the recursive PE estimators.

In general, the recursive PE algorithm is constructed, by analogy with recursive least squares and in accordance with the theoretical requirements for convergence, to take the following form,

$$\left.\begin{aligned} \hat{\underline{\theta}}_k &= \hat{\underline{\theta}}_{k-1} + P_k \underline{\Psi}_k \hat{e}_k && (1) \\ P_k &= P_{k-1} - P_{k-1}\underline{\Psi}_k[\sigma^2 + \underline{\Psi}_k^T P_{k-1}\underline{\Psi}_k]^{-1}\underline{\Psi}_k^T P_{k-1} && (2) \end{aligned}\right\} \quad (XX)$$

where $\underline{\theta}$ is a vector composed of all the unknown parameters and $\underline{\Psi}_k$ is a vector based on the negative of the gradient of $\hat{e}_k$ with respect to $\hat{\underline{\theta}}$, where $\hat{\underline{\theta}}$ is the estimate of $\underline{\theta}$, i.e.,

$$\underline{\Psi}_k = -\left[\frac{\partial \hat{e}}{\partial \hat{\underline{\theta}}}\right]_k = -\nabla_{\hat{\underline{\theta}}}[\hat{e}_k]$$

The motivation for this algorithm is clear when it is compared with both the SA algorithm (3.23) and the recursive least squares algorithm III of Chapter 4: it is simply a matrix gain SA or stochastic gradient algorithm, whose matrix gain is generated by analogy with the recursive least squares theory and whose gradient term, $\underline{g}_k = \underline{\Psi}_k\hat{e}_k$, can be interpreted as being proportional to the negative gradient of the *instantaneous* least squares cost function J_I, i.e.

$$J_I = \hat{e}_k^2$$

$$-\nabla_{\hat{\underline{\theta}}} J_I = -2\left[\frac{\partial \hat{e}}{\partial \hat{\underline{\theta}}}\right]_k \cdot \hat{e}_k = 2\underline{\Psi}_k\hat{e}_k$$

Note that to conform with convention in the PE literature we have defined $\hat{e}_k$ in XX (1) in accordance with the definition (9.7) rather than as usual in the rest of this book, where we use the negative of this error; as a result, the correction term in XX (1) is positive rather than negative. Of course, the resultant algorithms perform identically and the reader should not be confused by this slight change of definition.

In the case of the TF model (8.1) i.e.

$$y_k = \frac{B}{A}u_k + \frac{D}{C}e_k,$$

$\underline{\hat{\theta}} = [\hat{a}_1,\ldots, \hat{a}_n, \hat{b}_o,\ldots, \hat{b}_n, \hat{c}_1,\ldots, \hat{c}_n, \hat{d}_1,\ldots \hat{d}_n]$ and the PE innovations $\hat{e}_k$ is given by (9.6). The gradient vector $\underline{\Psi}_k$ can be evaluated quite simply in this case by differentiating $\hat{e}_k$ with respect to each parameter in turn. Thus, since

$$\hat{e}_k = \frac{\hat{C}}{\hat{D}\hat{A}}[\hat{A}y_k - \hat{B}u_k] = \frac{\hat{C}}{\hat{D}} y_k - \frac{\hat{B}\hat{C}}{\hat{D}\hat{A}} u_k$$

we obtain

$$\left.\begin{aligned}
\frac{\partial}{\partial \hat{a}_i}(\hat{e}_k) &= \frac{\hat{C}}{\hat{D}\hat{A}} \cdot \frac{\hat{B}}{\hat{A}} u_{k-i} = \frac{\hat{C}}{\hat{D}\hat{A}} \hat{x}_{k-i} = \hat{x}^*_{k-i} \\
\frac{\partial}{\partial \hat{b}_i}(\hat{e}_k) &= - \frac{\hat{C}}{\hat{D}\hat{A}} u_{k-i} = -u^*_{k-i} \\
\frac{\partial}{\partial \hat{c}_i}(\hat{e}_k) &= \frac{1}{\hat{D}}(y_{k-i} - \frac{\hat{B}}{\hat{A}} u_{k-i}) = \frac{1}{\hat{D}}\hat{\xi}_{k-i} = \hat{\xi}^*_{k-i} \\
\frac{\partial}{\partial \hat{d}_i}(\hat{e}_k) &= -\frac{1}{\hat{D}}[\frac{\hat{C}}{\hat{D}} y_{k-i} - \frac{\hat{C}\hat{B}}{\hat{D}\hat{A}} u_{k-i}] = -\hat{e}^*_{k-i}
\end{aligned}\right\} \quad (9.8)$$

where, as in the last chapter, the star superscript (*) denotes that the variable is filtered by a prefilter (assumed to be adaptive in some manner, as in refined IVAML) with transfer function $\hat{C}/\hat{D}\hat{A}$, while the star superscript on $\hat{\xi}_k$ and $\hat{e}_k$ indicates adaptive prefiltration by $1/\hat{D}$. In this manner, $\underline{\Psi}_k$ is defined as follows:

$$\underline{\Psi}_k = [-\hat{x}^*_{k-1},\ldots, -\hat{x}^*_{k-n}, u^*_k,\ldots, u^*_{k-n}, -\hat{\xi}^*_{k-1},\ldots, -\hat{\xi}^*_{k-n}, \hat{e}^*_{k-1},\ldots, \hat{e}^*_{k-n}]^T \quad (9.9)$$

The similarity between the algorithm with $\underline{\Psi}_k$ defined as in (9.9) and the refined IVAML algorithms XV and XVI of Chapter 8 is clear on inspection: the only major difference is that, whereas here all elements of $\underline{\theta}$ (i.e. a_i, b_i, c_i and d_i) are estimated *simultaneously* in a *single recursive algorithm* with a P_k matrix of dimension (4n+1) x (4n+1), in the refined IVAML case the parameters are estimated in two *separate but coordinated* algorithms; i.e. refined IV with $\hat{P}_k$ of dimension (2n+1) x (2n+1) and refined AML with P_k^n of dimension 2n x 2n. Clearly, therefore, the refined IVAML approach is computationally simpler than the recursive PE equivalent XX, which was first suggested by Ljung (1979a). But, given the powerful theoretical justification of the PE algorithm, we might ask whether the simpler refined IVAML approach can be fully justified. To consider this question, it is necessary to consider the asymptotic statistical properties of the estimates obtained from the PE algorithm XX.

Statistical Properties of the PE Estimates for the TF Model

In the case of the TF model, maximum likelihood estimation implies the minimization of the least squares cost function

$$J = \frac{1}{T} \sum_{k=1}^{T} \hat{e}_k^{\,2}$$

with respect to $\hat{\underline{\theta}}$. If we introduce an asymptotic cost function $J(\hat{\underline{\theta}})$ for the sample size T tending to infinity, then

$$J(\hat{\underline{\theta}}) = E\{\hat{e}_k^{\,2}\}$$

In this asymptotic situation, it is interesting to investigate the nature of the error covariance matrix for the parameter estimates. It is well known that this covariance matrix is, for large samples, given by the inverse of the so-called *information matrix* (Whittle, 1953; Fisher, 1956) and we can examine the covariance properties by reference to this information matrix $I(\underline{\theta})$, defined in the following manner

$$I(\underline{\theta}) = \frac{T}{2J(\underline{\theta})} \nabla^2 J(\underline{\theta}) = \frac{T}{2\sigma^2} \nabla^2 J(\underline{\theta}) \qquad (9.10)$$

where $\nabla^2 J(\underline{\theta})$ is the second partial derivative of $J(\hat{\underline{\theta}})$ with respect to $\hat{\underline{\theta}}$ evaluated at $\hat{\underline{\theta}} = \underline{\theta}$ where $\underline{\theta}$ is the *true* value of the parameter vector. In the present case,

$$\nabla^2_{\underline{\theta}} J(\theta) = E\{\nabla_{\underline{\theta}}[e_k]\cdot\nabla_{\underline{\theta}}^T[e_k]\} = E\{\underline{\Psi}_k\underline{\Psi}_k^T\} \qquad (9.11)$$

where $\underline{\Psi}_k$ is as defined in (9.9), with the prefilters set up in accordance with the *actual* parameter values.

If we now define

$$\hat{\underline{x}}_k^{*} = [-\hat{x}_{k-1}^{*}, \ldots, -\hat{x}_{k-n}^{*}, u_k^{*}, \ldots, u_{k-n}^{*}]^T \qquad (9.12)$$

and

$$\underline{m}_k = [-\xi_{k-1}^{*}, \ldots, -\xi_k^{*}, e_{k-1}^{*}, \ldots, e_{k-n}^{*}]^T \qquad (9.13)$$

then the information matrix can be written as

$$I(\underline{\theta}) = \frac{T}{2\sigma^2}\left[\begin{array}{c|c} E\{\hat{\underline{x}}_k^* \hat{\underline{x}}_k^{*T}\} & E\{\hat{\underline{x}}_k^* \underline{m}_k^T\} \\ \hline E\{\underline{m}_k \hat{\underline{x}}_k^{*T}\} & E\{\underline{m}_k \underline{m}_k^T\} \end{array}\right] \tag{9.14}$$

The error covariance matrix is given by the inverse of $I(\underline{\theta})$ in (9.14). But, from the definition of $\hat{\underline{x}}$ as an instrumental variable vector, we know that it is statistically independent of the noise terms ξ_k and e_k and is, therefore, independent of the vector $\underline{m}_k$. Consequently, the off-diagonal sub-matrices $E\{\underline{m}_k \hat{\underline{x}}_k^{*T}\}$ and $E\{\hat{\underline{x}}_k^* \underline{m}_k^T\}$ in the partitioned matrix $I(\underline{\theta})$ are identically zero and the estimate of the noise parameter vector $\underline{c}$ is asymptotically independent of the system parameter vector $\underline{a}$. This is, of course, precisely the result of Pierce (iii) given in Chapter 8; indeed the above analysis can be viewed as a simplistic proof of his theorem (although the interested reader should be directed to Pierce's paper for his more rigorous proof).

The asymptotic independence of the noise and system parameter estimates has important implications on both the PE and refined IVAML estimation procedures. We see that, for large T at least, little is accomplished by estimating the two sets of parameter estimates, $\hat{\underline{a}}$ and $\hat{\underline{c}}$ simultaneously, since the off diagonal blocks of the P_k matrix in algorithm XX will be close to zero. Indeed, from the numerical standpoint, it will not only be inefficient to compute these off diagonal blocks but it could be dangerous and lead to numerical problems, with the numbers composing these blocks being insignificantly different from zero and possibly going negative unless the computationally more expensive square-root formulation (Gelb *et al.*, 1974) is utilized.

For lower sample size, it could be argued that these off-diagonal terms *may* be significant and that their retention will improve the estimation. It is difficult to prove theoretically that this will be the case since, as we have indicated previously, no small sample theory has been developed for this problem. Recourse to Monte Carlo simulation could provide an indication of the relative importance of these terms as regards estimation accuracy but no thorough evaluation of algorithm XX in the TF model case has yet taken place. Initial simulation results do suggest, however, that there is little to choose between the PE and refined IVAML algorithms as regards estimation performance. On the other hand, the PE algorithm is obviously more complex in computational terms and this appears to lead to a greater number of the Monte Carlo realizations failing to converge. This appears partly due to the greater complexity of the PE algorithm and partly to the fact that the non-symmetric gain, refined IVAML algorithm appears more robust because of its IV-like structure (see Chapter 8). At this point in time, therefore, it is felt that the potential user will find the refined IVAML approach both simpler and easier to use than the PE approach (see Jakeman and Young, 1981 a, b).

Finally, the above discussion will reveal the similarities and differences between the PE minimization approach to the design of time-series estimation algorithms and the alternative OGEE approach discussed in the previous chapter. In general, we can conclude that the PE approach is somewhat more satisfactory in purely theoretical terms than the more heuristic OGEE approach, but it can lead to algorithmic designs which are less efficient computationally and tend to be less robust (e.g. Poljak and Tsypkin, 1980) in practical applications. In particular, the OGEE approach leads naturally to instrumental variable-like algorithms which are known to be very robust, in the sense that consistent and relatively efficient estimates are obtained even when the noise does not conform to the usual statistical assumptions.

9.2 The Extended Kalman Filter

The two other parameter estimation procedures discussed in this Chapter are both based on a state space (SS) representation and derive from a conceptualization of the estimation problem in Kalman filter-estimation terms. The first approach has, in fact, been termed the "Extended Kalman Filter" (EKF) and it is based on a simple modification of the Kalman filter equations (see e.g. Gelb *et al.*, 1974). Here both the state variables and any unknown parameters are estimated *simultaneously* using a recursive algorithm which closely resembles the normal Kalman filter and can be applied to both linear and nonlinear systems. But, even in the case of linear systems, the state variables and parameters to be estimated occur in product form, so that the estimation problem is *always* one of direct nonlinear estimation with its attendant practical difficulties. In order to allow for the nonlinearity, a form of linearization (sometimes termed re-linearization) of the filter-estimation equations is used, in which the covariance matrix (P_k^*; equation X (4) of Chapter 5)is based on the system equations *as linearized about the latest recursive estimates.*

Because the EKF is based on the Kalman filter, it can be implemented in completely digital (i.e. discrete-time model, discrete-time estimation, or DD) terms, in a purely analog (continuous-time model, continuous-time estimation or CC) form or, most usefully in practice, as a hybrid, continuous-discrete (CD) mechanization (see Young, 1979a). This latter formulation allows for the estimation of states in a continuous-time model of a stochastic dynamic system based on discrete time (sampled) data. In common with all EKF algorithms suggested in the literature so far, however, it does seem to have rather poor parameter estimation error variance, with the statistical efficiency usually rather low for any sample size in comparison to alternative, albeit less flexible, methods such as ordinary or refined IVAML (e.g. Neethling, 1974).

The EKF has been described in many texts (e.g. Jazwinski, 1970; Gelb *et al.*, 1974) but perhaps the best way of introducing the approach to the general reader is to look at a particular example in which the EKF is used in its CD form to estimate the states and parameters of a simple model for water quality in a non-tidal river system. Preparatory to this, let us consider a continuous-time system described by the following state-space differential equation

$$\dot{\underline{x}} = A\underline{x} + B\underline{u} + \underline{\eta} \tag{9.15}$$

where the dot notation indicates differentiation with respect to time, $\underline{x} = [x_1 x_2 \ldots\ldots x_n]^T$ is an n dimensional state vector, $\underline{u} = [u_1 u_2 \ldots\ldots u_q]^T$ is a q dimensional vector of deterministic forcing functions and $\underline{\eta} = [\eta_1 \eta_2 \ldots\ldots \eta_n]^T$ is an n dimensional vector of zero mean "white noise" disturbances to the system. A and B are, respectively, n x n and n x q matrices that characterise the system.

Suppose that some of the elements of A and B are unknown parameters and that we wish to estimate both the state $\underline{x}$ *and* the unknown parameters on the basis of our knowledge of the input forcing vector $\underline{u}$ and *discrete* observations of the system behaviour made by means of an observation process similar to that given earlier in equation (5.43), i.e.,

$$\underline{y}_k = C^* \underline{x}_k + \underline{\xi}_k \tag{9.16}$$

where $\underline{y}_k = [y_1 y_2 \ldots\ldots y_p]_k^T$ is the pth order observation vector at the kth time instant, and C^* is a p x n observation matrix.

One approach to this problem is to augment the state vector $\underline{x}$ with a parameter vector $\underline{a}$ composed of the unknown parameters and then hypothesise additional state equations for this parameter vector: e.g., if the parameters are all assumed time-invariant, a suitable state representation would be simply

$$\dot{\underline{a}} = 0$$

If we now define the augmented state vector $\underline{X}$ as

$$\underline{X} = \begin{bmatrix} \underline{x} \\ --- \\ \underline{a} \end{bmatrix} \tag{9.17}$$

the augmented state equations will be of the following general form,

$$\dot{\underline{X}} = \underline{f}\{\underline{X},\underline{u}\} + \underline{\eta} \tag{9.18}$$

with an observation equation

$$\underline{y}_k = C\underline{X} + \underline{\xi}_k \;;\; C = [C^* \,\vdots\, 0] \tag{9.19}$$

In equation (9.18) $\underline{f}\{\underline{X},\underline{u}\}$ is a *non-linear* vector function in $\underline{X}$ and $\underline{u}$ where the non-linearities arise from the product terms involving elements of $\underline{x}$ and $\underline{a}$.

A simple example of this kind of system representation can be obtained by considering the following equation for biochemical oxygen demand (BOD) at the output of a reach in a non-tidal river system (Young and Beck, 1974; Beck and Young, 1975, 1976)

$$\dot{x}_1 = -(k_1 + \frac{Q_f}{V_m})\, x_1 + \frac{Q_f}{V_m} u_1 + L_A + \eta_1 \qquad (9.20)$$

Here x_1 is the BOD, which can be considered as an aggregate or macro-measure of the oxygen absorbing potential of substances in the stream (such as decaying organic material from effluent discharges); it is defined as the oxygen absorbed (mg l^{-1}) over a 5 day period by a sample of the river water, in the absence of light, and at a constant temperature of 20°C. Other terms in equation (9.20) are defined as follows:

- k_1 is the BOD decay rate constant (day^{-1});
- Q_f is the volumetric flow rate in the stream (m^3 day^{-1});
- V_m is the mean volume of the reach (m^3);
- u_1 is the input BOD from the *previous* upstream reach of the river system (mg l^{-1});
- L_A is the rate of addition of BOD to the reach by local run-off from land adjoining the river (mg l^{-1} day^{-1});
- η_1 is the lumped effect of all stochastic disturbances to the BOD in the system (mg l^{-1} day^{-1}).

Suppose that the parameter k_1 and the variable L_A are to be estimated on the basis of noisy, daily observations of the BOD. In this situation, the observation equation (9.19) will be simply a scalar equation of the form

$$y_{1k} = x_{1k} + \xi_{1k}$$

where ξ_{1k} is the measurement noise of the kth day. In line with the suggested formulation, k_1 and L_A can be considered as additional states x_2 and x_3, i.e.

$$x_2 = k_1; \quad x_3 = L_A$$

If it is now assumed that k_1 is time invariant while L_A, which may vary, is a *continuous random walk* (i.e. the continuous-time equivalent of the random walk model (5.20) discussed in Chapter 5), then the augmented state equations can be written in the form (9.18) with

$$\left.\begin{aligned} \underline{f}(\underline{x},\underline{u}) &= \begin{bmatrix} -(x_2 + \frac{Q_f}{V_m})x_1 + \frac{Q_f}{V_m}u_1 + x_3 \\ 0 \\ 0 \end{bmatrix} \\ \underline{\eta} &= \begin{bmatrix} \eta_1 \\ 0 \\ \eta_2 \end{bmatrix} \end{aligned}\right\} \qquad (9.21)$$

Here $\underline{x} = [x_1 x_2 x_3]^T$ and η_2 is a 'white noise' disturbance introduced so that the final equation, i.e.

$$\dot{x}_3 = \eta_2$$

represents the required continuous random walk process for L_A.

The basic idea of the continuous-discrete EKF can be explained most easily by referring back to the ordinary discrete Kalman filter equations X in Chapter 5. In the present situation, the *a priori* prediction of the state between samples can be generated by replacing equation X (1) by an equivalent expression obtained by reference to equation (9.21). Since $\underline{\eta}$ is assumed to be a vector of serially uncorrelated 'white noise' disturbances, this new prediction equation will take the form

$$\hat{\underline{x}}_{k/k-1} = \hat{\underline{x}}_{k-1} + \int_{t_{k-1}}^{t_k} \underline{f}\{\hat{\underline{x}}_{k-1}, \underline{u}\}\, dt \tag{9.22}$$

where t_{k-1} and t_k are the times associated with the (k-1)th and kth samples, respectively. Equation (9.22) is simply the integration of the non-linear differential equation

$$\dot{\underline{x}} = \underline{f}\{\hat{\underline{x}}, \underline{u}\} \tag{9.23}$$

between t_{k-1} and t_k, from initial conditions defined by the *current* estimate $\hat{\underline{x}}_{k-1}$ at the (k-1)th instant. And, analogous to the purely discrete case, $\hat{\underline{x}}_{k/k-1}$ is the *a priori* prediction of $\underline{x}$ at the kth instant, given the estimate $\hat{\underline{x}}_{k-1}$ at the (k-1)th instant and knowledge of the state equations (9.18).

The remaining equations in the Kalman filter algorithm X all require a linear system of discrete-time state equations and would appear, at first sight, to be of little use in the present situation. But, if we assume that the estimation algorithm is to yield recursive estimates of the state vector $\underline{x}$ and parameter vector $\underline{a}$ *throughout* the observational interval, it seems reasonable to assume that these estimates provide a *nominal* trajectory for the system, about which the nonlinear equations can be *linearized* to produce *small perturbation linear equations* for the system. This process of 'statistical linearization' or 'relinearization' then yields, at *each* sampling instant, a set of linear, discrete-time state equations that can be used to define the remaining relationships required by the algorithm.

Considering the river water quality example, the prediction of the augmented state vector estimates between observations is obtained by the solution of the following equations between sampling instants, as indicated in equation (9.22)

$$\left.\begin{aligned}\dot{\hat{x}}_1 &= -(\hat{x}_2 + \frac{Q_f}{V_m})\hat{x}_1 + \frac{Q_f}{V_m}u_1 + \hat{x}_3 \\ \dot{\hat{x}}_2 &= 0 \\ \dot{\hat{x}}_3 &= 0\end{aligned}\right\} \quad (9.24)$$

The 'transition matrix' $A_k = A(k, k-1, \hat{\underline{x}}_{k-1}, \underline{u}_{k-1})$ for the equivalent discrete time, small perturbation equations is obtained by reference to the *Jacobian* matrix $F = F(\hat{\underline{x}},\underline{u})$ for the system (9.24) whose elements f_{ij} are defined in the following manner (see e.g. Dorf, 1965; Sage, 1968)

$$f_{ij} \stackrel{\Delta}{=} \left[\frac{\partial f_i(\hat{\underline{x}},\underline{u})}{\partial \hat{x}_j}\right]_{\hat{\underline{x}} = \hat{\underline{x}}_{k-1}}; \quad \text{for } i,\ j=1,\ 2,\ 3$$

Here, f_1, f_2 and f_3 are obtained by reference to equation (9.24) as

$$f_1 = -(\hat{x}_2 + \frac{Q_f}{V_m})\hat{x}_1 + \frac{Q_f}{V_m}u_1 + \hat{x}_3; \quad f_2 = 0; \; f_3 = 0$$

while the nomenclature $[.]_{\hat{\underline{x}}=\hat{\underline{x}}_{k-1}}$ indicates that the partial differential coefficients are evaluated at $\hat{\underline{x}}=\hat{\underline{x}}_{k-1}$.

In the present case, therefore, F is given by[†]

$$F = \begin{bmatrix} -([\hat{x}_2]_{k-1} + \frac{[Q_f]_{k-1}}{V_m}) & -[\hat{x}_1]_{k-1} & 1 \\ 0 & 0 & 0 \\ 0 & 0 & 0 \end{bmatrix}$$

Under certain minor and, in practice, non-limiting assumptions, A_k in X can now be derived as follows, using standard state-space techniques (see, for example, Dorf, 1965; Ogata, 1967) i.e.,

$$A_k = A(\hat{\underline{x}}_{k-1}) \stackrel{\Delta}{=} e^{F\Delta t}; \quad \Delta t = t_k - t_{k-1} \qquad (9.25)^{\dagger\dagger}$$

where $e^{F\Delta t}$ is the 'matrix exponential' function which can be evaluated in various ways, for example by taking sufficient terms in the infinite expansion

---

[†] The flow Q_f is a variable quantity and so its sampled value at the (k-1)th instant is used in the evaluation of F. Nominally V_m is also time variable but is assumed constant here for simplicity.

[††] In this example, we are not required to estimate any parameters in the coefficient of the input variable u_1 so A_k is independent of u_1.

$$e^{F\Delta t} = e^{M} = I + M + \frac{M^2}{2} + \frac{M}{3}\left[\frac{M^2}{2!}\right] + \ldots.\frac{M}{n}\left[\frac{M^{n-1}}{(n-1)!}\right] + \ldots.$$

Having obtained a discrete-time, linearized representation of the system at the (k-1)th instant, it is now simple to substitute in the remaining equations X (2) to X (4) of algorithm X to yield $P^{*}_{k/k-1}$, $\hat{x}_k$ and P^{*}_{k}, respectively. By analogy with this simple example, the general continuous-discrete time EKF algorithm can be written in the following form

$$\text{prediction:} \quad \hat{\underline{x}}_{k/k-1} = \hat{\underline{x}}_{k-1} + \int_{t_{k-1}}^{t_k} f(\hat{\underline{x}}_{k-1}, \underline{u})dt \qquad (1)$$

$$P^{*}_{k/k-1} = A_k P^{*}_{k-1} A_k^{T} + Q \qquad (2)$$

$$\text{correction:} \quad \hat{\underline{x}}_k = \hat{\underline{x}}_{k/k-1} + P^{*}_{k/k-1} C^T [R + CP^{*}_{k/k-1} C^T]^{-1} \{y_k - C\hat{\underline{x}}_{k/k-1}\} \qquad (3)$$

$$P^{*}_{k} = P^{*}_{k/k-1} + P^{*}_{k/k-1} C^T [CP^{*}_{k/k-1} C^T + R]^{-1} CP^{*}_{k/k-1} \qquad (4)$$

XXI

where it is understood that the transition matrix $A_k = A(\hat{\underline{x}}_{k-1}, \underline{u}_{k-1})$ is *re-evaluated* at each recursive step by reference to the small perturbation linearized equations derived on the basis of the *currently available* estimates obtained at the (k-1)th instant.

In algorithm XXI, Q is the covariance matrix of the discrete equivalent $\underline{\eta}_k$ of the white noise disturbance vector $\underline{\eta}$, while R is the covariance matrix of the measurement noise vector $\underline{\xi}_k$; i.e. both of the stochastic disturbance vectors are assumed to have the same kind of statistical properties as in the ordinary KF and, in this case, are assumed constant (i.e. the statistical properties of the noise are assumed stationary). Clearly the system equations must be defined such that these assumptions are reasonably valid, although in practice strict adherence to these requirements is rarely necessary.

In this latter connection, it is important to realize that, while algorithm XXI resembles the ordinary KF algorithm, it is of course a *non-linear* estimation procedure: thus unlike the KF, even if the theoretical assumptions are fully satisfied, there is no guarantee of convergence, nor is the P^{*}_{k} matrix *necessarily* any accurate indication of the covariance matrix of the estimation errors. Nevertheless, if care is taken in its utilization and provided that the statistical linearization procedure is valid within the range of confidence associated with the estimates at each recursive step, then the algorithm appears to work well in practice, as we shall see in the example at the end of this section.

Finally, it is instructive to consider the block diagram representation of the EKF algorithm XXI shown in Figure 9.5. This indicates that the EKF can be considered as a statistical version of the output error (OE) method of system estimation. Here the minimization of a quadratic form in the OE, (i.e. the difference between the observed and model predicted output) is accomplished recursively by a heuristic modification of the Kalman filter. Since this is nominally a statistical approach (although the statistical implications are not too clear because of the heuristic derivation of the algorithm), the reader might wonder whether the EKF can be considered in PE terms. This has been discussed by Ljung (1979b), who shows that it is only an approximate PE method because linearization is applied to the normal stochastic state-space representation rather than the innovations or Kalman filter representation (see Chapter 6), as required in the PE case.

This latter point probably accounts, in part, for the rather poor performance of the EKF in certain circumstances: problems of convergence and low statistical efficiency are often reported in the literature. But these problems have not deterred users and the EKF has been popular in applications work over the past fifteen years. Other related approaches such as "quasilinearization" (Bellman and Kalaba, 1965) and "invariant imbedding" (Bellman *et al.*, 1960; Bellman and Kalaba, 1964) have also received considerable attention. Typical of the applications of the EKF and related procedures is that of Detchmendy and Sridhar (1966). See also Sage (1968).

More recently, Beck and Young (1976) have suggested that the limitations of the EKF in statistical efficiency terms, coupled with its inherent flexibility, particularly when used in a CD mode with non-uniform sampling intervals, make it more appropriate for use in model structure identification studies than in final parameter estimation. Here, the EKF is used merely to test the dynamic behaviour of the continuous-time model (which is, as we can see in Figure 9.5, inbedded within the algorithm) and to evaluate structures which lead either to constant parameter representations or representations with variable parameters that have direct physical significance. In these applications the EKF is, in effect, being used as a sophisticated simulation/identification/estimation procedure and human interaction is essential. Having identified a satisfactory model structure, however, some more statistically efficient method is often necessary for final parameter estimation.

Ljung (1979b) has suggested modifications to the EKF to make it a PE-like algorithm. He expresses hopes that this new EKF algorithm will perform much better than the basic EKF alternative, although no thorough evaluation and comparative study has yet been attempted. Early simulation results, however, indicate that while the new algorithm is superior to the basic EKF, it still does not yield as good statistical efficiency as might have been expected from the theory.

As we see below the EKF, even in its basic form, is a rather useful and flexible tool and it probably warrants further research to see if a more efficient EKF-like algorithm can be evolved. One possibility emerges from consideration of the recursive-iterative approach used in refined IVAML: is it possible, for example, to decompose

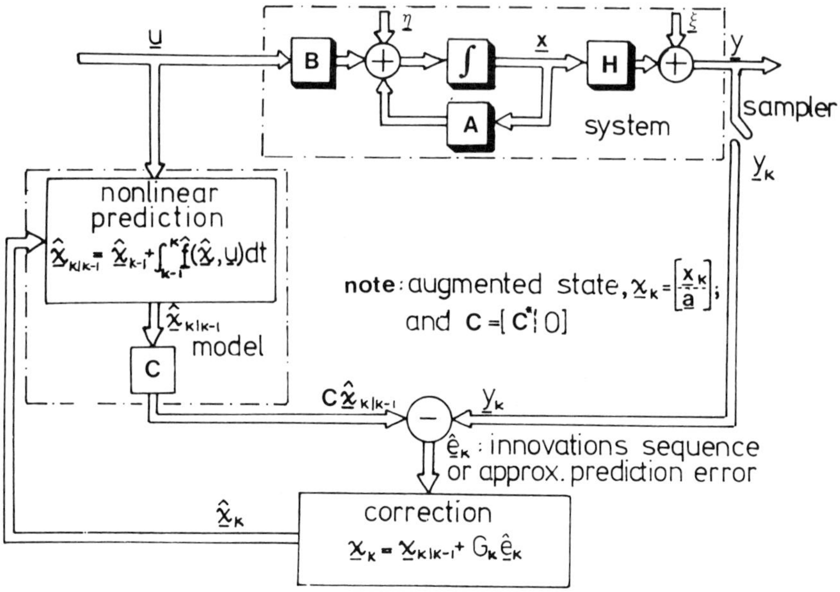

Figure 9.5 The EKF Algorithm

the estimation problem into a sub-problem which estimates the states and a second, coordinated sub-problem which estimates the parameters given the estimates obtained from the state estimation iteration? This seems viable and early results suggest that it does yield improved estimation performance. A fully recursive approach of this same type, which does not use iteration to enhance the estimation accuracy, has been suggested by Todini (1978) and this also seems to yield quite good results.

A practical example: A stochastic dynamic model for water quality in a non-tidal river system

One of the major objects of the Bedford-Ouse Study mentioned in Chapter 7 was to construct a dynamic model for water quality in the 55 km stretch of the Bedford-Ouse. As a pilot study for this project, water quality data were collected from a 4.7 km reach of the river Cam between Bait's Bite Lock and Bottisham Lock (Beck, 1974) in the summer of 1972. In all, a continuous 80 days of records were obtained including measurements of Biochemical Oxygen Demand (BOD - see earlier), Dissolved Oxygen (DO), flow, temperature and hours of sunshine.

The final mathematical model obtained from this pilot study is an expanded version of the BOD model outlined previously. This model, which includes an additional differential equation relating the BOD and DO, takes the form of two coupled, first order differential equations and can be written as follows

$$\text{BOD:} \qquad \dot{x}_1 = -(k_1 + \frac{Q_f}{V_m})\, x_1 + \frac{Q_f}{V_m} u_1 + L_A + a_1(I_k - \bar{I}) + \eta_1 \qquad (9.26)$$

$$\text{DO:} \qquad \dot{x}_2 = -k_1 x_1 - (k_2 + \frac{Q_f}{V_m})\, x_2 + \frac{Q_f}{V_m} u_2 + k_2 C_s + D_B + a_2(I_k - \bar{I}) + \eta_2 \qquad (9.27)$$

Here x_1, k_1, Q_f, V_m, u_1, L_A and ξ_1 are as given previously for equation (9.20) while the other terms are defined as follows:

x_2 is the DO in the reach (mg l^{-1});

k_2 is the DO reaeration rate constant (day^{-1});

u_2 is the input DO from the previous upstream reach of the river system (mg l^{-1});

C_s is the saturation concentration of DO (mg l^{-1});

D_B is the net rate of removal of DO from the reach by mud deposits etc. (mg l^{-1} day^{-1})

$a_i(I_k-\bar{I})$ are terms required to account for the effect of sustained sunlight on algal populations in the reach; a_1 and a_2 are unknown coefficients to be estimated while $(I_k - \bar{I})$ is the difference between a "sustained sunlight effect" I_k and a sunlight threshold $\bar{I}$ (see Beck and Young, 1976);

η_2 is the lumped effect of all stochastic disturbances to the DO in the system (mg l^{-1} day^{-1}).

This model was the outcome of a thorough identification stage in the analysis, in which various model structures were assessed using a statistical selection procedure based around the repeated use of the EKF algorithm XXI together with associated statistical and other tests (e.g. Young, Jakeman and McMurtrie, 1980). The EKF was used in exactly the same manner as that indicated earlier in connection with the simpler BOD model: in a final estimation run for the model given by equations (9.26) and (9.27) the states and parameters were estimated as shown in Figures 9.6 and 9.7. Here the parameters k_1, k_2, a_1 and a_2 were assumed to be constant over the period but the bias terms D_B and L_A were expected to vary and were modeled as random walks. Figure 9.6 compares the estimated BOD and DO with the measured values obtained over the 80-day experiment. Figure 9.7 give the recursive estimates of D_B, L_A, k_1 and k_2: as expected, sensibly constant estimates were obtained for the parameters k_1 and k_2, but the estimates of D_B and L_A show significant non-zero values at certain periods with the 80 days. This is not surprising since these terms are introduced to allow for lumped residual effects whose physical/biological meaning is not well understood at this time.

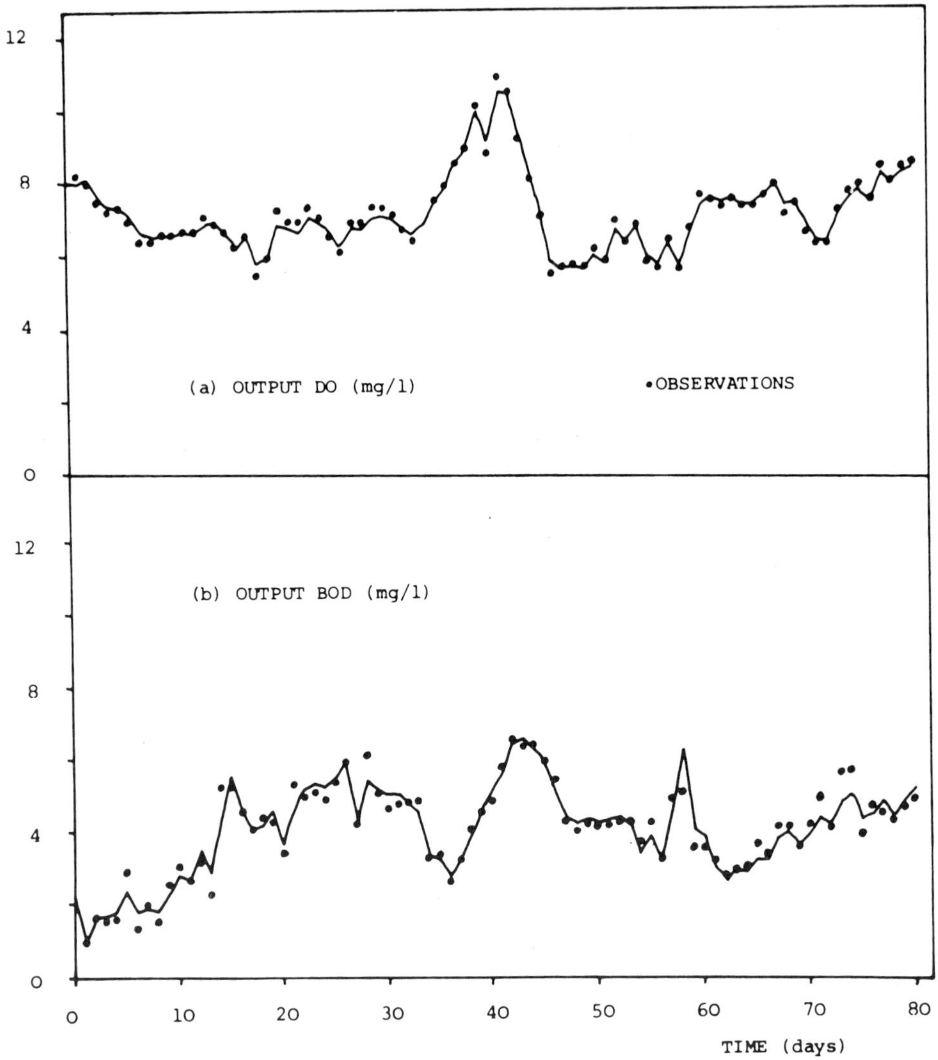

Figure 9.6 BOD-DO Model Results Obtained from EKF Estimation of Water Quality Data from River Cam, U.K.

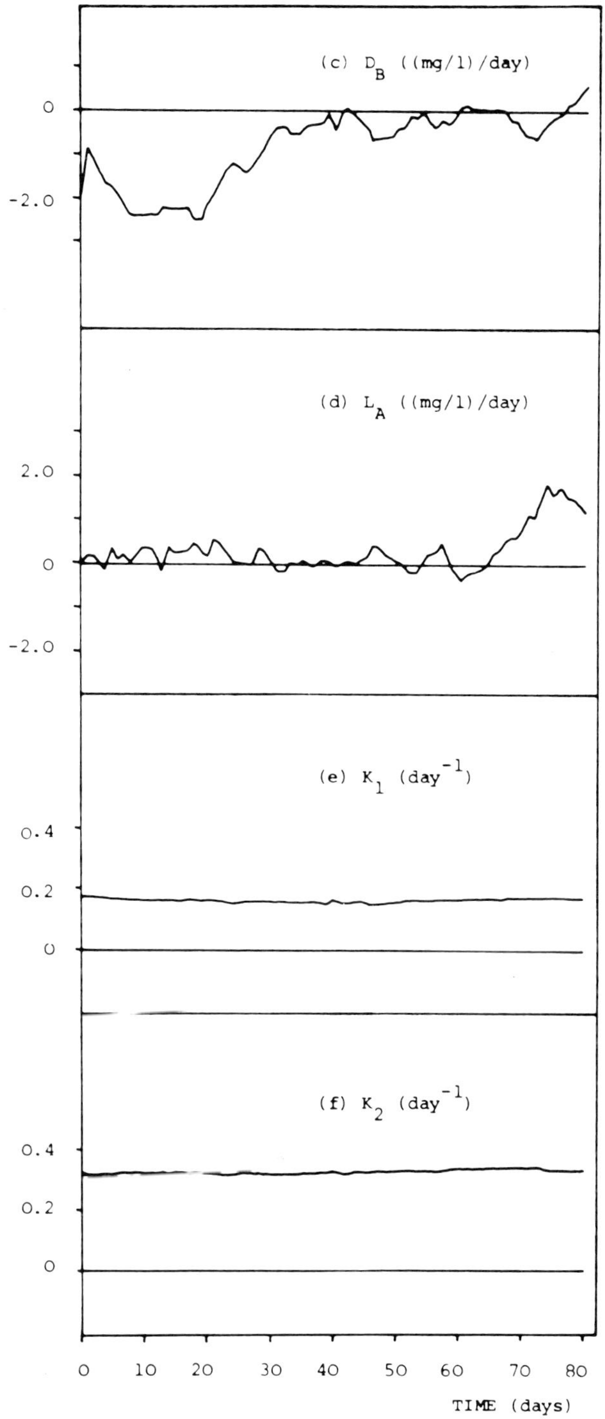

Figure 9.7 Recursive EKF Estimates of Parameters in BOD-DO Model for River Cam, U.K.

9.3 Maximum Likelihood Estimation in the State Space

In parameter estimation terms, a more satisfactory approach to the SS parameter/state estimation problem than the EKF is the ML method proposed by Stepner and Mehra (1973) and Astrom and Kallstrom (1973) (see also Kallstrom *et al.*, 1976; Mehra and Tyler, 1973). This is a logical extension both of the statistical OE approach of Rosenbrock and Storey (1966) and later, but similar, technique used by Taylor and Illiff (1972). It is also related to the ML approach discussed in theoretical terms by Balakrishnan (1973) and implemented by Illiff (1974) for the estimation of aircraft model parameters. Both Stepner *et al.* and Astrom *et al.* suggest solving the ML problem when there are both measurement and system noise inputs within a SS model formulation. In their CD mechanization, the problem is formulated in terms of the negative log-likelihood function and this results in a cost function J, of the form

$$J = \sum_{i=1}^{T} \{ ||\underline{\hat{e}}_i||^2_{B_{i|i-1}} + \ln|B_{i|i-1}| \} \qquad (9.28)$$

where $\underline{\hat{e}}_i$ and $B_{i|i-1}$[†] are, respectively, the innovation sequence and its covariance matrix *as obtained from a Kalman filter/estimation algorithm.* In relation to the system (9.15) this algorithm takes the form,

Prediction Equations:

$$\frac{d\underline{\hat{x}}(t|k-1)}{dt} = A\underline{\hat{x}}(t|k-1) + B\underline{u}(t) \qquad (1)$$

$$\frac{dP^*(t|k-1)}{dt} = AP^*(t|k-1) + P^*(t|k-1)A^T + Q \qquad (2)$$

Correction or Update Equations: XXII

$$\underline{\hat{x}}_k = \underline{\hat{x}}_{k|k-1} + G_k \{\underline{y}_k - H\underline{\hat{x}}_{k|k-1}\} \qquad (3)$$

where

$$G_k = P^*_{k|k-1}H^T[R + HP^*_{k|k-1}H^T]^{-1}$$

and

$$P^*_k = P^*_{k|k-1} - G_k HP^*_{k|k-1} \qquad (4)$$

[†] $B_{i|i-1}$ should not be confused with B in XXII (1)

In simple terms, these equations can be considered as the recursive CD Kalman Filter equations for the case of a stochastically variable, unknown state vector $\underline{x}(t)$ described by equations (9.15) and (9.16). Different nomenclature is utilized in the prediction equations and the correction equations to emphasize the CD nature of the algorithm in this case, where equations (1) and (2) are solved in continuous-time, so that the algorithm is based on continuous-time prediction and discrete-time correction. Thus at sampling instant k, $\hat{\underline{x}}(t|k-1) \equiv \hat{\underline{x}}_{k|k-1}$ and $P^*(t,k-1) \equiv P^*_{k|k-1}$ provide the *a priori* estimates of the state and its covariance, respectively, which are converted to *a posteriori* estimates $\hat{\underline{x}}_k$ and P^*_k by the correction equations XXII (3) and (4).†

The Bayesian interpretation is clear from this description. In (3) $\hat{\underline{y}}_k = H\hat{\underline{x}}_{k|k-1}$ is the conditional mean estimate, and $\hat{\underline{e}}_k = \underline{y}_k - \hat{\underline{y}}_k$ is the recursive residual or innovations sequence. $B_{k|k-1}$, the covariance of e_k, is obtained simply as

$$B_{k|k-1} = R + HP^*_{k|k-1}H^T \qquad (5) \qquad \text{XXII}$$

Since the minimization of (9.28) with the above definitions is a *very* difficult problem, Stepner and Mehra (1973) suggest a simplification by assuming that the Kalman filter gain G_k and the covariance $B_{k|k-1}$ have reached their asymptotic values G_∞ and B_∞, respectively. They then propose that the estimates are obtained by the use of a modified Newton-Raphson or quasilinearization algorithm. Astrom and Kallstrom suggest a similar approach but utilize a Fletcher-Powell method of hill-climbing.

The advantages of the comprehensive ML method are its generality and flexibility: as Stepner and Mehra point out, it can be applied to general nonlinear systems with the Kalman filter equation XXII replaced by equivalent nonlinear state estimation algorithms, such as those discussed by Jazwinski (1970).†† And, in theory, it can be used in CC, CD and DD models of mechanization. It is, however, an extremely complex approach to the problem: Astrom and Kallstrom, for instance, conclude that, "The actual computations are far from straightforward. A significant amount of manual interference like restarting, choice of (integration) step lengths etc., are required ... it is extremely tedious to compute the gradient of the likelihood function in this particular case". And later Sastry and Gauvrit (1978) say that "the algorithm requires excessive memory and computation time and hence should be limited to fairly small systems". Nevertheless, Stepner and Mehra's results (1973) are impressive and they give numerous examples of how the procedure has produced sensible estimates of aircraft system parameters from real data.

---

† Equations (1) and (2) of algorithm XXII are the prediction equations of the Kalman Filter in continuous time and were first developed by Kalman and Bucy (1961). For the present purposes, they can be considered simply as the continuous time equivalents of the discrete prediction equations (1) and (2) in Kalman Filter algorithm X.

†† Actually Mehra suggests the use of the EKF but it is applied in a purely state estimation role rather than as a state/parameter estimator, as discussed previously.

Of later work in this general ML area, the DD approach of Sastry and Gauvrit is probably the most notable. They follow the Stepner/Mehra ML formulation by considering the problem within a Bayesian context and including *a priori* information on the parameters: their cost function is then of the form

$$J = \sum_{i=1}^{N} \{||\hat{\underline{e}}_i||^2_{B_{i|i-1}} + \ln|B_{i|i-1}|\} + ||\underline{a} - \hat{\underline{a}}_0||_{P_0^{*-1}} \tag{9.29}$$

where $\hat{\underline{a}}_0$ is the *a priori* estimate of $\underline{a}$ and P_0^* is its associated error-covariance matrix. A diagrammatic representation of the algorithm is shown in Figure 9.8. Sastry and Gauvrit also discuss several methods of simplifying the computations, show how the off-line algorithm can be converted to a particular on-line form, and apply their method to the analysis of flight test data from the Concorde SST. While their analysis is carried out totally in discrete-time (DD) terms, the same approach could clearly be used in a CD context.

Even with the attendant simplifications of Sastry and Gauvrit, the ML method is computationally a very demanding approach which requires quite considerable manual intervention in truly practical applications. A similarly motivated but apparently simpler recursive-iterative solution to the ML problem with only output noise has been suggested by Sidar (1976) and, if the author's claims are correct, this seems to provide a potentially useful compromise between comprehensive formulation and practical utility in the SS situation. But only time will tell.

Finally, as Stepner and Mehra (1973) point out, the general ML formulation reduces to the simpler stochastic OE and EE formulations in conditions of no system noise ($\underline{w}(t) \equiv 0$) and complete observation of the state vector (i.e. $p = n$), respectively. Also, as we can see from the cost functions (9.28) and (9.29) which are posed in terms of the innovations sequence, the ML and Bayesian approaches are, as indicated previously, interpretable in PE terms.

9.4 Summary

In this chapter we have discussed briefly three of the major alternatives to the recursive methods of time-series analysis described in previous chapters of the book.

The first of these alternatives, the prediction error (PE) approach, is closely related to the optimal generalized equation-error (OGEE) approach discussed in Chapter 8. Indeed, the PE algorithm in the transfer function (TF) model case, is very similar to the refined IVAML algorithm: it is similarly posed in the observation space and it possesses the same asymptotic statistical properties. Also it is very flexible in the sense that, like refined IVAML, it can be modified to handle other candidate time-series model forms, such as the ARMAX and DA representations (e.g. Jakeman and Young, 1981a,b).

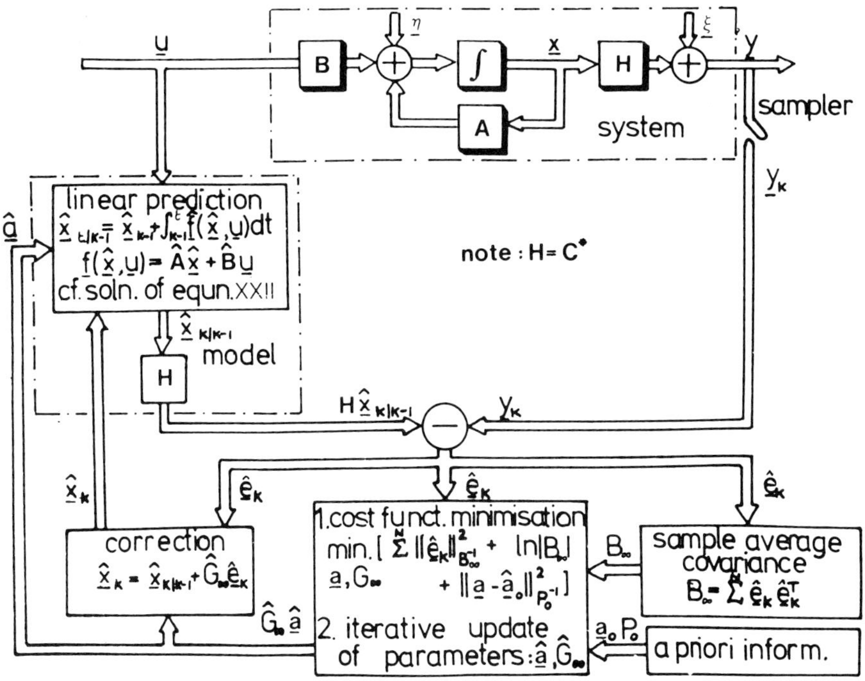

Figure 9.8 The ML Algorithm

In its current published form (Ljung, 1979a), however, the PE algorithm for the TF model has some disadvantages in relation to refined IVAML. First, it is a much less efficient algorithm computationally since it estimates the system and noise model parameters simultaneously and retains off diagonal blocks in the recursive algorithm for the P_k^* covariance matrix. Second, it is very much an unknown quantity in practical terms - so far it has only been described theoretically (Ljung, 1979a) and has not been evaluated either by stochastic simulation or by application to real time-series data.

The major advantage of the PE approach at this time is its strong and elegant theoretical background, which is helpful not only in assessing the convergence and asymptotic statistical properties of PE inspired algorithms in either the observation or state-space, but also in similarly assessing OGEE algorithms of the refined IVAML type (e.g. Solo, 1980).

The extended Kalman filter (EKF) and maximum likelihood (ML) methods both tackle the problem of time-series analysis from a state-space standpoint. The EKF is computationally straightforward but its performance is sub-optimal and can be problem dependent; in particular it is known to generate parameter estimates which can have low statistical efficiency. In this sense, it may have a more important role to play in the model structure identification stage of time-series analysis than in parameter estimation *per se* (see Beck and Young, 1976).

The ML method, in contrast to the EKF, is a true prediction error procedure; potentially it has very good statistical properties and will yield minimum variance estimates if convergence of the hill-climbing iterations is achieved. But it is highly complex in computational terms and can require manual intervention to obtain such convergence. In addition, it is not easy to program in general terms and does not really have much on-line potential; indeed, in its present form (e.g. Mehra and Tyler, 1973), it would have to be classed as strictly a recursive-iterative procedure, with maximum utility in those situations where the model structure is well defined and the parameters characterizing this structure are time-invariant over the observation interval.

There are, of course, other alternative recursive methods of time-series analysis which we have not attempted to describe in detail. But it is felt that refined IVAML and the other three methods discussed in this chapter provide what are currently, at least, the most attractive recursive approaches, both in theory and practice, to the *general* time-series estimation problem. This does not mean, however, that the other approaches, such as recursive least squares, generalized least squares and correlation methods (see e.g. Isermann *et al.*, 1973), may not have an important role to play in time-series analysis and the reader is urged to supplement his reading of this book by acquainting himself with these other alternatives, which may have advantages in particular circumstances.

10 Recursive Estimation: A General Tool in Data Analysis and Stochastic Model Building

In this book, we have attempted to provide an elementary introduction to recursive estimation; an introduction which has concentrated on the development of algorithms and has indicated how these algorithms can be useful in practical applications. Because of its introductory nature, however, much has had to be omitted either to protect the reader from overly esoteric theory at too early a stage in his exposure to the subject or to limit the size of the book in sympathy with the aims of an introductory text. In this final chapter, therefore, we mention some other, related, recursive estimation procedures that have been developed recently and which appear to have good potential for practical application. Together with the recursive algorithms discussed in previous chapters, these additional procedures provide the systems analyst with valuable additions to his 'tool bag' of techniques; additions which have utility in all phases of time-series analysis from data processing, through model structure identification and state/parameter estimation, to stochastic model building for linear and nonlinear systems in both well defined and poorly defined situations.

10.1 Pre-processing of Time-Series Data

Prior to the identification, estimation and validation stages of time-series analysis (see e.g. Box and Jenkins, 1970) it is often necessary to process the raw time-series data into a form appropriate to the implementation of the later procedures. For example, if linear time-series models are to be relevant, then it is likely that non-stationarity in the form of trends will need to be removed so that the data describe the 'small peturbation' behaviour about these trends.

An example of typical non-stationary data is shown in Figure 10.1, which shows two time-series for the sale of two competing products (Jenkins, 1979): both have clear rising time-trends and some form of 'seasonal' behaviour. Figure 10.2 shows the results of removing the trends from the data: the seasonal behaviour is now made much clearer and takes place around a zero mean value. The utility of such 'de-trending' of data is quite obvious; not only are linear models likely to be more appropriate to the analysis of the data in Figure 10.2 than the original data in 10.1, but also we see very clearly that the seasonal components are out of phase and, therefore, possibly the result of inherent feedback behaviour. This insight can be very valuable in the subsequent analysis of the data (see Young and Sirakoff, 1981).

The algorithm used in the analysis shown in Figure 10.2 has been developed recently by Young and Jakeman (1979c) using a particular "smoothing" form of the Kalman filter (see Chapter 5). This algorithm can also be used for the solution of other data processing problems such as data differentiation (or differencing), removal of

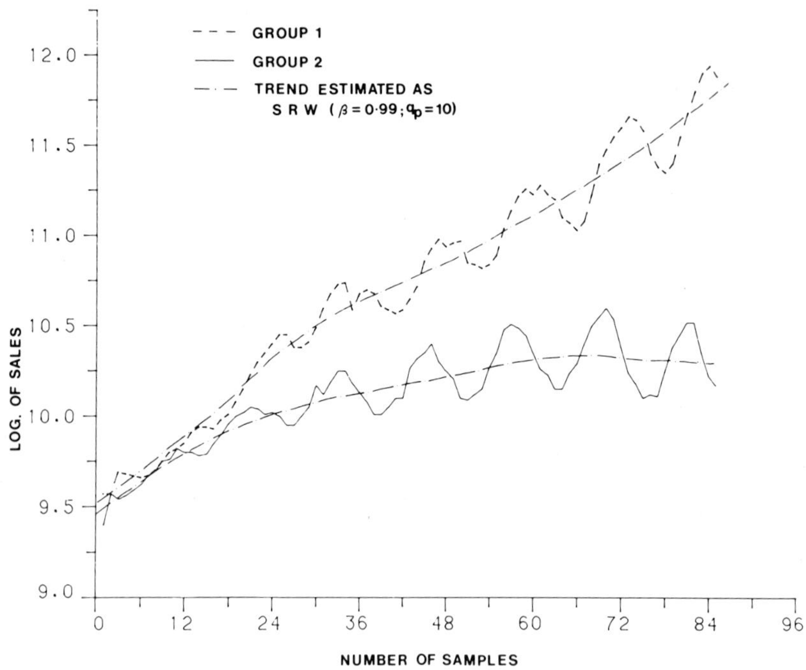

Figure 10.1 Sales Data for Two Competing Products (after Jenkins, 1979)

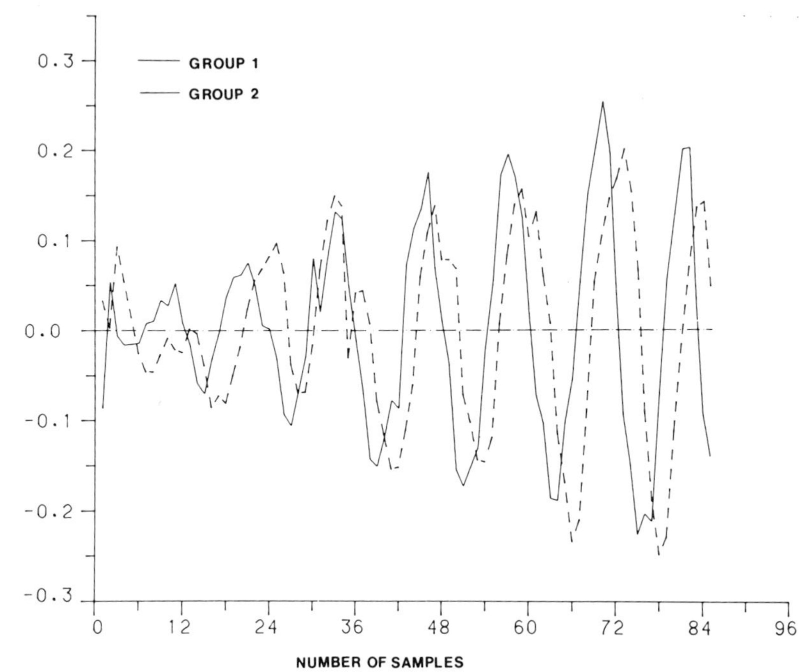

Figure 10.2 Jenkins' Sales Data with Trends Removed

seasonality, and the estimation of system input signals from noisy output measurements; all of which are examples of ill-posed problems (Jakeman and Young, 1980b). The smoothing solution effectively solves many of the noise amplification problems introduced by alternative procedures, such as trend removal by differencing or seasonality removal by seasonal differencing, as proposed by Box and Jenkins (1970).

10.2 Model Structure Identification

Following any data pre-processing, it is necessary to 'identify' an appropriate model structure such that subsequent parameter estimation will yield a unique and well defined set of parameter estimates.

In the case of badly defined systems (Young, 1978), such as those met in environmental and socio-economic systems analysis, this can pose quite difficult problems. Here the analyst is often confronted by 'passively observed' data obtained without the luxury of properly planned experiments. He cannot assume, therefore, that there is sufficient information in the data to allow for the estimation of all the parameters in some pre-specified 'theoretical' model structure (see e.g. Maciejowski, 1978). Rather he needs to identify a model structure which is appropriate to the explanation of the data and which is not over-parameterized. Over-parameterization can be particularly dangerous, leading in many cases to poorly defined (high variance) parameter estimates and surplus unvalidated content in the resulting model (Young, 1978; 1981).

Recursive methods cannot solve all the problems of badly defined system modelling but they can be extremely useful in identifying a model structure which is not over-parameterized. Sometimes such analysis will also reveal nonlinear behaviour which must be accounted for in some manner, as in the case of the rainfall-runoff model in Chapter 7. The recursive algorithms can then be useful in the later estimation phase of the analysis, as discussed by Priestley (1980) and Young (1978, 1981). Alternatively, if the appropriate model is either wholly or partly linear in dynamic terms (again as in the rainfall-runoff model, where the relationship between 'effective' rainfall and flow proved to be linear) then model structure identification can be assisted by a statistical identification procedure based on the results obtained by instrumental variable estimation (Young *et al.*, 1980).

In simple terms, the IV identification procedure examines the $\hat{P}_k$ (or $\hat{P}_k^*$) matrix obtained from the ordinary (or refined) IV algorithm discussed in Chapters 7 and 8 to see if they reveal evidence of high estimation error variance when the model complexity (the order of the A and B polynomials in the system Model) is increased. Alternative procedures, such as those of Akaike (1974), can be used to identify appropriate orders for the noise model polynomials C and D. These various order identification procedures can be considered as systematic and more discerning alternatives to the correlation methods of Box and Jenkins (1970), as demonstrated by Young *et al.* (1980).

10.3 Model Parameter Estimation

We have dealt with the estimation phase of time-series analysis in great detail in previous chapters. However, most attention has been concentrated on the estimation of parameters in discrete-time models from the analysis of discrete-time data (DD estimation). To complete the tool bag of the systems analyst, therefore, it is necessary to include some methods of continuous-time system parameter estimation.

The EKF and ML approaches to continuous-time model estimation have been reviewed in the previous chapter and provide what are probably the most flexible of the currently available approaches to the problem. It is worth noting, however, that recursive IV methods can be used in a continuous-time model estimation role using either a CC or CD mode of solution (see Chapter 9). Such procedures were first suggested in the mid-nineteen sixties (Young, 1965b; 1969a, b, c, 1970) and interest in them was revived with the development of the refined IVAML methods in the late nineteen seventies (Young and Jakeman, 1980). In simple terms, the prefilters, which we have seen are an inherent part of the refined IVAML algorithm, are utilized as "state variable filters" to generate estimates of the time derivatives of the system input and output signals. These estimates then replace the sampled, discrete time variables (i.e. $y_k, y_{k-1}, \ldots, u_k, u_{k-1}, \ldots, \hat{x}_k, \hat{x}_{k-1}, \ldots$) in the definition of the data vectors ($\underline{z}_k$; $\hat{\underline{x}}_k$) and allow for the estimation of the parameters in a *differential equation* model of the system.

Such an estimation procedure can be based on either the purely analog outputs of the filters and the use of continuous-time IV algorithms (i.e. CC estimation); or it can utilize *discrete-time data* obtained by sampling the continuous-time output of the filters, and then employ a discrete-time IV algorithm analogous to those discussed in Chapters 7 and 8 (i.e. CD estimation). These approaches appear to be useful in practice (Whitehead *et al.*, 1978) but are somewhat less robust than the DD alternatives.

The IV type methodology can also be extended to the estimation of multi-input, multi-output (MIMO; multivariable; or 'vector') models, as well as the estimation of time-variable parameters in time-series models. Multivariable estimation is discussed in detail by Jakeman and Young (1979a), where it is shown that the single input-single output (SISO) procedures carry over completely to the MIMO case and are just as effective. The problems of MIMO model structure identification and algorithmic complexity are not trivial, however, and the utility of such algorithms has to be assessed in practice with this in mind. Nevertheless, MIMO estimation can be useful in certain practical applications if used with care(e.g. Young and Whitehead, 1977; Young and Sirakoff, 1981). Alternatively, the equivalent MISO algorithms of Chapter 8 can provide a useful and simpler approach to multivariable systems analysis.

The modification of the refined IVAML algorithm to allow for the estimation or 'tracking' of time-variable parameters (TVP) is reasonably obvious from the discussion of such procedures for both regression models and ordinary IV estimation in Chapters 5 and 7, respectively. But refined IVAML has an improved theoretical basis which helps to justify the TVP extensions to the algorithm. In particular, the interpretation of the P matrix (algorithm XV, Chapter 8) as an estimate of the parametric error covariance matrix allows the heuristic extensions to be made on a much firmer and more reliable basis.

If we assume that the parameter variations can be described by the following discrete-time, Gauss-Markov model (with the usual definitions; Chapter 5)

$$\underline{a}_k = \Phi\underline{a}_{k-1} + \Gamma\underline{\eta}_k$$

then the TVP estimation algorithm in the refined IV case takes the form,

Prediction:

$$\hat{\underline{a}}_{k|k-1} = \Phi\hat{\underline{a}}_{k-1} \qquad (1)$$

$$\hat{P}_{k|k-1} = \Phi\hat{P}_{k-1}\Phi^T + \Gamma Q\Gamma^T \qquad (2)$$

Correction:

$$\hat{a}_k = \hat{a}_{k|k-1} - \hat{P}_{k|k-1}\hat{\underline{x}}_k^*[\hat{\sigma}^2 + \underline{z}_k^{*T}\hat{P}_{k|k-1}\hat{\underline{x}}_k^*]^{-1}\{\underline{z}_k^{*T}\hat{\underline{a}}_{k|k-1} - y_k^*\} \qquad (3)$$

$$\hat{P}_k = \hat{P}_{k|k-1} - \hat{P}_{k|k-1}\hat{\underline{x}}_k^*[\hat{\sigma}^2 + \underline{z}_k^{*T}\hat{P}_{k|k-1}\hat{\underline{x}}_k^*]^{-1}\underline{z}_k^{*T}\hat{P}_{k|k-1} \qquad (4)$$

(1)–(4): XXIII

This algorithm, or simpler versions of it, can be useful in nonlinear or TVP model identification and estimation; and it can also provide a basis for on-line adaptive forecasting and control (see below). A similar algorithm can be developed for refined AML.

10.4 State Estimation

Recently Young (1979b) has shown how the estimates $\hat{x}$ of the state $\underline{x}$ of a stochastic, discrete-time, SISO system can be obtained as a linear function of the outputs of the adaptive prefilters used in the refined IV algorithm when it is applied to the following ARMAX model (see (6.18) Chapter 6)

$$y_k = \frac{B(z^{-1})}{A(z^{-1})}u_k + \frac{D(z^{-1})}{A(z^{-1})}e_k \qquad (10.1)$$

In particular, it can be shown that the state estimate $\hat{\underline{x}}_k$ is generated theoretically from a relationship of the following form,

$$\hat{\underline{x}}_k = Z_k\underline{p} \tag{10.2}$$

where

$$Z_k = [N_1\underline{\zeta}_{1k} \mid \cdots \mid N_n\underline{\zeta}_{1k} \mid N_1\underline{\zeta}_{2k} \mid \cdots \mid N_n\underline{\zeta}_{2k}] \tag{10.3}$$

and

$$\underline{p}^T = [\underline{d}^T - \underline{a}^T \,\vdots\, \underline{\beta}^T] \tag{10.4}$$

Here $\underline{\zeta}_{1k}$ and $\underline{\zeta}_{2k}$ are the following vectors of prefiltered variables

$$\underline{\zeta}_{1k} = [y_{k-1}^*, \ldots, y_{k-n}^*]^T; \quad \underline{\zeta}_{2k} = [u_{k-1}^*, \ldots, u_{k-n}^*] \tag{10.5}$$

while N_i are n x n matrices, i = 1, 2, ..., n, composed of the numerator coefficients of $[I - Fz^{-1}]^{-1}\underline{\delta}_i$, where $\underline{\delta}_i$ is the ith unit vector and F is the following matrix

$$\begin{bmatrix} -d_1 & 1 & 0 \ldots 0 \\ -d_2 & 0 & 1 \ldots 0 \\ \cdot & & \cdot \\ \cdot & & \cdot \\ \cdot & & \cdot \\ \cdot & & 1 \\ -d_n & 0 & 0 \ldots 0 \end{bmatrix} \tag{10.6}$$

Finally, the vectors $\underline{a}$, $\underline{\beta}$ and $\underline{d}$ in (18) are defined as

$$\underline{a} = [a_1, \ldots, a_n]^T; \qquad \underline{\beta} = [b_1, \ldots, b_n]^T$$

and

$$\underline{d} = [d_1, \ldots, d_n]^T \tag{10.7}$$

The state estimation filter (10.2) can be implemented in practice by replacing $\underline{p}$ by its estimate $\hat{\underline{p}}$ obtained from a recursive IVAML algorithm and using the outputs of the prefilters in the same algorithm to define Z_k.

If the fully recursive IV solution is utilized in the above fashion then the overall procedure constitutes an optimal adaptive Kalman filter, in the sense that the asymptotically optimal state estimates are obtained as a by-product of an asymptotically efficient parameter estimation scheme. In the off-line recursive-iterative implementation, use of the algorithm in this fashion represents a Kalman filter design procedure in which the 'asymptotic gain' Kalman (or Wiener) filter represented by equation (10.2) is synthesized directly from the system data.† This should be

†This would seem to satisfy Kalman's requirement (1960), that "the two problems (parameter and state estimation) should be optimized jointly if possible".

particularly useful in practice because it obviates the need for specifying noise statistics and solving the covariance matrix (Riccati) equations X (2) and (4) discussed in Chapter 5. Note also that this approach to state estimation can also be applied in the purely stochastic situation where no input variables u_k are present: here the refined AML algorithm would provide the source of parameter estimates and prefiltered variables.

An extension of the above approach to continuous-time systems with deterministic inputs is fairly obvious but involves some technical problems. The expression for the estimate of the continuous-time state vector $\hat{\underline{x}}(t)$ is of the same basic form as (10.2), i.e.

$$\hat{\underline{x}}(t) = Z(t)\underline{p} \tag{10.8}$$

where, theoretically, $Z(t)$ is the continuous equivalent of Z_k. But the noise model in the continuous-time equivalent of equation (10.1) has equal order numerator and denominator polynomials which introduces estimation problems (see, e.g. Phillips, 1959; Phadke and Wu, 1974).

We will not discuss these problems here but will merely note that, in this situation, a straightforward yet clearly suboptimal approach is to use a more arbitrary "state variable filter" ($1/\hat{D}$). For example, the choice of D could be based on the heuristic notion that its passband should encompass the passband of the system under study (e.g. $\hat{D} = \hat{A}$). In this manner, the filter will pass frequency components of interest in the estimation but attenuate high frequency noise. The resultant IV algorithm is then identical to that suggested by Young (1970), and the state estimate obtained from (10.8) with $\underline{p}$ replaced by $\hat{\underline{p}}$, although not optimal in any minimum variance sense, should provide useful procedure with reasonable practical potential.

The IV algorithm in this latter case can be considered as a stochastic equivalent of the "adaptive observer" suggested by Kreisselmeier (1977) which is based on the deterministic equivalent of equation (10.8) and uses a continuous-time deterministic estimation algorithm with constant gains. But unlike the Kreisselmeier observer, this stochastic "state reconstruction filter" will function satisfactorily, albeit non-optimally, in the presence of even high levels of noise. A typical example of the performance of the state estimator (10.2) for a discrete-time second order system is given in Young and Jakeman (1980).

Finally, with the utilization of suitable multivariable canonical forms, it is possible to extend the arguments in this section to multivariable systems (Jakeman and Young 1979b). Such extensions will, however, suffer from the disadvantages of complexity associated with all multivariable 'black box' methods (see Jakeman and Young, (1979a) and they will need to be considered in detail before their practical potential can be evaluated.

10.5 Self-Adaptive (or Self-Tuning) Estimation, Forecasting and Control

Perhaps the prime motivation for the development of recursive estimation algorithms during the early nineteen sixties was their potential use in self-adaptive control. Of late, this early stimulus has been revived with the development of the "self-tuning regulator" (STR) based on recursive least squares (RLS) estimation (e.g. Aström and Wittenmark 1973, Clarke and Gawthrop 1975). In the STR the effect of the noise induced asymptotic bias on the RLS parametric estimates is cleverly neutralized by embedding the algorithm within a closed adaptive loop which automatically adjusts the estimates and the control law to yield either minimum variance regulation or some other objective, such as closed loop pole assignment (Wellstead *et al.*, 1979).

The concept of the STR can be compared with the earlier concept of self adaptive control by "identification and synthesis" (SAIS), where the objective is to explicitly obtain unbiased parameter estimates and then to separately synthesize the control law using these estimates (e.g. Kalman 1958; Young 1965a,b; Young 1979c).

While the STR seems to possess good practical potential, there are situations where the alternative of SAIS will have some advantages. For example, the stability of the adaptive loop in the STR is not easy to ensure *a priori* because of the close integration between the recursive algorithm and the control law, and the highly non-linear nature of the resulting closed loop system. On the other hand, the separation of estimation and synthesis in the SAIS system means that the question of convergence and stability is largely one of ensuring the identifiability of the system under closed loop control. This will always be possible provided an external command input is present which is both 'sufficiently exciting' and statistically independent of the noise in the closed loop. The requirement for such an input can be problematical, however, in the pure regulatory situation, where the STR clearly comes into its own. And whilst the ordinary or refined AML algorithms could be utilized in such a situation they would not, of course, be exploiting the advantages of the IV method as regards robustness in relation to the definition of noise characteristics.

In cases where the SAIS procedure seems advantageous, the refined IV algorithm provides the best, currently available recursive estimation strategy: it is robust, can be applied in continuous or discrete-time and its results can be used for either deterministic or stochastic control system design. The efficacy of such an SAIS strategy is demonstrated in the self adaptive autostabilization system for a missile mentioned in Chapter 5. As we can see from Figure 5.8, the adaptive control system performs very well with the simulated missile closely following the normal acceleration commands despite the large and rapid changes in the system dynamic characteristics. Such tight control is achieved by using the recursive parameter estimates to synthesize an adaptively updated control system based on closed loop pole assignment using state variable feedback (see Young, 1979c).

In the case of adaptive, stochastic control by SAIS, the present paper has shown that the refined IV approach provides an added bonus: the single IV-AML algorithm yields not only the estimates of the model parameters but also estimates of the state variables, which can then be used in state variable feedback control. And in the discrete-time, linear case, such an SAIS system could be considered optimally adaptive, since the state variable estimates would then, as we have seen, be optimal in a Kalman sense.

Of course, in adaptive control applications, the recursive algorithm need not be as complicated as those discussed in detail in this book. For example, simple stochastic approximation (SA) versions of the algorithms with scalar gains may be appropriate in certain circumstances. Such simple SA algorithms cannot be recommended for general applications, however, since their rates of convergence can be intolerably low (see, e.g. Ho and Blaydon, 1966). But it is possible to consider somewhat more complicated algorithms which represent a compromize between the simplicity of basic SA and the complexity of the fully recursive matrix-gain algorithms. A simple example would be the following modification to the refined IV algorithm developed in Chapter 8

$$\hat{\underline{a}}_k = \hat{\underline{a}}_{k-1} - \gamma_k \hat{P}_k^D \hat{\underline{x}}_k^* \{\underline{z}_k^{*T} \hat{\underline{a}}_{k-1} - y_k^*\} \qquad \text{XXIV}$$

Here $\hat{P}_k^D$ is a 2n + 1, diagonal matrix with elements $(\hat{x}_{k-i}^*)^{-2}$, i = 1, 2, ..., n, and $(u_{k-j}^*)^{-2}$, j = 0, 1, ..., n; while γ_k is a SA sequence, say $\gamma|k$. In other words, the gain matrix P_k is replaced by a purely diagonal 'approximation' $\gamma_k \hat{P}_k^D$, so that the computational burden is proportional to n rather than n^2 for the full refined IV algorithm.

Algorithms such as XXIII seem to work reasonably well (see, e.g. Kumar and Moore 1979). As might be expected, their performance seems to fall somewhere between that of the full algorithm and the scalar gain equivalent. In general, however, the simpler algorithms should only be used when this is necessitated by computational limitations, as for example in on-line applications using special low storage capacity microprocessors.

Finally, it should be noted that algorithms such as refined IVAML can be used for adaptive forecasting of time-series. We have seen how optimal state estimates can be obtained in section 10.4 and a simple extension of this approach can be used to obtain optimal forecasts of the state variables. Alternatively, the TF model can be used as a basis for forecasting the output variable y_k as discussed, for example, by Box and Jenkins (1970).

10.6 Summary

This final chapter has done little more than introduce the reader to the many ways in which recursive estimation can prove useful in all aspects of time-series analysis from data pre-processing, through model structure identification and parameter/state estimation, to adaptive estimation, forecasting and control. Nevertheless we hope that it has helped to stimulate an interest in recursive estimation methodology and its potential for practical application. Certainly the present book will have achieved its objective if it persuades the reader to pursue the subject of recursive estimation in more detail both in theory, which has been kept to a minimum here, and in practice, where most of the challenge still resides. In this sense, the present book provides a primer for a forthcoming graduate text which will consider all of the topics discussed here in more detail and will also review other areas of time-series analysis where recursive methods have particular importance, including the modelling of poorly defined systems and econometric analysis (Young, 1984).

11 Epilogue

This book is intended to introduce the reader to aspects of the theory and practice of recursive estimation, including its use in time-series analysis. A number of other excellent books, dealing at least in part with recursive estimation, have been available for some time (e.g. Sage, 1968; Bryson and Ho, 1969; Jazwinski, 1970; Gelb, 1974 etc.), but these tend to be orientated towards state estimation and Kalman filtering, rather than parameter estimation. Others (e.g. Tsypkin, 1971; Mendel, 1973; Eykhoff, 1974; Sorenson, 1980 etc.) deal in part with recursive parameter estimation but do not consider, in depth, the problems of recursive analysis applied to time-series models.

The current situation is, then, that no single text does real justice to the topic of recursive estimation and time-series analysis. The present book will, hopefully, help a little to introduce a wider audience to the subject but, because of time and space limitations, it still lacks the breadth of coverage and theoretical content that would be required for a really comprehensive text. The author is currently preparing a book (Young, 1984) which covers these other areas (and will hopefully provide an advanced text which will be complementary to the present introductory one) but it is unlikely that it will be available for some time. In the meantime, a number of publications have appeared recently, or will be appearing shortly, which may help the readers of the present book to extend their appreciation of recursive estimation and time-series analysis in both theoretical and practical terms.

1. First, there are two important books on parameter estimation from the "Swedish-School" of research workers. These follow on from that excellent general stochastic control text written by the originator of the school, Karl Aström (Aström, 1970). However, they are concerned more specifically with parameter estimation and recursive methodology. One of these (Soderstrom and Stoica, 1983) is complementary to the present book since it deals, in some depth, with the topic of *Instrumental Variable Methods for System Identification* (using "identification" to include parameter estimation, as is the norm in the control field). The treatment is, however, more theoretical and by concentrating on the topic of IV estimation they are able to treat, in a more rigorous manner, topics such as asymptotic properties and optimality. In this regard, it is interesting to note that the book, which only deals in part with recursive methods, discusses an 'optimum' IV method on page 105 et seq. which, not surprisingly, is almost identical to the refined IV-AML algorithm discussed in Chapter 8 of the present book. Soderstrom and Stoica also provide some interesting practical examples in the latter part of their book, including another analysis of the Box-Jenkins gas furnace data (see Chapter 8 of the present book; Young et al., 1971; and Young and Shellswell, 1972). In these examples the authors concentrate on the DA form of model, as given in equation (6.38) of the present book.

The other text is by Ljung and Soderstrom (1983) and is entitled *The Theory and Practice of Recursive Estimation*. Although it has not been possible to review this book up to now, it is likely to follow the content and style of the many excellent publications by both authors in recent years, and should be most comprehensive in theoretical terms (see the main reference list). In this sense, the book should again prove complementary to the present text and should provide a very suitable sequel to it.

2. A treatment of recursive estimation and time-series analysis very similar to that given here, although with somewhat wider but less detailed coverage, appears in the book *Computer Aided Modelling and Simulation* by Spriet and Vansteenkiste (1982). A large part of Chapter 3 in this book (pages 56-203) is based on lecture notes given by the present author in a course at the University of Ghent in January 1980 (see Acknowledgements). Much of the material touched on in Chapter 10 of the present book is, however, considered in more detail in Spriet and Vansteenkiste's book and many readers should find it useful in this respect. They should also find that the rest of the book makes interesting background reading.

3. Another forthcoming publication on recursive estimation is the Chapter entitled "Recursive Identification, Estimation and Control" contributed by the author to the book *Time Series in the Time Domain*, edited by E.J. Hannan et al. (1984). This chapter outlines and unifies some of the topics discussed in the present book, with the aim of introducing them to a wider audience in the statistics and econometrics areas. It also demonstrates how the recursive methods can be useful in the development of computer-aided procedures for time-series analysis, by discussing the MICROCAPTAIN computer program (see Preface). This includes the presentation of a number of practical examples not discussed in the present text, including the modelling of environmental, business and (with a little 'tongue-in-cheek') macro economic data. These examples include exercises in both univariate (purely stochastic AR and ARMA models) and multivariable (vector) time-series analysis.

4. The practical utility of recursive estimation in environmental systems analysis is demonstrated in three other recent publications by the author. The first is a chapter entitled "The Validity and Credibility of Models for Badly Defined Systems" which appears in the book *Uncertainty and Forecasting of Water Quality* edited by Beck and van Straten (1983). This includes an example where recursive estimation is used to identify the nonlinear dynamics of an estuary in Western Australia. The second is a chapter entitled "Time-Series Methods and Recursive Estimation in Hydrological Systems Analysis" which will appear in the book *Real-Time Flow Forecasting* edited by Kraijenhoff and Moll (1984). The book is based on lecture notes given by various well known authors to a postgraduate course of the same name held in Wageningen, the Netherlands, in July 1983. The third publication is again a Chapter from a book; this time one entitled "Mining and the Natural Environment: Systems Analysis and Mathematical Modelling" which appears in *Social and Environmental Choice: The Impact of Mining in the Northern Territory*, edited by S. Harris (1980). This chapter discusses certain aspects of modelling water flow and quality in a remote

area of Northern Australia where large deposits of uranium have been discovered. Both this chapter and the one in the Beck/van Straten book have benefitted by research on hydrology and water quality carried out by the author and his colleagues during the past few years (Beer and Young, 1981; Beer et al., 1982; Humphries et al., 1981).

5. Finally, and from a somewhat different standpoint, there is a forthcoming tutorial paper entitled "A Unified View of Statistical Forecasting Procedures" by the well known econometrician Andrew Harvey (1984), which will appear in the Journal of Forecasting. This stimulating paper concentrates on fairly simple forecasting procedures, but nicely unifies them within the context of the Kalman filter. The suggested approach has much in common with the methods of time-variable parameter estimation based on random-walk model assumptions, as discussed in Chapter 5 of the present book (pages 63-75). Young and Jakeman (1979c) have shown how similar procedures, using optimal smoothing algorithms, can be used for locally smooth trend estimation. They can also be extended to handle problems such as Stein-type shrunken estimators, ridge regression and numerical differentiation, all of which are discussed by Akaike (1979) in relation to equivalent non-recursive Likelihood and Bayes procedures. Harvey's paper will provide the readers of the present book with a good introduction to time-series forecasting and will expose them to various aspects of the statistical and econometric literature, including the interesting work on recursive estimation by Harrison and Stevens (1976). Harvey's book *Time-Series Models* (1981) also has much to commend it as a general text on the subject.

Recent References (not appearing in the main Bibliography)

AKAIKE, H. (1979) Likelihood and the Bayes procedure. Int. Mtg on Bayesian Statistics, Valencia, Spain.

BEER, T., YOUNG, P.C., HUMPHRIES, R.B., and BURGESS, J.S. (1982) *Environmental Water Quality: A Systems Study of Tuggeranong Creek and Kambah Pool*, CRES Monograph 5, Australian National University Press: Canberra (176 pages).

HARRISON, P.J., and STEVENS, C.F. (1976) Bayesian forecasting, *Jnl. Royal Stat. Soc., Series B*, 38, 205-247.

HARVEY, A.C. (1981) *Time-Series Models*, Phillip Allan: Oxford.

HARVEY, A.C. (1984) A unified view of statistical forecasting procedures, to appear *Jnl. of Forecasting*.

HUMPHRIES, R.B., YOUNG, P.C., and BEER, T. (1981) *Systems analysis of an Estuary*, Bulletin 100, Western Australian Dept. of Conservation and Environment: Perth.

LJUNG, L., and SODERSTROM, T. (1983) *Theory and Practice of Recursive Identification*, MIT Press: Cambridge, Mass.

MENDEL, J.M. (1973) *Discrete Techniques of Parameter Estimation*, Marcel Dekker: New York.

SODERSTROM, T., and STOICA, P.G. (1983) *Instrumental Variable Methods for System Identification*, Springer-Verlag: Berlin.

SORENSON, H.W. (1980) *Parameter Estimation*, Marcel Dekker: New York.

SPRIET, J.A., and VANSTEENKISTE, G.C. (1982) *Computer Aided Modelling and Simulation*, Acad. Press: New York.

YOUNG, P.C. (1980) Mining and the natural environment: systems analysis and mathematical modeling, in S.F. Harris (ed.) *Social and Environmental Choice: The Impact of Uranium Mining in the Northern Territory*, CRES Monograph 3, Australian National University Press: Canberra.

YOUNG, P.C. (1983) Validity and credibility of models for badly defined systems, in M.B. Beck and G. van Straten (eds.) *Uncertainty and Forecasting of Water Quality*, Springer-Verlag: Berlin.

YOUNG, P.C. (1984) Recursive identification, estimation and control, in E.J. Hannan, P.R. Krishnaiah and M.M. Rao (eds.) *Time-Series in the Time Domain*, North Holland: Amsterdam (to appear).

YOUNG, P.C. (1984) Time-series methods and recursive estimation in hydrological systems analysis, in Kraijenhoff and Moll (eds.) *Real-Time Flow Forecasting*, to appear.

Appendix 1 Relevant Mathematical and Statistical Background Material

The reader of these lecture notes should have access to standard texts in mathematics, statistics and dynamic systems to aid in his understanding of the mathematical analysis. As a further aid, this Appendix highlights certain results in these areas which are particularly relevant to the analysis.

A.1.1 Matrix Algebra

1. Matrices

A *matrix* is defined as a rectangular array of elements arranged in rows and columns; in this book it is denoted by a capital letter, e.g.

$$A = \begin{bmatrix} a_{11} & a_{12} & \cdots & a_{1n} \\ a_{21} & a_{22} & \cdots & a_{2n} \\ \vdots & \vdots & & \vdots \\ a_{m1} & a_{m2} & \cdots & a_{mn} \end{bmatrix}$$

Often A is alternatively denoted by $[a_{ij}]$ to indicate that it is characterized by elements a_{ij}, $i = 1, 2, \ldots, m$; $j = 1, 2, \ldots, n$. If it has m.n elements arranged in m rows and n columns, then it is said to be of order m by n, usually written m x n.

The following should be noted in relation to matrices:

(i) a *null* matrix has all of its elements set to zero, i.e. $a_{ij} = 0$ for all i, j;

(ii) a *symmetric* matrix is a square matrix in which $a_{ij} = a_{ji}$; i.e. it is symmetric about the diagonal elements;

(iii) the *trace* of a square n x n matrix, denoted by Tr., is the sum of its diagonal elements i.e.

$$\text{Tr}.A = a_{11} + a_{22} + \ldots. + a_{nn}$$

(iv) a *diagonal* matrix is a square matrix with all its elements *except those on the diagonal* set to zero i.e.

$$A = \begin{bmatrix} a_{11} & 0 & \cdots & 0 \\ 0 & a_{22} & & \vdots \\ \vdots & & \ddots & \vdots \\ 0 & \cdots & & a_{nn} \end{bmatrix}$$

(v) an n x n diagonal matrix with elements set to *unity* is denoted by I_n and termed the *identity* (or unit) matrix of order n, e.g. for a 3 x 3 identity matrix

$$I_3 = \begin{bmatrix} 1 & 0 & 0 \\ 0 & 1 & 0 \\ 0 & 0 & 1 \end{bmatrix};$$

sometimes the subscript is omitted if the order is obvious.

(vi) an *idempotent* matrix is a square matrix such that

$$A^2 = AA = A$$

i.e. it remains unchanged when multiplied by itself.

2. Vectors

A matrix of order m x 1 contains a single column of m elements and is termed a *column vector* (or sometimes just a *vector*); in this book, it is denoted by a lower case letter with an underscore i.e. for a vector $\underline{b}$

$$\underline{b} = \begin{bmatrix} b_1 \\ b_2 \\ \cdot \\ \cdot \\ \cdot \\ b_n \end{bmatrix}$$

3. Matrix Addition (or Subtraction)

If two matrices A and B are of the same order then we define A + B to be a new matrix C where

$$c_{ij} = a_{ij} + b_{ij}$$

In other words, the addition of the matrices is accomplished by adding *corresponding* elements. A - B is defined in an analogous manner.

4. Matrix or Vector Transpose

The *transpose* of a matrix A is obtained from A by interchanging the rows and columns; in this book, it is denoted by a superscript capital T; e.g. for A defined in 1., above,

$$A^T = \begin{bmatrix} a_{11} & a_{21} & \cdots & a_{m1} \\ a_{12} & a_{22} & \cdots & a_{m2} \\ \cdot & & & \\ \cdot & & & \\ \cdot & & & \\ \cdot & & & \\ a_{1n} & a_{2n} & \cdots & a_{mn} \end{bmatrix}$$

The transpose of a column vector $\underline{b}$, denoted by $\underline{b}^T$ is termed a *row vector*, e.g. for $\underline{b}$ in 2., above,

$$\underline{b}^T = [b_1 \; b_2 \; .... \; b_n]$$

Note that

(i) in the case of a symmetric matrix $A^T = A$

(ii) $[A^T]^T = A$

(iii) $[A+B]^T = A^T + B^T$

5. Matrix Multiplication

If A is of order m x n and B is of order n x p then the product AB is defined to be a matrix of order m x p whose (ij)th element c_{ij} is given by

$$c_{ij} = \sum_{k=1}^{n} a_{ik} \, b_{kj}$$

i.e. the (ij)th element is obtained by, in turn, multiplying the elements of the ith row of the matrix A by the jth column of the matrix B and summing over all terms ($\therefore$ the number of elements (n) in each row of A must be equal to the number of elements in each column of B). Note that, in general, the commutative law of multiplication which applies for scalars does *not* apply for matrices i.e.

$$AB \neq BA$$

so that *pre-multiplication* of B by A does not, in general, yield the same as *post-multiplication* of B by A. However, pre-multiplying or post-multiplying by the identity matrix leaves the matrix unchanged i.e.

$$AI_n = I_nA = A$$

Note also that for A of order m x n, B of order n x p and C of order p x q the following results apply

(i) $(AB)C = A(BC)$

(ii) $A(B+C) = AB+AC$ with orders m, n, p and q chosen appropriately

(iii) $(B+C)A = BA+CA$

(iv) for A, B and C, the multiplication by a scalar λ yields a corresponding matrix with *all* its elements multiplied by λ, i.e. $\lambda A = [\lambda a_{ij}]$

(v) $[AB]^T = B^TA^T$

(vi) $[ABC]^T = C^TB^TA^T$ (since $[ABC]^T = [(AB)C]^T = C^T[AB]^T = C^TB^TA^T$ from (v))

Finally, it should be observed that, for a vector $\underline{x} = [x_1 \; x_2 \; ... \; x_n]^T$, the *inner product* $\underline{x}^T\underline{x}$ yields a scalar quantity which is the sum of the squares of the elements of $\underline{x}$, i.e.

$$[x_1 \; x_2 \; ... \; x_n] \begin{bmatrix} x_1 \\ x_2 \\ \vdots \\ x_n \end{bmatrix} = x_1^2 + x_2^2 + ... \, x_n^2$$

The product $\underline{x}\,\underline{x}^T$, on the other hand yields a symmetric square matrix of order n x n, whose elements are the squares (on the diagonal) and cross products (elsewhere) of the $\underline{x}$ elements, i.e.

$$\begin{bmatrix} x_1 \\ x_2 \\ \cdot \\ \cdot \\ \cdot \\ x_n \end{bmatrix} [x_1 \; x_2 \; \cdots \; x_n] = \begin{bmatrix} x_1^2 & x_1x_2 & \cdots & x_1x_n \\ x_2x_1 & x_2^2 & \cdots & x_2x_n \\ \cdot & & & \cdot \\ \cdot & & & \cdot \\ \cdot & & & \cdot \\ x_nx_1 & \cdots\cdots & & x_n^2 \end{bmatrix}$$

Both products are of importance in the present text.

6. Determinant of a Matrix

The *determinant* of a square n x n matrix A is a scalar quantity, denoted by |A| or det.[A], obtained by performing certain systematic operations on the matrix elements. In particular, if the *cofactors* c_{ij} of A are defined as follows

$$c_{ij} = (-1)^{i+j}|A_{ij}|, \tag{A.1.1}$$

where $|A_{ij}|$ is the determinant of the submatrix obtained when the ith row and jth column are deleted from A, then the determinant of A can be defined as follows in terms of the elements of the ith row or their co-factors.

$$|A| = a_{i1}c_{i1} + a_{i2}c_{i2} + \ldots + a_{in}\, c_{in} \tag{A.1.2}$$

|A| may be similarly expanded in terms of the elements of any row or column.

Note that, for a matrix of order greater than 2, it is necessary to nest the operations (A.1.1) and (A.1.2) and apply them repeatedly until A_{ij} is reduced to a scalar, in which case the determinant is equal to the scalar. The following example demonstrates this process:

$$A = \begin{bmatrix} a_{11} & a_{12} & a_{13} \\ a_{21} & a_{22} & a_{23} \\ a_{31} & a_{32} & a_{33} \end{bmatrix}$$

then,

$$|A| = a_{11} \begin{vmatrix} a_{22} & a_{23} \\ a_{32} & a_{33} \end{vmatrix} - a_{12} \begin{vmatrix} a_{21} & a_{23} \\ a_{31} & a_{33} \end{vmatrix} + a_{13} \begin{vmatrix} a_{21} & a_{22} \\ a_{31} & a_{32} \end{vmatrix}$$

so that, applying (A.1.1) and (A.1.2) again to the sub-determinants, we obtain,

$$|A| = a_{11}(a_{22}a_{33} - a_{32}a_{23}) - a_{12}(a_{21}a_{33} - a_{31}a_{23}) + a_{13}(a_{21}a_{32} - a_{31}a_{22})$$

For further discussion on determinants see, for example, Johnston (1963).

7. Partitioned Matrices

Since a matrix is a rectangular array of elements, we may divide it up by means of horizontal and vertical dotted lines into smaller rectangular arrays of sub-matrices e.g.

$$A = \left[\begin{array}{ccc|c} a_{11} & a_{12} & a_{13} & a_{14} \\ a_{21} & a_{22} & a_{23} & a_{24} \\ \hline a_{31} & a_{32} & a_{33} & a_{34} \end{array}\right]$$

has been divided in this manner into 4 sub-matrices

$$A_{11} = \begin{bmatrix} a_{11} & a_{12} & a_{13} \\ a_{21} & a_{22} & a_{23} \end{bmatrix} ; \quad A_{12} = \begin{bmatrix} a_{14} \\ a_{24} \end{bmatrix}$$

$$A_{21} = [a_{31} \; a_{32} \; a_{33}] ; \quad A_{22} = a_{34}$$

So that A_{11} is a 2 x 3 submatrix, A_{12} is a 2 x 1 column vector, A_{21} is a 1 x 3 row vector, and A_{22} is a scalar. As a result A can be denoted by

$$A = \left[\begin{array}{c|c} A_{11} & A_{12} \\ \hline A_{21} & A_{22} \end{array}\right]$$

The basic operations for addition, multiplication and transposition apply for partitioned matrices but the matrices must be partitioned conformably to allow for such operations. A multiplicative example is

$$AB = \left[\begin{array}{c|c} A_{11} & A_{12} \\ \hline A_{21} & A_{22} \end{array}\right] \left[\begin{array}{c} B_{11} \\ \hline B_{21} \end{array}\right] = \left[\begin{array}{c} A_{11}B_{11} + A_{12}B_{21} \\ \hline A_{21}B_{11} + A_{22}B_{21} \end{array}\right]$$

The results of such operations will be the same as would be obtained by multiplying the unpartitioned matrices element by element (as in 5., above) but the partitioning approach may be extremely useful in simplifying the analysis.

One theorem for partitioned matrices that is useful in the context of the book (see Chapter 7) concerns the determinant of a partitioned matrix A where

$$A = \left[\begin{array}{c|c} A_{11} & A_{12} \\ \hline A_{21} & A_{22} \end{array}\right]$$

It can be shown (e.g. Gantmacher, 1960; Dyrymes, 1970) that

$$|A| = |A_{22}| \cdot |A_{11} - A_{12} A_{22}^{-1} A_{21}|$$

or alternatively,

$$|A| = |A_{11}| \cdot |A_{22} - A_{21} A_{11}^{-1} A_{12}|$$

where A_{11}^{-1} and A_{22}^{-1} are, respectively, the "inverses" of the matrices A_{11} and A_{22}, respectively, as defined in 8. below.

8. Inverse of a Matrix

If a matrix A^{-1} exists such that

$$AA^{-1} = A^{-1}A = I$$

where I is an appropriately ordered identity matrix, then A^{-1} is termed the *inverse* (or reciprocal) of A by analogy with the scalar situation.

The inverse of a square matrix A of order n x n is obtained from A by means of the formula,

$$A^{-1} = \frac{1}{|A|} [\text{Adj}.A]$$

$$= \begin{bmatrix} \frac{c_{11}}{|A|} & \frac{c_{21}}{|A|} & \cdots & \frac{c_{n1}}{|A|} \\ \vdots & & & \vdots \\ \frac{c_{1n}}{|A|} & \cdots & \cdots & \frac{c_{nn}}{|A|} \end{bmatrix}$$

where Adj. A denotes the *adjoint* of the matrix A and is obtained as the *transpose* of an n x n matrix C with elements c_{ij} which are the *co-factors* of A, *as defined by (A.1.1) in 6., above,* i.e.

$$c_{11} = |A_{11}|, \; c_{12} = -|A_{12}|, \; c_{22} = |A_{22}| \text{ etc.}$$

Note that, by definition, the inverse will only exist if $|A| \neq 0$; otherwise the matrix is non-invertible or *singular*. A non-singular matrix is, therefore, invertible.

Several theorems on inverse matrices are useful, e.g.

(i) $[AB]^{-1} = B^{-1}A^{-1}$

(ii) $[AB][B^{-1}A^{-1}] = A[BB^{-1}]A^{-1} = AIA^{-1} = AA^{-1} = I$

(iii) $[ABC]^{-1} = C^{-1}B^{-1}A^{-1}$

(iv) $[A^T]^{-1} = [A^{-1}]^T$

(v) $|A^{-1}| = 1/|A|$

One of the most common uses of the inverse matrix is in solving a set of algebraic, simultaneous equations such as,

$$X\underline{a} = \underline{b} \tag{A.1.3}$$

where X is a *known* n x n matrix, $\underline{a}$ is an n x 1 vector of *unknowns*, and $\underline{b}$ is a *known* n x 1 vector. The reader can easily verify that this represents a set of simultaneous equations in the elements of $\underline{a}$, where $\underline{a} = [a_1\, a_2 \ldots a_n]^T$, by defining $X = [x_{ij}]$ and $\underline{b} = [b_1\;\; b_2 \ldots b_n]^T$. Premultiplying both sides of (A.1.3) by X^{-1} we obtain

$$X^{-1}X\underline{a} = X^{-1}\underline{b} \text{ or } I\underline{a} = X^{-1}\underline{b}$$

so that

$$\underline{a} = X^{-1}\underline{b} = \frac{1}{|X|}[\text{adj}.X]\,\underline{b}$$

which is the required solution for $\underline{a}$ and is an alternative to other methods of solution such as pivotal elimination. For further discussion on matrix inverses see, for example, Johnston (1963).

9. Quadratic Forms

A *quadratic form* in a vector $\underline{e} = [e_1\; e_2\; \ldots\; e_n]^T$ is defined as

$$\underline{e}^T Q\, \underline{e}$$

where Q is a *symmetric* matrix of order n x n. The reader can verify that, for $Q = [q_{ij}]$ with off diagonal elements $q_{ij} = q_{ji}$, $\underline{e}^T Q\, \underline{e}$ is a scalar given by

$$\begin{aligned}\underline{e}^T Q\, \underline{e} &= q_{11}\, e_1^2 + 2q_{12}\, e_1\, e_2 + \ldots + 2q_{1n}\, e_1\, e_n \\ &\quad + q_{22}\, e_2^2 + \ldots + 2q_{2n}\, e_2\, e_n \\ &\quad + \ldots + q_{nn}\, e_n^2\end{aligned} \tag{A.1.4}$$

Note that if Q is *diagonal*, then this reduces to (cf inner product)

$$\underline{e}^T Q\, \underline{e} = q_{11}\, e_1^2 + q_{22}\, e_2^2 + \ldots q_{nn}\, e_n^2$$

A quadratic form such as (A.1.4) is sometimes termed the *weighted Euclidian Squared Norm* of the vector $\underline{e}$ and is denoted by

$$||\underline{e}||_Q^2 \tag{A.1.5}$$

As we see, it represents a very general or weighted (by the elements of Q) "sum of squares" type operation on the elements of $\underline{e}$. It proves particularly useful as a cost (or criterion function) if $\underline{e}$ represents a vector of errors (or lack of fit) associated with some model (see Chapters 3, 5, 8 and 9).

10. Positive Definite or Semi-Definite Matrices

A symmetric matrix A is said to be *positive definite* (p.d.) if

$$\underline{x}^T Q \underline{x} > 0$$

where $\underline{x}$ is any non-null vector. It is termed positive semi-definite (p.s.d.) if

$$\underline{x}^T Q \underline{x} \geqslant 0$$

For an n x n p.d. matrix A, $a_{ii} > 0$, i=1, 2, ... , n; for a p.s.d. matrix $a_{ii} \geqslant 0$, i=1, 2, ... , n.

Note that if A is p.d. then A is *non-singular* and can be inverted; if A is p.s.d. (but not p.d.) then A is singular (see Dhrymes, 1970).

11. The Rank of a Matrix

The *rank* of a matrix is the order of its largest sub-matrix that is non-singular and so has a non-zero determinant. Thus for a square n x n matrix the rank must be n (i.e. the matrix must be *full rank*) for the matrix to be non-singular and invertible. For further discussion on the rank of a matrix see, for example, Johnston (1963).

12. Differentiation of Vectors and Matrices

The differentiation of vectors and matrices is most important in optimization and statistical analysis. The main result concerns the differentiation of an inner product of two vectors with respect to the elements of one of the vectors.

Consider the inner product of two (n x 1) vectors $\underline{x}$ and $\underline{a}$, i.e.

$$\underline{x}^T \underline{a} = [x_1 \; x_2 \; \dots \; x_n] \begin{bmatrix} a_1 \\ a_2 \\ \cdot \\ \cdot \\ \cdot \\ a_n \end{bmatrix}$$

$$= x_1 a_1 + x_2 a_2 + \dots x_n a_n$$

It is clear that for all i, i=1, 2, ... , n, the partial differentials with respect to a_i are given by

$$\frac{\partial(\underline{x}^T \underline{a})}{\partial a_i} = x_i$$

As a result, if the partial differentials are arranged in order of their subscripts as a vector, then this vector is simply $\underline{x}$. Thus it is convenient to refer to the process of *vector differentiation* in shorthand as

$$\frac{\partial(\underline{x}^T\underline{a})}{\partial\underline{a}} = \underline{x}$$

or

$$\frac{\partial(\underline{x}^T\underline{a})}{\partial\underline{a}^T} = \underline{x}^T$$

The analogy with scalar differentiation is apparent from the above result. A particularly important example of vector differentiation which occurs in this book (e.g. Chapter 3 *et seq.*) is concerned with the differentiation of a least squares cost function J_2 which, in its simplest form, is defined as

$$\sum_{i=1}^{k} e_i^2$$

where $e_i = \underline{x}_i^T\hat{\underline{a}} - y_i$ is an error measure based on a vector of estimated coefficients or parameters $\hat{\underline{a}}$. In order to obtain the estimate $\hat{\underline{a}}$, it is necessary to differentiate J_2 with respect to all of the elements $\hat{a}_i$, i=1, 2, ... n, of $\hat{\underline{a}}$. Using the above results, we see that since

$$J_2 = \sum_{i=1}^{k} [(\underline{x}_i^T\hat{\underline{a}})^2 - 2\underline{x}_i^T\hat{\underline{a}}\,y_i + y_i^2]$$

then

$$\frac{\partial J_2}{\partial\hat{\underline{a}}} = \sum_{i=1}^{k} [2\underline{x}_i\underline{x}_i^T\hat{\underline{a}} - 2\underline{x}_i y_i]$$

$$= 2\sum_{i=1}^{k} \underline{x}_i\underline{x}_i^T\hat{\underline{a}} - \underline{x}_i y_i \qquad \text{(A.1.6)}$$

which, when set to zero in the usual manner, constitutes a set of n simultaneous equations in the n unknowns $\hat{a}_i$, i=1, 2, ... , n; the normal equations

Alternatively, we can proceed by forming the k x n matrix X with rows defined by $\underline{x}_i^T$, i=1, 2, ... , k. The reader can then verify that the vector $\underline{e} = [e_1\ e_2\ \dots\ e_k]^T$ is defined by

$$\underline{e} = X\hat{\underline{a}} - \underline{y}$$

where $y = [y_1\ y_2\ \dots\ y_k]^T$, and so,

$$J_2 = [X\hat{\underline{a}} - \underline{y}]^T[X\hat{\underline{a}} - \underline{y}]$$

$$= \hat{\underline{a}}^T X^T X\hat{\underline{a}} - 2\hat{a}^T X^T\underline{y} + \underline{y}^T\underline{y}$$

since $\underline{y}^T X \hat{\underline{a}}$ is a scalar and so equal to its transpose $\hat{\underline{a}}^T X^T \underline{y}$. It now follows straightforwardly that

$$\frac{\partial J_2}{\partial \hat{\underline{a}}} = 2X^T X\hat{\underline{a}} - 2X^T \underline{y}$$

which will be seen as identical to (A.1.6) by substituting for X in terms of $\underline{x}_i$.

If J_2 is replaced by the more general weighted least squares cost function (see 9., above) i.e.

$$J_2 = [X\hat{\underline{a}} - \underline{y}]^T Q[X\hat{\underline{a}} - \underline{y}] \qquad (A.1.7)$$

$$= ||X\hat{\underline{a}} - \underline{y}||_Q^2$$

where Q is a symmetric p.d. weighting matrix, then it is straightforward to show that

$$\frac{\partial J_2}{\partial \hat{\underline{a}}} = 2X^T QX\hat{\underline{a}} - 2X^T Q\underline{y}$$

A.1.2 Statistics and Probability

1. Discrete Random Variables

A *discrete-valued random variable* x is defined as a discrete valued function x(j) with a probability of occurence of the jth value given by p(j): p(j) is the *probability mass function* of the random variable x(j). For simplicity x(j) and p(j) are denoted by x and p(x).

The random variable x can be characterised approximately in probabilistic terms by specifying a finite number of moments of p(j). The first two moments are

(i) the *mean value* or *first moment of p(x)* which is defined as the *expected value of x*, denoted by $\bar{x}$, i.e.

$$E\{x\} = E_x = \bar{x} \triangleq \sum_j x(j)\, p(j)$$

(ii) the *variance* or *second central moment of p(x)* which is defined as the expected value of the square of the difference between x(j) and its mean value $\bar{x}$, i.e.

$$E\{(x-\bar{x})^2 = \sigma^2 \triangleq \sum_j \{x(j) - \bar{x}\}^2\, p(j)$$

2. Discrete Random Vectors

A *random vector* is a column vector $\underline{x}$ whose elements are discrete random variables, e.g. x_i, i=1, 2, ... , n. If each component x_i can take on a discrete set of values $x_i(j_i)$ where $j_i = 1, 2, \ldots, m_i$ then there are $m_1.m_2 \ldots m_n$ possible vectors. The *joint probability mass function* $p(j_1, j_2, \ldots, j_n)$ is the probability that x_1 has its j_1th value, x_2 has its j_2th value, etc. For simplicity, the joint probability mass function is usually written $p(\underline{x}) = p(x_1, x_2, \ldots, x_n)$. *The marginal probability mass function* $p(j_1)$ is the probability that x_1 takes on its j_1th value while $x_2, \ldots, x_n$ take on any possible values, i.e. in general

$$P(j_i) = \sum_{j_2=1}^{m_2} \sum_{j_3=1}^{m_3} \ldots \sum_{j_n=1}^{m_n} p(j_1, j_2, \ldots, j_n)$$

As in the scalar case, it is possible to characterise $\underline{x}$ approximately by specifying moments of $p(\underline{x})$, i.e.

(i) *the mean of* $\underline{x}$:

$$E_{\underline{x}} = \bar{\underline{x}} \triangleq \sum_{j_1=1}^{m_1} \ldots \sum_{j_n=1}^{m_n} \begin{bmatrix} x_1(j_1) \\ \cdot \\ \cdot \\ \cdot \\ \cdot \\ x_n(j_n) \end{bmatrix} p(j_1, j_2, \ldots, j_n)$$

(ii) *the covariance of* $\underline{x}$: since $\underline{x}$ is a vector it has n variances and n(n-1)/2 *covariances* associated with it, where the covariances are defined as the expected value of the cross products of the elements with means removed; thus the covariance is specified by an n x n symmetric *covariance matrix* P defined by

$$P = P_{\underline{x}} = E\{[\underline{x}-\bar{\underline{x}}][\underline{x}-\bar{\underline{x}}]^T\} = E\left\{\begin{bmatrix} (x_1-\bar{x}_1)^2, & \ldots\ldots, & (x_1-\bar{x}_1)(x_n-\bar{x}_n) \\ \cdot & (x_2-\bar{x}_2)^2 & \cdot \\ \cdot & \cdot & \cdot \\ \cdot & \ddots & \cdot \\ \cdot & & \cdot \\ (x_n-\bar{x}_n)(x_1-\bar{x}_1) & \ldots\ldots\ldots & (x_n-\bar{x}_n)^2 \end{bmatrix}\right\}$$

$$= \sum_{j_1=1}^{m_1} \ldots \sum_{j_n=1}^{m_n} [\underline{x}-\bar{\underline{x}}][\underline{x}-\bar{\underline{x}}]^T p(j_1, j_2, \ldots, j_n)$$

Any such covariance matrix is at least positive semi-definite.

3. Conditional Probabilities

If a random vector $\underline{x}$ is characterised by a covariance matrix $P = [p_{ij}]$ with $p_{ij} \neq 0$ for $i \neq j$, then the elements of $\underline{x}$ are *correlated*. The elements of $\underline{x}$ are said to be *dependent* if knowledge of $p(x_1)$, $p(x_2)$, ... , $p(x_n)$ does not determine $p(x_1, \ldots, x_n)$ completely; if, on the other hand,

$$p(x_1, \ldots, x_n) = p(x_1).p(x_2) \ldots\ldots p(x_n)$$

for all possible values of $x_1, \ldots, x_n$, then the elements are said to be *independent*.

If two random vectors $\underline{x}$ and $\underline{y}$ are dependent and $\underline{x}$ takes on a particular value, it should be possible to predict $\underline{y}$ better than if this additional information was not available. This leads to the concept of the *conditional probability mass function* $p(\underline{y}|\underline{x})$, where

$$p(\underline{y}|\underline{x}) = p(\underline{y},\underline{x})/p(\underline{x}), \text{ for } p(\underline{x}) \neq 0$$

is the probability of $\underline{y}$ conditioned on a given value of $\underline{x}$. The conditional mean and covariance are defined in a similar manner to that shown in 2. above, with the joint probability mass function replaced by the conditional probability mass function. The conditional mean and covariance are random variables because they are a function of the conditioning random variable $\underline{x}$. Since

$$p(\underline{x}|\underline{y}) = p(\underline{y},\underline{x})/p(\underline{y})$$

then

$$p(\underline{y},\underline{x}) = p(\underline{x}|\underline{y}).p(\underline{y})$$

and so

$$p(y|x) = \frac{p(\underline{x}|\underline{y}).p(\underline{y})}{p(\underline{x})}$$

This is known as the Bayes Rule for conditional probabilities and is a most important concept in recursive estimation theory (see e.g. Bryson and Ho, 1969; Young, 1982). If we consider that $p(\underline{y})$ is the *a priori* probability of $\underline{y}$ without any knowledge of $\underline{x}$ then $p(\underline{y}|\underline{x})$ can be considered the *a posteriori* probability of $\underline{y}$ given that $\underline{x}$ has taken on a certain value. Of course if $\underline{x}$ and $\underline{y}$ are *independent* then $p(\underline{y}|\underline{x}) = p(\underline{y})$, so that knowledge of $\underline{x}$ is not useful in predicting the value of $\underline{y}$.

4. Continuous Random Variables and Vectors

The concepts described in the above sections can be extended to continuous random variables and vectors (see Bryson and Ho, 1969). So, for example,

$$\bar{\underline{x}} = \int_{-\infty}^{+\infty} \cdots \int_{-\infty}^{+\infty} \begin{bmatrix} x_1 \\ \vdots \\ x_n \end{bmatrix} p(x_1, \ldots, x_n) dx_1 \ldots dx_n$$

$$P = \int_{-\infty}^{+\infty} \cdots \int_{-\infty}^{+\infty} [\underline{x}-\bar{\underline{x}}][\underline{x}-\bar{\underline{x}}]^T p(x_1, \ldots, x_n) dx_1 \ldots dx_n$$

where the function $p(x_1, \ldots, x_n)\, dx_1 \ldots dx_n$ is the *probability density function* and is defined as the probability that the random vector $\underline{x}$ will lie in the *differential volume* $dx_1, \ldots, dx_n$ with centre at $(x_1, \ldots, x_n)$.

5. The Normal or Gaussian Density Function

A *normally distributed random variable (scalar)* has an amplitude density function p(x) defined by

$$p(x) = \frac{1}{(2\Pi)^{\frac{1}{2}}\sigma} \exp. \frac{(x-\bar{x})^2}{2\sigma^2}$$

so that

$$\int_{-\infty}^{+\infty} p(x)dx = 1.0$$

$$E\{x\} = \bar{x}$$

$$E\{(x-\bar{x})^2\} = \sigma^2$$

The distribution is, therefore, completely specified by its mean and variance, and so it is usual to summarise the distribution as follows

$$x \sim N(\bar{x},\sigma^2)$$

A *normally distributed random vector* $\underline{x} = [x_1\ x_2\ \ldots\ x_n]^T$ has a multivariate normal density function defined by

$$p(\underline{x}) = \frac{1}{(2\Pi)^{n/2}|P|^{\frac{1}{2}}} \exp. \{-\tfrac{1}{2}[\underline{x}-\bar{\underline{x}}]P^{-1}[\underline{x}-\bar{\underline{x}}]^T\}$$

so that

$$\int_{-\infty}^{+\infty} \cdots \int_{-\infty}^{+\infty} p(\underline{x})dx_1 \ldots dx_n = 1$$

$$E\{\underline{x}\} = \underline{\bar{x}}$$

$$E\{[\underline{x}-\underline{\bar{x}}][\underline{x}-\underline{\bar{x}}]^T\} = P$$

and $p(\underline{x})$ is completely characterized by its mean $\underline{\bar{x}}$ and covariance matrix P. Once again, it is usual to summarize the distribution as

$$\underline{x} \sim N(\underline{\bar{x}},P)$$

The concept of the normally distributed random variable or vector is most useful in analytical terms: often the density functions for random variables or vectors can be considered as *approximately* normally distributed and so characterised almost completely by their mean and variance or covariance matrix properties; as a result mathematical analysis can be made much more straightforward by making the normal distribution assumption.

6. Properties of Estimators

Suppose a model is characterized by a single unknown parameter θ and we need to find an *estimate* $\hat{\theta}$ of θ based on T observations $y_1, \ldots, y_T$. The rule or algorithm for processing the observations is termed the *estimator* of θ and the estimate $\hat{\theta}$ is a function of the observations. A reasonable estimator should produce estimates for different sample sizes that is reasonably close, in some sense, to the true value θ. Such an estimator is said to be *unbiased* if

$$E\{\hat{\theta}\} = \theta \quad \text{for all } \theta$$

The estimate is sometimes said to be *asymptotically unbiased* if the estimate $\hat{\theta}_k$ based on k samples is unbiased for $k\to\infty$.† Clearly unbiasedness is a more desirable property than asymptotic unbiasedness but the latter may often be acceptable if sample sizes are reasonably high.

The *mean square error* of this estimate is defined simply as

$$MSE(\hat{\theta}) = E\ \{(\hat{\theta}-\theta)^2\}$$

and we see in 8. below that we can design estimates that produce unbiased estimates which attain the lowest possible or minimum value of the MSE. Such estimators are termed *minimum variance unbiased estimators.*

A consistent estimator is one which produces estimates $\hat{\theta}_k$ which become more accurate, in the sense that the probability of its being close to the true value θ increases as

† A more rigorous definition is that $\hat{\theta}_k$ is an asymptotically unbiased estimate of θ if the mean of the limiting distribution of $\sqrt{k}(\hat{\theta}_k-\theta)$ is zero.

the sample size k increases. Mathematically this can be written

$$\lim_{k\to\infty} \Pr(|\hat{\theta}_k - \theta| > \varepsilon) = 0 \quad \text{for any } \varepsilon > 0$$

This is usually written more concisely as

$$\text{p.lim. } \hat{\theta}_k = \theta$$

or the *probability in the limit* of the sequence $\hat{\theta}_k$ is θ.

7. The Likelihood Function and Maximum Likelihood Estimation

Suppose that we have a set of observations on T random variables $y_1, y_2, \dots, y_T$ which compose a vector $y = [y_1, \dots y_T]^T$. It is possible to consider a *joint density function* $L(\underline{\theta}; y_1, y_2 \dots y_T)$ which depends upon the n unknown parameters in the vector $\underline{\theta} = [\theta_1, \dots, \theta_n]^T$ and which can be interpreted as the probability of obtaining particular values of $y_1, \dots, y_T$. Once a sample has been taken then $y_1, \dots, y_T$ becomes a set of fixed numbers and the expression for L can be re-interpreted as a function of $\hat{\underline{\theta}}$ where $\hat{\underline{\theta}}$ is *any admissable value* of the parameter vector rather than the *true* value. In this sense L can be considered as a means of assessing the relative merits of different values of $\hat{\underline{\theta}}$, given the sample. L is, therefore, termed the *Likelihood Function* and denoted by $L(\underline{\theta})$ or $L(\underline{\theta}, \underline{y})$.

The maximum likelihood (ML) approach to the problem of estimating θ is one of investigating which value of $\hat{\underline{\theta}}$ is most likely given the observations: the *ML estimate* is then given by $\underline{\theta}_0$ where

$$L(\underline{\theta}_0) \geqslant L(\hat{\underline{\theta}})$$

where $\hat{\underline{\theta}}$ is any other admissable estimate of $\underline{\theta}$.

The classical theory of maximum likelihood (see e.g. Kendall and Smart, 1961) is based on the situation in which the T observations are drawn independently of each other from the same distribution so that

$$L(\hat{\underline{\theta}}; y_1, \dots, y_T) = \prod_{k=1}^{T} p(y_k; \underline{\theta})$$

where $\prod_{k=1}^{T}$ denotes the multiplication operator. If, for example, we consider the likelihood function for a sample of T independent observations from a normal distribution with mean $\bar{x}$ and variance σ^2 then

$$\log_e L(\bar{x}, \sigma^2; y_1, \dots, y_T) = -\frac{T}{2}\log_e 2\Pi - \frac{T}{2}\log_e \sigma^2 - \frac{1}{2\sigma^2}\sum_{k=1}^{T}(y_k - \bar{x})^2$$

where, in this case, $\theta = [\bar{x}, \sigma^2]^T$. Note that, as is usual in this kind of analysis, the natural logorithm of L is considered here so that the analysis is easier; this is quite allowable since if $\underline{\theta}_0$ satisfies $L(\underline{\theta}_0) \geqslant L(\hat{\underline{\theta}})$, it also satisfies $\log_e L(\underline{\theta}_0) \geqslant \log_e L(\hat{\underline{\theta}})$. The maximum likelihood estimates are, therefore, obtained by finding those estimates of $\bar{x}$ and σ^2 which simultaneously maximise L. These can be obtained in the usual manner by differentiating L with respect to $\bar{x}$ and σ^2 in turn and setting the resultant expression to zero, i.e.

$$\frac{\partial \log_e L}{\partial \underline{\theta}} = \nabla_{\underline{\theta}} \log_e L = 0$$

where $\nabla_{\underline{\theta}}$ denotes the partial differential with respect to each element of $\underline{\theta}$ in turn. In the present example, this yields,

$$\frac{\partial L}{\partial \bar{x}} = \frac{1}{\sigma^2} \Sigma(y_k - \bar{x}) = 0$$

and

$$\frac{\partial L}{\partial \sigma^2} = \frac{T}{2\sigma^2} + \frac{1}{2\sigma^4} + \Sigma(y_k - \bar{x})^2 = 0$$

As a result,

$$\bar{x}_0 = \sum_{k=1}^{T} y_k / T$$

and (A.1.8)

$$\sigma_0^2 = \sum_{k=1}^{T} (y_k - \bar{x})^2 / T$$

which are the ML expressions for the sample mean and variance of a random variable. We see that, in maximising the likelihood function, these estimates also minimize the sum of the squares function (see Chapter 2 *et seq*). Also the reader can verify easily that the estimates do indeed maximise L, since the matrix of second partial derivatives $H(\underline{\theta})$ (the "Hessian" of the log-likelihood function $\log_e$ L) given by

$$H(\underline{\theta}) = \nabla_{\underline{\theta}}^2 \log_e L$$

$$= \begin{bmatrix} \dfrac{\partial^2 \log_e L}{\partial \bar{x}^2} & \dfrac{\partial^2 \log_e L}{\partial \bar{x} \partial \sigma^2} \\ \\ \dfrac{\partial^2 \log_e L}{\partial \sigma^2 \partial \bar{x}} & \dfrac{\partial^2 \log_e L}{\partial \sigma^2} \end{bmatrix}$$

$$= \begin{bmatrix} -\dfrac{T}{\sigma^2} & -\dfrac{1}{\sigma^4} \Sigma(y_k - \bar{x}) \\ \\ -\dfrac{1}{\sigma^4} \Sigma(y_k - \bar{x}) & \dfrac{T}{2\sigma^4} - \dfrac{1}{\sigma^6} \Sigma(y_k - \bar{x})^2 \end{bmatrix}$$

Consequently, substituting from (A.1.8)

$$H(\underline{\theta}_o) = \begin{bmatrix} -\frac{T}{\sigma_o^2} & 0 \\ 0 & -\frac{T}{2\sigma_o^4} \end{bmatrix}$$

which is negative definite because $\sigma_o^2 > 0$ for a non-zero random variable.

8. The Cramer-Rao Lower Bound

If the vector $\underline{\theta}$ is of order one, so that the unknown parameter is a scalar θ, then the amount of *information* in the sample is defined by

$$I(\theta) = - E[H(\theta)] = - E[\nabla^2 \log_e L]$$

$$= - \quad H(\theta).L(\theta; y_1, \ldots , y_T)dy_1 \ldots dy_T$$

The value $1/I(\theta)$ is termed the *minimum variance bound* (MVB) for any unbiased estimate $\hat{\theta}$ of θ; in other words, the variance of $\hat{\theta}$ must be greater than or equal to $1/I(\theta)$ or

$$E\{(\hat{\theta}-\theta)^2\} \geqslant 1/I(\theta)$$

which is known as the *Cramer-Rao inequality*. This concept of minimum variance estimation can be extended to the vector situation, as discussed briefly in Chapter 9, but $I(\underline{\theta})$ is now a matrix termed the *information matrix*.

9. Time-Series

If we consider a simple time-series of a random variable x_i, $i = -\infty, \ldots ,$ $-1, 0, 1, \ldots \infty$ where the subscript i denotes the sampled value of the variable x at the ith instant of time, then the mean, variance and covariance are defined as follows if x_i is stationary,

(i) mean : $E\{x_i\} = \bar{x}$

(ii) variance : $E\{(x_i-\bar{x})^2\} = \sigma^2$

(iii) covariance : $E\{(x_i-\bar{x})(x_j-\bar{x})\} = \mu(i-j)$

$$= \mu_\tau, \quad \tau = i-j$$

Note that the covariance is defined as the expected value of the random variable multiplied by itself lagged by a given number of time instants; this is sometimes termed the *autocovariance* and non-zero values for $\tau \neq 0$ indicates that the variable is *autocorrelated* in time. A *white noise variable* is defined here as one which is *serially uncorrelated* in time, i.e. $\mu_\tau = 0$ for all $\tau \neq 0$.

For a vector of time-series variables $\underline{x} = [x_1, \dots, x_n]^T$, it is necessary to allow for the possibility of the *serial correlation* in time of the individual elements and *cross correlation* between elements at different lag values. A *white noise vector* is one whose elements are *serially uncorrelated* in time but may be correlated with other elements of the vector *at the same instant of time.* The *covariance matrix* of such a white noise vector is usually defined as

$$R_{\underline{x},\underline{x}} = E\{[\underline{x}_i - \bar{\underline{x}}][\underline{x}_j - \bar{\underline{x}}]^T\}$$

$$= Q\delta_{ij}$$

where δ_{ij} is the so-called Kronecker delta function, which is equal to unity if i=j and zero if i≠j. Often, where the mean value $\bar{\underline{x}}$ is equal to zero it is omitted from the definition. If Q is a diagonal matrix, then the elements are *mutually uncorrelated white noise variables*. A vector of time-series variables $\underline{e}_i$ with zero mean and covariance matrix Q i.e.

$$E\{\underline{e}_i\} = \underline{0}$$

$$E\{\underline{e}_i\ \underline{e}_j^T\} = Q\delta_{ij}$$

provides a useful source of random variables in the mathematical description of stochastic dynamic systems (see 10. below): in effect the system is seen to "process" the vector in some manner to yield other vectors composed of correlated random variables (of greater or less dimension than $\underline{e}_i$) which will, in general, be composed of "coloured noise" components; i.e. each element will be serially correlated in time and cross-correlated with all other elements of the vector at all instants of time and all lags.

The *autocorrelation* ρ_τ of a time-series variable x_i at lag τ is simply the normalized autocovariance of the variable, where normalization is based on the autocovariance at lag zero, μ_0, i.e.

$$\rho_\tau = \frac{\mu_\tau}{\mu_0} \quad ; \quad \tau = 0, 1, 2 \dots$$

so that ρ_0, the instantaneous autocorrelation is normalized to unity.

10. Gauss-Markov Random Sequences

To describe a random time-series vector (or scalar) sequence $\underline{x}_i$, i=1, 2, ... , T, completely, the joint probability density function

$$p(\underline{x}_T, \underline{x}_{T-1}, \dots , \underline{x}_o)$$

of *all* the elements in the sequence must be specified. Although this involves an enormous amount of information in general terms, it is possible to simplify the situation by assuming that the sequence is a *Markov sequence* where the *conditional (or transition) probability density function* has the special property

$$p(\underline{x}_k|\underline{x}_{k-1}, \underline{x}_{k-2}, \dots , \underline{x}_o) = p(\underline{x}_k|\underline{x}_{k-1})$$

for all k. In other words, the probability density function of $\underline{x}_k$ depends *only* on knowledge of $\underline{x}_{k-1}$ at the previous instant and not on any *previous* values $\underline{x}_{k-\tau}$, τ=2, 3 ... . The knowledge of $\underline{x}_{k-1}$ can be either deterministic, in which the exact value of $\underline{x}_{k-1}$ is known, or probabilistic, where only $p(\underline{x}_{k-1})$ is known.

The *joint probability density function* of a Markov random sequence can be described completely by specifying its initial density function $p(\underline{x}_o)$ and the transition density functions $p(\underline{x}_k|\underline{x}_{k-1})$ i.e.

$$p(\underline{x}_T, \dots , \underline{x}_o) = p(\underline{x}_T|\underline{x}_{T-1}).p(\underline{x}_{T-1}|\underline{x}_{T-2}) \dots p(\underline{x}_1|\underline{x}_o).p(\underline{x}_o)$$

A purely random (or white noise) sequence is defined by the property that

$$p(\underline{x}_k|\underline{x}_{k-1}) = p(\underline{x}_k), \quad \text{for all k.}$$

A Gauss-Markov random sequence is a Markov random sequence with the additional requirement that $p(\underline{x}_k)$ and $p(\underline{x}_k|\underline{x}_{k-1})$ are Gaussian probability density functions for all k. The density function for a Gauss-Markov random sequence is, in this manner, described completely by the mean value vector $\bar{\underline{x}}_k = E\{\underline{x}_k\}$ and covariance matrix $P_k = E\{[\underline{x}_k-\bar{\underline{x}}_k][\underline{x}_k-\bar{\underline{x}}_k]^T\}$.

A Gauss-Markov random sequence of nth order vectors $\underline{x}_k$ can always be represented by the following vector-matrix model

$$\underline{x}_k = \Phi_k\underline{x}_{k-1} + \Gamma_k\underline{e}_{k-1} \qquad (A.1.9)$$

where Φ_k is an n x n transition matrix, Γ_k is an n x m input matrix, and $\underline{e}_k$ is an mth order white noise vector with mean $\bar{\underline{e}}_k$ and covariance matrix Q i.e.

$$E\{\underline{e}_k\} = \bar{\underline{e}}_k$$

$$E\{[\underline{e}_j-\bar{\underline{e}}_j][\underline{e}_k-\bar{\underline{e}}_k]^T\} = Q\,\delta_{jk}$$

Such Gauss-Markov processes are discussed in the text (see Chapter 5 *et seq.*), usually with $\bar{\underline{e}}_k = 0$, but the reader is advised to consult Bryson and Ho (1969) for a more complete background.

A.1.3 Simple Deterministic Dynamic Systems

1. First Order Continuous-Time Linear Dynamic System

A simple deterministic, first order, linear dynamic system with input u(t) and output x(t) can be described by the following ordinary differential equation

$$\frac{dx(t)}{dt} + \alpha x(t) = \beta u(t) \qquad \text{(A.1.10)}$$

or

$$\frac{Tdn(t)}{dt} = -x(t) + \beta T u(t)$$

where $T = 1/\alpha$ is the *time constant* of the system. This system responds in a very simple manner to input stimuli u(t), e.g. in the case of a unit step (i.e. $u(t) = 0$ for $t<0$, $u(t) = 1.0$ for $t \geqslant 0$), x(t) is given by (see e.g. Takahashi *et al.*, 1972),

$$x(t) = \frac{\beta}{\alpha}(1-e^{-\alpha t})$$

Consequently for t = T (the time constant)

$$x(T) = \frac{\beta}{\alpha}(1-e^{-1.0}) \simeq 0.63\,\frac{\beta}{\alpha}$$

while for $t = \infty$, i.e. in the *steady state*

$$x(\infty) = \beta/\alpha$$

Therefore the system is said to have a *steady state gain* (SSG) of β/α and it reaches 0.63 of this steady state after a period of time equal to the time-constant.

2. A First Order Discrete-Time Linear Dynamic System

If the input signal u(t) can be assumed constant over a sampling period T_s time units (i.e. it is a staircase type function) then the continuous time system (A.1.10) can be represented *exactly* in discrete-time terms by the equation

$$x_k = ax_{k-1} + bu_{k-1} \tag{A.1.11}$$

where x_k is the value of x(t) at the kth sampling instant and u_k is the value of u(t) at the kth sampling instant. The parameters a and b are related to α and β by the following equations

(i)
$$a = e^{-\alpha T_s} \quad ; \quad \text{so that}$$

$$\alpha = -\frac{\log_e a}{T_s} \text{ and } T = -\frac{T_s}{\log_e a}$$

(ii)
$$b = \frac{\beta}{\alpha}(1-e^{-\alpha T_s})$$

$$= \frac{\beta}{\alpha}(1-a) \quad ; \quad \text{so that}$$

$$\beta = \frac{\alpha b}{1-a} = \frac{-(\log_e a)b}{T_s(1-a)}$$

Although, in general, the discrete-time solution (A.1.11) will not be exact for u(t) not constant over the sampling period T_s, any first order linear dynamic system can be represented by a model such as (A.1.11) although the relationships (i) and (ii) will not hold exactly. As a result, we can use it as a general deterministic representation in discrete-time terms. It is useful to refer back to the continuous time representation (A.1.10), however, so that we can compute easily the time constant of the system: this is because, while there is only *one* representation of (A.1.10) characterized by the parameters α and β (and a time constant $T = 1/\alpha$), there are infinitely many representations (A.1.11) with parameters a and b which depend upon the chosen sampling interval T_s. Note also that the time constant of the discrete-time system (A.1.11) *in sampling intervals* is given by

$$T = -\frac{1}{\log_e a}$$

3. <u>The Discrete-Time State-Space Representation of a Deterministic Dynamic System</u>

If we consider a sampled vector of random variables $\underline{x}_k$, then it can be represented by the vector-matrix analogue of (A.1.10) i.e.

$$\underline{x}_k = A\underline{x}_{k-1} + B\underline{u}_{k-1} \tag{A.1.12}$$

where $\underline{u}_k$ is an mth order vector of input variables, A is an nxn transition matrix and B is a nxm input matrix. This is, of course, simply the deterministic analogue of the Gauss-Markov model (A.1.9) discussed in Section A.1.2 previously, with the white noise vector $\underline{e}_k$ replaced by the deterministic input vector $\underline{u}_k$.

4. Transfer Function Representation of a Single Input, Single Output (SISO) Discrete Pyramid System

If the backward shift operator z^{-1} is introduced into (A.1.12), where $z^{-1}\underline{x}_k = \underline{x}_{k-1}$, then it can be represented by

$$[I - Az^{-1}]\underline{x}_k = Bz^{-1}\underline{u}_k$$

where I is the nxn unit matrix I_n.

For a single (scalar) input system, i.e. $\underline{u}_k = u_k$, the input matrix B becomes a vector $\underline{b}$. If A and $\underline{b}$ are now defined in the following special form,

$$A = \begin{bmatrix} -a_1 & 1 & 0 & \dots & 0 \\ -a_2 & 0 & 1 & \dots & 0 \\ \cdot & & & & \vdots \\ \cdot & & & & \\ \cdot & & & & 1 \\ -a_n & 0 & 0 & \dots & 0 \end{bmatrix}; \quad \underline{b} = \begin{bmatrix} b_1 \\ b_2 \\ \cdot \\ \cdot \\ \cdot \\ b_n \end{bmatrix}$$

and we define the output of the system as $x_k = (x_1)_k$ (i.e. the output is the first element of $\underline{x}_k$), then

$$x_k = \underline{c}^T[I - Az^{-1}]\underline{b}z^{-1}u_k$$

where $\underline{c}^T = [1\ 0\ 0\ \dots\ 0]$.

As a result,

$$x_k = [1\ 0\ 0\ \dots\ 0] \begin{bmatrix} (1+a_1z^{-1}) & -z^{-1} & 0 & \dots & 0 \\ +a_2z^{-1} & 1 & -z^{-1} & & 0 \\ \cdot & & & & \cdot \\ \cdot & & & & \cdot \\ \cdot & & & & \cdot \\ \cdot & & & & -z^{-1} \\ +a_nz^{-1} & 0 & 0 & \dots & 1 \end{bmatrix}^{-1} \begin{bmatrix} b_1 \\ b_2 \\ \\ \\ \\ \\ b_n \end{bmatrix} z^{-1}u_k$$

The reader can then verify that this yields the following transfer function (TF) representation,

$$x_k = \frac{B(z^{-1})}{A(z^{-1})}u_k = \frac{b_1z^{-1} + b_2z^{-2} + \dots + b_nz^{-n}}{1 + a_1z^{-1} + a_2z^{-2} + \dots a_nz^{-n}}u_k \tag{A.1.13}$$

where $B(z^{-1})/A(z^{-1})$ is termed the *rational transfer function* of the system. Cross-multiplying and converting back to a discrete-time equation form, we obtain

$$x_k + a_1 x_{k-1} + \dots + a_n x_{k-n} = b_1 u_{k-1} + \dots b_n u_{k-n} \qquad (A.1.14)$$

which is the nth order extension of (A.1.11). The reader will see that x_k is now a function of past values x_{k-i} and u_{k-i}, i=1, 2, ... n, of itself and the input variable, respectively. Consequently the response of x_k to input stimuli u_k is much more complex than in the first order case (see e.g. Box and Jenkins, 1970). However, if the system is *stable* in the sense that the roots of the equation

$$1 + a_1 z^{-1} + \dots + a_n z^{-n} = 0$$

lie *outside* the unit circle in complex plane (or conversely the roots of $z^n + a_1 z^{n-1} + \dots + a_n = 0$ lie *inside* the unit circle) then x_k will reach a *steady state value* if u_k is chosen as a unit step function ($u_k = 0$ for $k < 0$; $u_k = 1.0$ for k = 0, 1, 2 ...). This steady state value, which is obtained simply by setting $u_k = 1.0$ and $z^{-1} = 1.0$ (i.e. $x_k = x_{k-1}$ at steady state) in (A.1.13), provides the *steady state gain* of the system i.e.

$$SSG = \frac{b_1 + b_2 + \dots + b_n}{1 + a_1 + \dots + a_n}$$

5. <u>The Infinite Dimensional Impulse Response Representation of a Linear SISO Discrete Dynamic System</u>

In the TF representation (A.1.13) the transfer function is defined as the ratio of two finite dimensional polynomials in the backward shift operator z^{-1}. If the numerator polynomial $B(z^{-1})$ is divided by the denominator polynomial $A(z^{-1})$ then, in general, we obtain an *infinite* polynomial $G(z^{-1})$ in the backward shift operator z^{-1}. Consequently (A.1.13) can be written alternately in the form,

$$\begin{aligned} x_k &= G(z^{-1})\, u_k \\ &= (g_1 z^{-1} + g_2 z^{-2} + \dots + g_\infty z^{-\infty}) u_k \end{aligned}$$

Once again, converting back to discrete-time equation form we see that

$$x_k = g_1 u_{k-1} + g_2 u_{k-2} + \dots + g_\infty u_{k-\infty}; \qquad (A.1.15)$$

in other words, we see that x_k is dependent on the input variations into the *infinite* past and the nature of this dependency is defined by the coefficients $g_1, g_2, \dots g_\infty$ of the $G(z^{-1})$ polynomial.

The reader will see that if u_k is defined as the unit impulse ($u_k = 0$, for $k < 0$; $u_k = 1$ for $k = 0$; $u_k = 0$ for $k > 0$) then, if $x_0 = 0$, the output x_k for k = 0, 1, 2 is defined by the coefficients of $G(z^{-1})$ i.e.

$$x_0 = 0$$
$$x_1 = g_1$$
$$x_2 = g_2$$
etc.

and we see that the infinite dimensional polynomial $G(z^{-1})$ defines the *impulse response* of the system. Equation (A.1.15) is, in fact, the discrete-time equivalent of the well known *convolution integral equation* in continuous time terms, i.e.

$$x(t) = \int_0^t g(\tau)\, u(t-\tau)\, d\tau$$

where $g(\tau)$ is the *continuous-time impulse response function*

Note that the TF representation (A.1.14) is parametrically much more efficient than the impulse response representation (A.1.15) requiring only 2n parameters (a_i, b_i, i=1, 2, ... , n) rather than an infinite number (g_i, i=1, 2, ... , ∞) to *completely* describe the system behaviour. It is, therefore, a more suitable form for parameter estimation purposes although, as we see in the text, it does pose certain parameter estimation problems because of the presence of the lagged terms in x_k, i.e. x_{k-i}, i=1, 2, ... , n. Box and Jenkins (1970) term (A.1.14) a "parsimonious" representation because of its parametric efficiency.

6. Differentiation of a TF with respect to a Given Parameter

When considering Maximum Likelihood estimation of the parameters of a TF model such as (A.1.14) in Chapter 8, it is necessary to differentiate expressions written in TF terms with respect to each coefficient in turn. This is accomplished quite easily by reference to the rule for differentiating a product or quotient. For example for i=1, 2, ... n

$$\frac{\partial}{\partial b_i}\left\{\frac{B(z^{-1})}{A(z^{-1})}u_k\right\} = \frac{\partial}{\partial b_i}\left\{\frac{b_1 z^{-1} + \ldots + b_n z^{-n}}{1 + a_1 z^{-1} + \ldots a_n z^{-n}}u_k\right\}$$

$$= \frac{z^{-i}}{A(z^{-1})}u_k$$

while

$$\frac{\partial}{\partial a_i}\left\{\frac{B(z^{-1})}{A(z^{-1})}u_k\right\} = \frac{A(z^{-1})\,\frac{\partial}{\partial a_i}\{B(z^{-1})\} - B(z^{-1})\,z^{-i}}{[A(z^{-1})]^2}$$

$$= -\frac{B(z^{-1})}{[A(z^{-1})]^2}\,z^{-i}$$

since $\frac{\partial}{\partial a_i}\{B(z^{-1})\} = 0$ as $B(z^{-1})$ has no terms in a_i.

Appendix 2 Gauss's Derivation of Recursive Least Squares

This Appendix is based on pages 53 to 55 of the book *Méthode des Moindres Carrés: Mémoires sur la Combinaison des Observations*, which is the French translation by J. Bertrand of Gauss's collected works on least squares (1803-1826) and was published in 1855, with the authorisation of Gauss. In Section 35, page 53, Bertrand's translation states:

"Nons traiterons particuliérement le problème suivant, tant à cause de son utilité pratique, que de la simplicité de la solution:

Trouver les changements que les valeurs les plus plausibles des inconnues subissent par l'adjonction d'une nouvelle équation, et assigner les poids de ces nouvelles determinations".

In this Appendix, we will reproduce the following two and a half pages of Bertrand's book, which constitute the main part of Gauss's derivation. To aid the reader both to appreciate the analysis and to draw comparisons with the equivalent vector-matrix analysis in the present book, we will add comments at various stages in the analysis. These appear in italics between horizontal lines and, in all cases, the vector matrix nomenclature is similar to that used in Chapters 3 and 4 of the main text. Note that additional equation numbers have been introduced into the analysis for clarity. An English translation of Bertrand's book has been produced by Hale F. Trotter (1957)

Gauss's analysis begins as follows:

Conservons les notations précédentes. Les équations primitives, réduites à avoir pour poids l'unité, seront

$$\nu = 0,\ \nu' = 0,\ \nu'' = 0, \ldots ;$$

on aura

$$\Omega = \nu + \nu'^2 + \nu''^2 + \ldots, \qquad \text{(A.2.1)}$$

ξ, η, ζ, etc., seront les dérivées partielles

$$\frac{d\Omega}{2dx},\ \frac{d\Omega}{2dy},\ \frac{d\Omega}{2dz}, \ldots, \qquad \text{(A.2.2)}$$

et enfin on aura, par l'élimination

$$\text{(1)}\quad \begin{aligned} x &= A + (\alpha\alpha)\xi + (\alpha\beta)\eta + (\alpha\gamma)\zeta + \ldots \\ y &= B + (\alpha\beta)\xi + (\beta\beta)\eta + (\beta\gamma)\zeta + \ldots \\ z &= C + (\alpha\gamma)\xi + (\beta\gamma)\eta + (\gamma\gamma)\zeta + \ldots \end{aligned} \qquad \text{(A.2.3)}$$

Comment:

Here Ω is the 'sum of squares' cost function, after a given number of observations, with the unknown parameters set to their true values. Gauss refers to these observations as "equations" since he associates each new set of observations with the latest equation of the system under study. Using the nomenclature of the present book, each equation is of the form,

$$e_{y_i} = \underline{x}_i^T \underline{a} + y_i \tag{A.2.4}$$

This is similar to equation (3.4) of Chapter 3 but with slightly different sign convention. Using the nomenclature of Chapter 4, the whole set of equations so constituted after k-1 observations can be written

$$(\underline{e}_y)_{k-1} = X_{k-1}\underline{a} + \underline{y}_{k-1} \tag{A.2.5}$$

Where we have assumed initially k-1 observations in order to facilitate comparison of later equations with equivalent ones in the main text.

The variables ξ, η, ζ, etc., are the partial derivatives (or gradients) of Ω with respect to the unknown parameters x, y, z, etc., as defined in equation (A.2.2). Equation (A.2.3) (which, in the original, is Gauss's equation (1)) is simply a statement of the solution to the normal equations of least squares. This becomes clear if we consider equation (4.4) of Chapter 4. With the different sign convention of equation (A.2.5), this equation yields a relationship between $\underline{a}$ and $\underline{\hat{a}}$ of the form

$$\begin{aligned} \underline{a} &= \underline{\hat{a}} + [X_{k-1}^T X_{k-1}]^{-1} X_{k-1}^T (\underline{e}_y)_{k-1} \\ &= \underline{\hat{a}} + [\sum_{i-1}^{k-1} x_i x_i^T]^{-1} \sum_{i=1}^{k-1} \underline{x}_i e_{y_i} \end{aligned} \tag{A.2.6}$$

In equation (A.2.3) x,y,z, etc. are the elements of the true parameter vector $\underline{a}$; A, B, C, etc., are the elements of the estimate vector $\underline{\hat{a}}$; $(\alpha\alpha)$, $(\alpha\beta)$, $(\alpha\gamma)$, etc., are the elements of the inverse matrix $[X_{k-1}^T X_{k-1}]^{-1} = P_{k-1}$; and ξ, η, ζ, etc., are associated with the gradient vector $\underline{g}_{k-1} = X_{k-1}^T(\underline{e}_y)_{k-1}$. We can see that this is the gradient vector by referring to the least squares cost function in this case, i.e.

$$\Omega = \sum_{i=1}^{k-1} e_{y_i}^{\ 2} = (\underline{e}_y)_{k-1}^T (\underline{e}_y)_{k-1} = [X_{k-1}\underline{a} + \underline{y}_{k-1}]^T [X_{k-1}\underline{a} + \underline{y}_{k-1}]$$

the gradient of which is (see Appendix 1),

$$\underline{g}_{k-1} = X_{k-1}^T [X_{k-1}\underline{a} + \underline{y}_{k-1}] = X_{k-1}^T (\underline{e}_y)_{k-1} = \sum_{i=1}^{k-1} \underline{x}_i e_{y_i} \qquad (A.2.7)$$

From this comparison, and noting that $[X_{k-1}^T X_{k-1}]^{-1}$ *is the* P_{k-1} *matrix of the main text, we see that equation (A.2.3) can be written, in the nomenclature of the present book, as,*

$$\begin{aligned} a_1 &= \hat{a}_1 + p_{11}g_1 + p_{12}g_2 + p_{13}g_3 + \ldots. \\ a_2 &= \hat{a}_2 + p_{12}g_1 + p_{22}g_2 + p_{23}g_3 + \ldots. \\ a_3 &= \hat{a}_3 + p_{13}g_1 + p_{23}g_2 + p_{33}g_3 + \ldots. \end{aligned}$$

where a_1, a_2, a_3, *etc., are the elements of the unknown parameter vector* $\underline{a}$*;* $\hat{a}_1$, $\hat{a}_2$, $\hat{a}_3$, *etc., are the elements of the estimate vector* $\hat{\underline{a}}_{k-1}$ *after* $k-1$ *samples;* p_{ij} *are the elements of* P_{k-1}*; and* g_1, g_2, g_3, *etc. are the elements of* $\underline{g}_{k-1}$.

Supposons maintenant que l'on ait une nouvelle equation approximative,

$$\overset{*}{\nu} = 0 \qquad (A.2.8)$$

dont nous supposerons le poids égal à l'unité. Cherchons les changements que subiront les valeurs les plus plausibles A, B, C, etc. et celles des coefficients $(\alpha\alpha)$, $(\beta\beta)$, etc.

Posons

$$\Omega + \overset{*}{\nu}{}^2 = \overset{*}{\Omega}, \qquad (A.2.9)$$

$$\frac{1}{2}\frac{d\overset{*}{\Omega}}{dx} = \overset{*}{\xi}, \quad \frac{1}{2}\frac{d\overset{*}{\Omega}}{dy} = \overset{*}{\eta}, \quad \frac{1}{2}\frac{d\overset{*}{\Omega}}{dz} = \overset{*}{\zeta}, \ldots.,$$

et soit

$$x = \overset{*}{A} + (\alpha\overset{*}{\alpha})\overset{*}{\xi} + (\alpha\overset{*}{\beta})\overset{*}{\eta} + (\alpha\overset{*}{\gamma})\overset{*}{\zeta} + \ldots. \qquad (A.2.10)$$

le résultat de l'élimination

Comment:

$\overset{*}{\Omega}$ *in equation (A.2.9) is the updated sum of squares with* $\overset{*}{\nu}$ *denoting the latest error squared term. The associated new gradients are given as* $\overset{*}{\xi}$, $\overset{*}{\eta}$, $\overset{*}{\zeta}$, *etc., and equation (A.2.10) is the new equation for* x *(with equations for* y, z, *etc. not shown). This follows directly from equation (A.2.3) but with the updated values for the variables denoted by the star superscripts. Gauss is simply pointing out that all* *the variables need to be updated on receipt of new information in order to obtain new estimates* A^*, B^*, C^*, *etc. This is the prelude to his development of the recursive equations, which now follows.*

Soit enfin

$$\nu^* = fx + gy + hz + \ldots + k, \tag{A.2.11}$$

qui deviendra, en ayant égard aux équations (1)

$$\nu^* = F\xi + G\eta + H\zeta + \ldots + K, \tag{A.2.12}$$

et posons

$$Ff + Gg + Hh + \ldots = \omega \tag{A.2.13}$$

K sera évidemment la valeur la plus plausible de la fonction ν^*, telle qu'elle résulte des équations primitives, sans avoir égard à la valeur o fournie par la nouvelle observation, et $\frac{1}{\omega}$ sera le poids de cette détermination.

Comment:

Equation (A.2.11) is Gauss's version of equation (A.2.4) with ν^* *denoting the latest error* (e_{y_k})*;* f, g, h, *etc., the regressors (the elements of* $\underline{x}_k$)*; and* k *the new observation of the dependent variable* (y_k). *He obtains (A.2.12) by substituting for* x, y, z, *etc. from equation (A.2.3), i.e.*

$$\begin{aligned}\nu^* &= f(A + (\alpha\alpha)\xi + (\alpha\beta)\eta + \ldots) + g(B + (\alpha\beta)\xi + (\beta\beta)\eta + \ldots) \\ &\quad + h(C + (\alpha\gamma)\xi + (\beta\gamma)\xi + \ldots) + \ldots + k \\ &= [f(\alpha\alpha) + g(\alpha\beta) + h(\alpha\gamma) + \ldots]\xi + [f(\alpha\beta) + g(\beta\beta) + h(\beta\gamma) + \ldots]\eta \\ &\quad + fA + gB + hC + k \\ &= F\xi + G\eta + \ldots + K\end{aligned}$$

where K *will be recognised as the latest recursive residual (innovations process). Using the nomenclature of the present text, the reader can verify that the equivalent vector matrix expression is*

$$e_{y_k} = \underline{x}_k^T P_{k-1}\underline{g}_{k-1} + \{\underline{x}_k^T \hat{a}_{k-1} + y_k\} \tag{A.2.14}$$

where Gauss's F, G, H, etc., are the elements of the vector $x_k^T P_{k-1}$; *and* K *is the latest recursive residual shown in curly brackets* $\{\cdot\}$. *Note also that* ω *defined by equation (A.2.13) is equivalent to* $\underline{x}_k^T P_{k-1} \underline{x}_k$.

Or nous avons

$$\xi^* = \xi + f\nu^*,\ \eta^* = \eta + g\nu^*,\ \zeta^* = \zeta + h\nu^*, \ldots, \tag{A.2.15}$$

et, par suite,

$$F\xi^* + G\eta^* + H\zeta^* + \ldots + K = \nu^* \ (1 + Ff + Gg + Hh + \ldots);$$

d'où l'on déduit:

$$\nu^* = \frac{F\xi^* + G\eta^* + H\zeta^* + \ldots + K}{1 + \omega} \tag{A.2.16}$$

Comment:

Here ξ^*, η^*, ζ^*, *etc. in (A.2.15) represent the updated gradient measures. In the vector terms used above, these equations are equivalent to*

$$\underline{g}_k = \underline{g}_{k-1} + \underline{x}_k e_{y_k} \tag{A.2.17}$$

which follows from equation (A.2.7). Using equations (A.2.15), Gauss now defines ν^* *in terms of the updated gradients, i.e.*

$$\begin{aligned}\nu^* &= F\xi + G\eta + H\zeta + \ldots + K \\ &= F(\xi^* + f\nu^*) + G(\eta^* + g\nu^*) + \ldots + K\end{aligned}$$

so that,

$$\nu^*[1 + Ff + Gg + Hh \ldots] = F\xi^* + G\eta^* + \ldots + K$$

and equation (A.2.16) for ν^* *follows because of the definition of* ω *in (A.2.13). The following vector-matrix equivalent of equation (A.2.16) is obtained straightforwardly by reference to equation (A.2.14) and the vector matrix definition of* $\omega = \underline{x}_k^T P_{k-1} \underline{x}_k$

$$e_{y_k} = \frac{\underline{x}_k^T P_{k-1} \underline{g}_k + \{\underline{x}_k^T \hat{\underline{a}}_{k-1} + y_k\}}{1 + \underline{x}_k^T P_{k-1} \underline{x}_k} \tag{A.2.18}$$

where $\underline{g}_k$ *is defined in (A.2.17).*

On a, en outre,

$$\begin{aligned} x &= A + (\alpha\alpha)\xi^* + (\alpha\beta)\eta^* + (\alpha\gamma)\zeta^* + \ldots. \\ &\quad - \nu^*[f(\alpha\alpha) + g(\alpha\beta) + h(\alpha\gamma) + \ldots] \\ &= A + (\alpha\alpha)\xi^* + (\alpha\beta)\eta^* + \ldots. - F\nu^* \\ &= A + (\alpha\alpha)\xi^* + (\alpha\beta)\eta^* + \ldots. - \frac{F}{1+\omega}(F\xi^* + G\eta^* + H\zeta^* + \ldots + K) \end{aligned} \tag{A.2.19}$$

Nous d'éduirons de là,

$$A^* + A - \frac{FK}{1+\omega} \tag{A.2.20}$$

qui sera la valeur la plus plausibles de x, deduite de *toutes* les observations.

On aura aussi

$$(\alpha\alpha^*) = (\alpha\alpha) - \frac{F^2}{1+\omega} ; \tag{A.2.21}$$

par conséquent,

$$\frac{1}{(\alpha\alpha) - \frac{F^2}{1+\omega}}$$

sera le poids de cette determination.

On trouvera de la mème manière, pour valeur la plus plausible de y, déduite de *toutes* observations,

$$B^* = B - \frac{GK}{1+\omega}$$

le poids de cette détermination sera

$$\frac{1}{\frac{1}{(\beta\beta)} - \frac{G^2}{1+\omega}}$$

et ainsi de suite.

Le problème est donc résolu.

<u>*Comment:*</u>

The above equations (A.2.20) and (A.2.21) for A^* *and* $(\alpha\alpha^*)$*, respectively, constitute the recursive least squares update equations for the first unknown parameter and the associated diagonal element of the inverse matrix* (P_k). *The associated equations for all other parameter estimates and the elements of the* P_k *matrix follow in a similar manner to provide, finally, the complete recursive algorithm. Gauss does not continue further, however, since the subsequent derivation is obvious.*

Equation (A.2.19) follows by substituting from equations (A.2.15) into equation (A.2.3) in the following manner,

$$x = A + (\alpha\alpha)[\xi^* - f\nu^*] + (\alpha\beta)[\eta^* + g\nu^*]+\dots$$

$$= A + (\alpha\alpha)\xi^* + (\alpha\beta)\eta^*+\dots -\nu^*[f(\alpha\alpha) + g(\alpha\beta)+\dots];$$

and then noting that $F = [f(\alpha\alpha) + g(\alpha\beta)+\dots]$, *while* ν^* *is defined in terms of* ξ^*, η^*, ζ^*, *etc. by equation (A.2.16).*

Equation (A.2.21) which, taken together with similarly derived equations for $(\alpha\beta^*)$, $(\alpha\gamma^*)$, *etc. constitutes the equivalent of the matrix inversion lemma can be obtained quite straightforwardly but with rather lengthy algebraic manipulation. It is not clear from Gauss's reported analysis, however, exactly how he obtained these relationships since he does not include the details: in the classic phrase "one has also" he parallels the over-used present day phrase "it can be shown" and leaves the reader to his own devices. Alas poor reader, we will do the same!*

It remains to note that the vector-matrix equivalent of the above equations are, of course, the recursive least squares equations of algorithm II in Chapter 3, with the minor sign difference in the recursive residual arising because of Gauss's sign convention (see equation (A.2.4)). In other words,

$$\hat{\underline{a}}_k = \hat{\underline{a}}_{k-1} - \frac{P_{k-1}\underline{x}_k}{1 + \underline{x}_k^T P_{k-1}\underline{x}_k}\{\underline{x}_k^T\hat{\underline{a}}_{k-1} + y_k\}$$

$$P_k = P_{k-1} - \frac{P_{k-1}\underline{x}_k\underline{x}_k^T P_{k-1}}{1 + \underline{x}_k^T P_{k-1}\underline{x}_k}$$

The above analysis, carried out at the beginning of the nineteenth century, serves to illustrate yet again the enormous contributions Gauss made to science and mathematics. While it may be arguable that Gauss and Lagrange evolved the method of least squares independently and at about the same time, it is clear that only Gauss was responsible for the development of the theory in its most elegant, recursive form. And while the development of the recursive form is fairly straightforward in these days of the digital computer and matrix analysis, we can only marvel at Gauss's derivation relying, as it had to, on the use of scalar algebra. Finally, it is nice to note that Gauss did not develop the method for its own sake (although he too was surely impressed by the elegance of the algorithmic form), but because it solved a very real practical problem. Gauss's practicality is demonstrated later in the analysis when he concludes (page 58 of Bertrand's book):

"If, after the calculation is finished, several new equations should be adjoined to the original, or if the weights attributed to several of them were in error, the calculations of the corrections would become very complicated and it would be better to begin all over again".

Of course, had he had access to the modern digital computer, he would not have needed to worry.

Appendix 3 The Instantaneous Cost Function Associated with the Recursive Least Squares Algorithm

The derivation of the recursive least squares regression algorithm used in the main text does not directly address the situation at the start of the algorithm, when it is necessary to choose the initial estimate $\underline{\hat{a}}_0$ of the parameter vector $\underline{a}$ and the associated matrix P_0 or P_0^*. In this connection, it is interesting to consider the following instantaneous cost function at the kth sampling instant (Young, 1965c, Rauch et al, 1965),

$$J = \frac{1}{\sigma^2}[\underline{x}_k^T\underline{\hat{a}}_k - y_k]^2 + [\underline{\hat{a}}_k - \underline{\hat{a}}_{k-1}]^T(P_{k-1}^*)^{-1}[\underline{\hat{a}}_k - \underline{\hat{a}}_{k-1}] \qquad \text{(A.3.1)}$$

As usual, the conditions for a minimum of J with respect to the unknown vector $\underline{\hat{a}}_k$, are

$$\frac{\partial J}{\partial \underline{\hat{a}}_k} = \nabla_{\underline{\hat{a}}} J = 0$$

Consequently, we can obtain $\underline{\hat{a}}_k$ from the solution of the following equation,

$$\frac{1}{\sigma^2}\underline{x}_k[\underline{x}_k^T\hat{a}_k - y_k] + (P_{k-1}^*)^{-1}\underline{\hat{a}}_k - (P_{k-1}^*)^{-1}\underline{\hat{a}}_{k-1} = 0$$

or,

$$[\frac{1}{\sigma^2}\underline{x}_k\underline{x}_k^T + (P_{k-1}^*)^{-1}]\underline{\hat{a}}_k = \frac{1}{\sigma^2}\underline{x}_k y_k + (P_{k-1}^*)^{-1}\underline{\hat{a}}_{k-1}$$

so that,

$$\underline{\hat{a}}_k = \{\frac{1}{\sigma^2}\underline{x}_k\underline{x}_k^T + (P_{k-1}^*)^{-1}\}^{-1}\{\frac{1}{\sigma^2}\underline{x}_k y_k + (P_{k-1}^*)^{-1}\underline{\hat{a}}_{k-1}\}$$

Referring now to the matrix inversion lemma of equation II(1), this latter equation can be written in the form,

$$\underline{\hat{a}}_k = \{P_{k-1}^* - P_{k-1}^*\underline{x}_k[\sigma^2 + \underline{x}_k^T P_{k-1}^*\underline{x}_k]^{-1}\,\underline{x}_k^T P_{k-1}^*\}\{\frac{1}{\sigma^2}\underline{x}_k y_k + (P_{k-1}^*)^{-1}\underline{\hat{a}}_{k-1}\}$$

As a result, we can obtain the recursive algorithm for $\hat{a}_k$ in terms of $\underline{\hat{a}}_{k-1}$ by multiplying out the expression on the right hand side of this equation and re-arranging the terms, i.e.

$$\hat{\underline{a}}_k = \hat{\underline{a}}_{k-1} + P^*_{k-1}\underline{x}_k\frac{y_k}{\sigma^2} - P^*_{k-1}\underline{x}_k[\sigma^2 + \underline{x}_k^T P^*_{k-1}\underline{x}_k]^{-1}\underline{x}_k^T P^*_{k-1}\underline{x}_k\frac{y_k}{\sigma^2}$$

$$- P^*_{k-1}\underline{x}_k[\sigma^2 + \underline{x}_k^T P^*_{k-1}\underline{x}_k]^{-1}\underline{x}_k^T\hat{\underline{a}}_{k-1}$$

$$= \hat{\underline{a}}_{k-1} - P^*_{k-1}\underline{x}_k[\sigma^2 + \underline{x}_k^T P^*_{k-1}\underline{x}_k]^{-1}\{-[\sigma^2 + \underline{x}_k^T P^*_{k-1}\underline{x}_k]\frac{y_k}{\sigma^2}$$

$$+ \underline{x}_k^T P^*_{k-1}\underline{x}_k\frac{y_k}{\sigma^2} + \underline{x}_k^T\hat{\underline{a}}_{k-1}\}$$

$\therefore$

$$\hat{\underline{a}}_k = \hat{\underline{a}}_{k-1} - P^*_{k-1}\underline{x}_k[\sigma^2 + \underline{x}_k^T P^*_{k-1}\underline{x}_k]^{-1}\{\underline{x}_k^T\hat{\underline{a}}_{k-1} - y_k\}$$

which is the least squares regression equation III(1).

This analysis reveals that each step in the recursive least squares algorithm III can be considered as minimising the instantaneous cost function J defined in (A.3.1), with P^*_{k-1} and σ^2 defined as in the main text. Considering the situation at the beginning of the algorithm, therefore, we see that the initial estimates $\hat{\underline{a}}_0$ and their associated covariance matrix P_0^* appear in the cost function via the additive quadratic form (see Appendix 1 and Section 5.3, Chapter 5),

$$[\hat{\underline{a}}_1 - \hat{\underline{a}}_0]^T(P_0^*)^{-1}[\hat{\underline{a}}_1 - \hat{\underline{a}}_0] \qquad \text{(A.3.2)}$$

Thus, at its initiation, the algorithm is selecting $\hat{a}_1$ in order to minimise not only the normal least square cost term, i.e.

$$[\underline{x}_k^T\hat{\underline{a}}_k - y_k]^2 \; ; \; k = 1$$

but also a quadratic form in the difference between $\hat{\underline{a}}_1$ and the initial *a priori* estimate $\underline{a}_0$, weighted by the inverse of the associated *a priori* covariance matrix P_0^*.

From this simple analysis, we see that, if the analyst has little confidence in the *a priori* estimate $\hat{\underline{a}}_0$ and so chooses P_0^* to be large (e.g. 10^6 diagonal), then little notice will be taken of $\hat{\underline{a}}_0$ in determining $\hat{\underline{a}}_1$. On the other hand, if there is good prior knowledge of the parameter values, then the algorithm can be informed of this by the analyst choosing a suitably smaller P_0^* covariance matrix which reflects the increased confidence associated with his knowledge of $\hat{\underline{a}}_0$. In this manner, the second term (A.3.2) in the cost function will be given more weight and the estimate $\hat{\underline{a}}_1$ will be much more dependent upon $\hat{\underline{a}}_0$.

The Bayesian statistical interpretation of the above procedure is obvious, with the *a priori* information ($\hat{\underline{a}}_0$; P_0^*) playing an important role in the computation of the *a posteriori* estimate $\hat{\underline{a}}_1$. But the algorithm remains essentially the same without these statistical interpretations: in the deterministic, recursive least squares algorithm II, for example, we see that $\sigma^2 = 1$ and $P_0^* = P_0$, but the above algebraic results still apply. Consequently, on purely deterministic, numerical grounds, the P_0 matrix should be chosen by the analys so that P_0^{-1} suitably reflects the "weight" (to use the term favoured by Gauss) he wishes to associate with his *a priori* choice of the initial estimate vector $\hat{\underline{a}}_0$. And the choice of P_0 as a diagonal matrix with large elements is clearly consistent with the usual situation of low confidence in $\hat{\underline{a}}_0$, since then the diagonal elements of P_0^{-1} will be very small and the quadratic form (A.3.2) will play little part in the recursive update of $\hat{\underline{a}}_0$ to yield $\hat{\underline{a}}_1$.

Finally, it should be noted that the evaluation of the recursive least squares algorithm in the manner shown in this Appendix can be made much more general, since it applies to all recursive algorithms of the "least squares-like" form. For example, as in Chapter 5, we could consider the case of a regression function with vector measurements, $\underline{y}_k$, and replace (A.3.1) by the following cost function,

$$J = [\Pi_k\hat{\underline{a}}_k - \underline{y}_k]^T R_N^{-1}[\Pi\hat{\underline{a}}_k - \underline{y}_k] + [\hat{\underline{a}}_k - \hat{\underline{a}}_{k-1}](P_k^*)^{-1}[\hat{\underline{a}}_k - \hat{\underline{a}}_{k-1}],$$

minimisation of which gives rise to algorithm VIII with $W = R_N^{-1}$. Or again, we could consider the Kalman filter algorithm by repeating the analysis for a cost function

$$J = [C_k\hat{\underline{x}}_k - \underline{y}_k]R_k^{-1}[C_k\hat{\underline{x}}_k - y_k] + [\hat{x}_k - \hat{\underline{x}}_{k/k-1}]^T(P^*_{k/k-1})^{-1}[\hat{\underline{x}}_k - \hat{x}_{k/k-1}]$$

This yields the KF algorithm X if $\hat{x}_{k/k-1}$ and $P^*_{k/k-1}$ are suitably defined in X(1) and X(2) in relation to the state equations (5.43)

References

AKAIKE, H. (1974) A new look at statistical model identification, *IEEE Trans. Auto. Control* AC19, 716-722.

AOKI, M., and STALEY, R.M. (1970) On input signal synthesis in parameter identification, *Automatica*, 6, 431-440.

AOKI, M., and YUE, P.C. (1970) On certain convergence questions in system identification, *SIAM Jnl. Control*, 8, 239-256.

ASTROM, K.J. (1970) *Introduction to Stochastic Control Theory*, Acad. Press: New York.

ASTROM, K.J., and BOHLIN, T. (1966) Numerical identification of linear dynamic systems from normal operating records, in P.H. Hammond (ed.), *Theory of Self Adaptive Systems*, Plenum Press: New York.

ASTROM, K.J., and EYKHOFF, P. (1971) System identification - a survey, *Automatica*, 7, 123.

ASTROM, K.J., and KÄLLSTROM, C.G. (1973) Application of system identification techniques to the determination of ship dynamics, appears in P. Eykhoff (ed.), *Identification and System Parameter Estimation*, North Holland/American Elsevier: Amsterdam/New York.

ASTROM, K.J., and WITTENMARK, B. (1973) On self tuning regulators, *Automatica*, 9, 185-199.

ASTROM, K.J., BOHLIN, T., and WENSMARK, S. (1965) Automatic construction of linear stochastic dynamic models for stationary industrial processes with random disturbances using operating records, *IBM Nordic Lab. Report TP-18, 150.*

BALAKRISHNAN, A.V. (1973) *Stochastic Differential Equation Systems I*, Springer Verlag: New York.

BEAUMONT, C. (1980) Stochastic hydrology - an update, *Progress in Phys. Geog.*, 4, 549-556.

BECK, M.B. (1974) *Ph.D. Thesis, Dept. Eng., Univ. of Cambridge, England.*

BECK, M.B., and YOUNG, P.C. (1975) A dynamic model for DO-BOD relationship in a non-tidal stream, *Water Res.*, 9, 769-776.

BECK, M.B., and YOUNG, P.C. (1976) Systematic identification of DO-BOD model structure, *Proc. A.S.C.E., Jnl. Env. Eng. Div.*, 102,EE5, 909.

BEER, T., and YOUNG, P.C. (1981) On the characterisation of longitudinal dispersion in natural streams, *Centre for Resource and Environmental Studies, ANU, Rep. No. AS/R42 (1980).*

BELLMAN, R., KALABA, R.E., and WING, G.M. (1960) Invariant imbedding and mathematical physics I: particle processes, *Int. Math. Phys.*, 1, 280.

BELLMAN, R., and KALABA, R.E. (1964) Dynamic programming, invariant imbedding and quasilinearisation: comparisons and interconnections, in A. Balakrishnan and L. Neustadt (eds.) *Computing Methods in Optimization Problems*, Acad. Press: New York.

BELLMAN, R., and KALABA, R.E. (1965) *Quasilinearisation and Nonlinear Boundary Layer Problems*, American Elsevier: New York.

BENNETT, R.J. (1976) Non-stationary parameter estimation for small sample situations: a comparison of methods, *Int. Jnl. Systems. Sci.*, 7, 257-275.

BENNETT, R.J. (1977) Consistent estimation of non-stationary parameters for small sample situations: a Monte-Carlo study, *Int. Econ. Rev.*, 18, 489-502.

BENNETT, R.J. (1979) *Spatial Time-Series: Analysis, Forecasting and Control*, Pion: London.

BERTRAND, J. (1855) *Méthode des Moidres Carrés* translation into French of 'Memoirs on the Combination of Observations' by K.F. Gauss, published with authorization of Gauss, Mallet-Bachelier: Paris.

BLAKELOCK, J.H. (1965) *Automatic Control of Aircraft and Missiles*, Wiley: New York.

BLUM, J.A. (1954) Multidimensional stochastic approximation methods, *Ann. Math. Stat.*, 25, 737-744.

BODEWIG, E. (1956) *Matrix Calculus*, North Holland: Amsterdam.

BOHLIN, T. (1970) On the maximum likelihood method of identification, *IBM Jnl. Res. and Dev.*, 14, 41-51.

BOX, G.E.P., and JENKINS, G.M. (1970) *Time Series Analysis, Forecasting and Control*, Holden Day: San Francisco.

BRAY, J.W., HIGH, R.J., McCANN, A.D., and JEMMESON, H. (1965) On-line model making for chemical plant, *Trans. Soc. Inst. Tech.*, 17.

BROWNLEE, K.A. (1965) *Statistical Theory and Methodology in Science and Engineering*, John Wiley: New York.

BROWN, R.L., DURBIN, J., and EVANS, J.M. (1975) Techniques for testing the constancy of regression relationships over time, *Jnl. Royal. Stat. Soc., Series B*, 37, 149-192.

BRYSON, A.E., and HO, Y.C. (1969) *Applied Optimal Control*, Blaisdell: Mass.

CAINES, P.E., and LJUNG, L. (1976) Asymptotic normality and accuracy of prediction error estimators, *Res. Rep. No. 7602, Dept. of Electrical Eng., Univ. of Toronto* (also *JACC reprints*, 1976 and *Stochastics*, 3, 29-46).

CAREW, B., and BELANGÉR, P.R. (1973) Identification of optimum filter steady state gain for systems with unknown noise covariances, *IEEE Trans. Auto. Control*, AC-18, 582-587.

CHATFIELD, C. (1975) *The Analysis of Time-Series: Theory and Practice*, Chapman and Hall: London.

CHOW, G.C. (1960) A test for equality between sets of observations in two linear regressions, *Econometrica*, 28, 591-605.

CLARKE, D.W. (1967) Generalized least squares estimation of the parameters of a dynamic model, *paper 3.17 Int. Fed. Auto. Control (IFAC) Congress Preprints, Prague*.

CLARKE, D.W., and GAWTHROP, P.J. (1975) Self tuning controller, *Proc. Inst. Elect. Eng.*, 122, 929-934.

DETCHMENDY, D.M., and SRIDHAR, R. (1966) Sequential estimation of states and parameters in noisy, nonlinear, dynamical systems, *ASME Trans. Jnl. Bas. Eng.*, 880, 362.

DHRYMES, P.J. (1970) *Econometrics: Statistical Foundations and Applications*, Harper and Row: New York.

DORF, R.C. (1965) *Time-Domain Analysis and Design of Control Systems*, Addison Wesley: Reading, Mass.

DUNCAN, D.B., and HORN, S.D. (1972) Linear dynamic recursive estimation from the view-point of regression analysis, *Jnl. Am. Statist. Assoc.* 67, 815-821.

DURBIN, J. (1954) Errors in variables, *Rev. Int. Statist. Inst.*, 22, 23-32.

DURBIN, J. (1960) The fitting of time-series models, *Rev. Int. Stat. Inst.* 28, 233-43. (See also DURBIN, J. (1960) estimation of parameters in time-series regression models, *Jnl. Roy. Stat. Soc. Series B*, 22, 139-153.)

DVORETSKY, A. (1956) On stochastic approximation, *Proc. 3rd Berkeley Symp. Math. Statist. Prob.* J. Neyman (ed.), Univ. Calif. Press: Berkeley.

ELGERD, O.I. (1967) *Control Systems Theory*, McGraw Hill: New York.

EYKHOFF, P. (1974) *System Identification*, Wiley: New York.

FINIGAN, B.M., and ROWE, I. (1974) Strongly consistent parameter estimates by the introduction of strong instrumental variables, *IEEE Trans. Auto. Control*, AC-19, 825-831.

FISHER, R.A. (1956) *Statistical Methods and Scientific Inference*, Oliver and Boyd: Edinburgh.

FREEMAN, T.G. (1981) Introduction to the use of CAPTAIN for time-series analysis, *Commonwealth Scientific and Industrial Research Organisation (CSIRO), Australia, Div. of Computing Res., Computing Note 43* (learners guide to supplement User Manual by VENN, M.W., and DAY, B. (1977)).

FRÖBERG, C.E. (1970) *Introduction to Numerical Analysis*, Addison-Wesley: Reading, Mass.

GANTMACHER, F.R. (1960) *Matrix Theory, Vol. 1*, Chelsea: New York.

GAUSS, K.F. (1809) Theoria Motus corporum coelestium, *Werke 7*, Hamburg (First English translation 1857; recent translation C.H. DAVIS (1963), Dover Pub.: New York).

GAUSS, K.F. (1821, 1823, 1826) Theoria combinationis observationum erroribus minimis obnoxiae, Parts 1, 2 and supplement, *Werke 4*, 1-108 (French Translation J. BERTRAND (1855), English translation H.F. TROTTER (1957)).

GELB, A. (ed.) (1974) *Applied Optimal Estimation* MIT Press for The Analytic Sciences Corporation: Cambridge, Mass.

GOODWIN, G.C. (1969) Input synthesis for minimum covariance state and parameter estimation, *Elect. Letters*, 5, 539-540.

GOODWIN, G.C. (1971) Optimal input signals for nonlinear system identification, *Proc. Inst. Elect. Eng.*, 118, 922-926.

GOODWIN, G.C., MURDOCH, J.C., and PAYNE, R.L. (1973) Optimal test signal design for linear SISO system identification, *Int. Jnl. Control*, 17, 45-55.

GOODWIN, G.C., and PAYNE, R.L. (1977) *Dynamic System Identification: Experiment Design and Data Analysis*, Acad. Press: New York.

GRANGER, C.W.J., and NEWBOLD, P. (1977) *Forecasting Economic Time Series*, Acad. Press: New York.

GRANGER, C.W.J., and MORRIS, M.J. (1976) Time-series modelling and interpretation, *Jnl. Royal Stat. Soc., Series A*, 139, 246-257.

GRAYBILL, F.A. (1961) *An Introduction to Linear Statistical Models*, McGraw Hill: New York.

GRINDLEY, J. (1967) The estimation of soil moisture deficit, *Met. Mag.*, 96, 97-108.

GUSTAVSSON, I. (1971) Choice of sampling interval for parametric identification, *Lund Inst. of Tech., Div. Auto. Control, Rep. No. 7103*. (See ASTROM, K.J. (1968) On the choice of sampling rates in parametric identification of time series, *Lund Inst. of Tech., Div. Auto. Control, Rep. No. 6807*; also appears in *Inf. Sciences*, 1, 273-278, 1969).

HAMMERSLEY, J.M., and HANDSCOMB, D.C. (1964) *Monte-Carlo Methods*, Methuen: London.

HANNAN, E.J. (1970) *Multiple Time-Series*, Wiley: New York.

HANNAN, E.J. (1976) The convergence of some recursions, *Ann. Statist.*, 4, 1258-1270.

HANNAN, E.J., and TANAKA, K. (1976) ARMAX models and recursive calculations, in H. Myoken (ed.) *Proc. Conf. System Dynamics and Control in Quantitative Economics*, Nagoya City Univ.

HARVEY, A.C. (1976) An alternative proof and generalisation of a test for structural change, *American Statistician*, 30, 122-23.

HASTINGS-JAMES, R. (1970) *Ph.D. Thesis, Dept. Eng., Univ. of Cambridge, England.*

HASTINGS-JAMES, R., and SAGE, M.W. (1969) Recursive generalised least squares procedure for on-line identification of process parameters, *Proc. Inst. Elect. Eng.*, 116, 2057-2062.

HOLST, J. (1977) Adaptive prediction and recursive estimation, *Lund Inst. of Tech., Div. of Auto. Control, Rep. No. LUTF D2/(TRFT-1013)/1-206/(1977).*

HO, Y.C. (1962) On the stochastic approximation method and optimal filtering theory, *Jnl. Math. Anal. App.*, 6, 152.

HO, Y.C., and BLAYDON, C. (1966) On the abstraction problem in pattern classification, *Proc. Nat. Elect. Conf., U.S.A.*

ILLIFF, K.W. (1974) Identification of aircraft stability and control derivatives in the presence of turbulence, in *Parameter Estimation Techniques and Applications in Flight Testing,* NASA TN D-7647.

ISERMANN, R., BAUR, U., BAMBERGER, W., KNEPPO, P., and SIEBERT, H. (1973) Comparison of 6 on-line indentification and estimation methods, in R. Isermann (ed.) *Identification and System Parameter Estimation,* Pergamon: Oxford (also *Automatica,* 10, 81-103).

JAKEMAN, A.J., and YOUNG, P.C. (1979a) Refined instrumental variable methods of recursive time-series analysis, Part II: multivariable systems, *Int. Jnl. Control,* 29, 621-644.

JAKEMAN, A.J., and YOUNG, P.C. (1979b) Joint parameter/state estimation, *Electronics Letters,* 15, 582.

JAKEMAN, A.J., and YOUNG, P.C. (1980a) Towards optimal modeling of translocation data from tracer studies, *Proc. 4th Biennial Conf., Simulation Soc. of Australia,* 248-253.

JAKEMAN, A.J., and YOUNG, P.C. (1980b) Systems identification and estimation for convolution integral equations, in R.S. Anderssen *et al.* (eds.), *The Application and Numerical Solution of Integral Equations,* Sijthoff and Noordhoff: Netherlands.

JAKEMAN, A.J., and YOUNG, P.C. (1981a) On the decoupling of system and noise model parameter estimation in time-series analysis, CRES Rep. No. AS/R45(1981), see *Int. Jnl. Control, 34,* 423-431.

JAKEMAN, A.J. and YOUNG, P.C. (1981b) Statistically efficient methods of recursive time-series analysis, CRES Rep. No. AS/R46(1981), see *Int. Jnl. Control, 37,* 1291-1310.

JAKEMAN, A.J., STEELE, L.P., and YOUNG, P.C. (1980) Instrumental variable algorithms for multiple input systems described by multiple transfer functions, *IEEE Trans. Syst., Man and Cyb.* SMC-10, 593-602.

JAMES, P.N., SOUTER, P., and DIXON, D.C. (1972) Sub-optimal estimation of the parameters of discrete systems in the presence of correlated noise, *Electronics Letters,* 8, 411-412.

JAZWINSKI, A.H. (1969) Adaptive filtering, *Automatica,* 5, 475-485.

JAZWINSKI, A.H. (1970) *Stochastic Processes and Filtering Theory,* Acad. Press: New York.

JENKINS, G.M. (1979) Practical experiences with modelling and forecasting time-series, in O.D. Anderson (ed.) *Forecasting,* North Holland: Amsterdam.

JOHNSTON, J. (1963) *Econometric Methods,* McGraw-Hill: New York.

JOSEPH, P., LEWIS, J., and TOU, J. (1961) Plant identification in the presence of disturbances and application to digital adaptive systems, *AIEE Trans. App. Ind.,* 80, 18.

JURY, E.I. (1964) *Theory and Application of the z-Transform Method,* Wiley: New York.

KAILATH, T., and FROST, P. (1968) An innovations approach to least squares estimation: I, *IEEE Trans. Auto. Control,* AC-13, 646-655.

KALLSTROM, C.G., ESSEBO, T., and ASTROM, K.J. (1976) A computer program for maximum likelihood identification of linear, multivariable, stochastic systems, *Proc. 4th IFAC Symp. on Identification and System Parameter Estimation, Tblisi, USSR.*

KALMAN, R.E. (1958) Design of a self optimizing control system, *A.S.M.E. Trans., Jnl. Basic Eng.,* 80-D, 468-478.

KALMAN, R.E. (1960) A new approach to linear filtering and prediction problems, *ASME Trans., Jnl. Basic. Eng.*, 83-D, 95-108.

KALMAN, R.E. (1979) A system theoretic critique of dynamic economic models, paper presented at Economics Seminar, Univ. of Cambridge, England (appears in *Int. Jnl. Policy Anal. and Inf. Syst.*, 4, 3-22).

KALMAN, R.E., and BUCY, R.S. (1961) New results in linear filtering and prediction theory, *ASME Trans., Jnl. Basic Eng.*, 83-D, 95.

KENDALL, M.G., and STUART, A. (1961) *The Advanced Theory of Statistics, Vol.2*, Griffin: London.

KIEFER, J., and WOLFOWITZ, J. (1952) Stochastic estimation of the maximum of a regression function, *Ann. Math. Stat.*, 23, 462-466.

KOPP, R.E., and ORFORD, R.J. (1963) Linear regression applied to system identification for adaptive control systems, *AIAA Jnl.*, 1, 2300.

KREISSELMEIER, G. (1977) Adaptive observers with exponential rate of convergence, *IEEE Trans. Auto. Control*, AC-22, 2-8.

KUMAR, R., and MOORE, J.B. (1979) Inverse state and decorrelated state stochastic approximation, to appear *Automatica* (appears originally as Dept. of Elect. Eng., Univ. of Newcastle, N.S.W., Australia, Rep. No. 7808, 1978).

LANDAU, I.D. (1976) Unbiased recursive identification using model reference adaptive techniques, *IEEE Trans. Auto. Control*, AC-21, 194-202.

LASDON, L.S., MITTER, S.K., and WAREN, A.D. (1967) The conjugate gradient method for optimal control problems, *IEEE Trans. Auto. Control*, AC-12, 132-138.

LEE, R.C.K. (1964) *Optimal Identification, Estimation and Control*, M.I.T. Press: Cambridge, Mass.

LEGENDRE, A.M. (1805) Sur la methode des moindres carrés, Appendix in Legendre, A.M. *Nouvelle Méthodes pour la Determination des Orbites des Cometes*, Paris (English translation by H.A. Ruger and H.M. Walker see Smith, D.E. *A Source Book in Mathematics*, Dover: New York). See also Legendre, A.M. (1810) Méthode des moindres carrés pour trouver le milieu le plus probable entre les resultats de différent observations, *Mem. Inst. de France*, 149-154.

LEVIN, M.J. (1963) Estimation of system pulse transfer function in the presence of noise, *Proc. Joint Automatic Control Conference*, 452-458 (also *IEEE Trans. Auto. Control*, AC-9, 229-235 and 214-215).

LJUNG, L. (1976) On the consistency of prediction error methods, in R.K. Mehra and D.G. Lainiotis (eds.) *System Identification: Advances and Case Studies*, Acad. Press: New York.

LJUNG, L. (1977a) On positive-real functions and convergence of some recursive schemes, *IEEE Trans. Auto. Control*, AC-22, 539.

LJUNG, L. (1977b) Analysis of recursive stochastic algorithms, *IEEE Trans. Auto Control*, AC-22, 551.

LJUNG, L. (1978) Convergence analysis of parametric identification methods, *IEEE Trans. Auto. Control*, AC-23, 770-783.

LJUNG, L. (1979a) Convergence of recursive estimators, in R. Isermann (ed.) *Identification and System Parameter Estimation*, Pergamon: Oxford 131-144.

LJUNG, L. (1979b) Asymptotic behaviour of the Extended Kalman filter as a parameter estimator for linear systems, *IEEE Trans. Auto. Control*, AC-24, 36-50.

LJUNG, L., SODERSTROM, T., and GUSTAVSSON, I. (1975) Counter examples to the general convergence of a commly used recursive identification method, *IEEE Trans. Auto. Control*, AC-20, 643-652.

LOEVE, M.M. (1963) *Probability Theory*, Von Nostrand: New York.

MACIEJOWSKI, J.M. (1978) *The Modelling of Systems With Small Observational Sets*, Lecture Notes in Control and Information Sciences, No. 10., Springer-Verlag: Berlin, New York.

MANN, H.B., and WALD, A. (1943) On the statistical treatment of linear stochastic difference equations, *Econometrica*, 11, 173-220.

MARDEN, M. (1949) The geometry of the zeros of a polynomial in a complex variable, *Trans. Amer. Math. Soc.:* New York, 152.

MEHRA, R.K. (1970) Maximum likelihood estimation of aircraft parameters, *Proc. Joint Automatic Control Conf., Atlanta, Georgia, U.S.A.*

MEHRA, R.K. (1971) On-line identification of linear dynamic systems with applications to Kalman filtering, *IEEE Trans. Auto. Control*, AC-16, 12-21.

MEHRA, R.K., and TYLER, J.S. (1973) Case studies in aircraft parameter identification, in P. Eykhoff (ed.) *Identification and System Parameter Estimation*, North-Holland/American Elsevier: Amsterdam/New York.

MENDEL, J.M., and FU, K.S. (1970) *Adaptive Learning and Pattern Recognition Systems*, Acad. Press: New York.

MOORE, R.J., and CLARKE, R.T. (1979) Some properties of variance reduction techniques where hydrological extremes are estimated by Monte-Carlo analysis, *Water Resources Res.*, 14, 55-61.

NARENDRA, K.S. (1976) Stable identification schemes, appears in R.K. Mehra and D.G. Lainiotis, *System Identification: Advances and Case Studies*, Acad. Press: New York.

NEETHLING, C. (1974) *Ph.D. Thesis, Dept. of Engineering, Univ. of Cambridge, England.*

NEETHLING, C., and YOUNG, P.C. (1974) Comments on "Identification of optimum filter steady state gain for systems with unknown noise covariances", *IEEE Trans. Auto. Control*, AC-19, 623-5.

NORTON, J.P. (1975) Optimal smoothing in the identification of linear time-varying systems, *Proc. Inst. Elect. Eng.*, 122, 663-668.

NORTON, J.P. (1977) Initial convergence of recursive maximum likelihood identification algorithms, *Electronics Letters*, 13, 621-2.

OGATA, K. (1967) *State Space Analysis of Control Systems*, Prentice Hall: N.J.

OGATA, K. (1970) *Modern Control Engineering*, Prentice Hall: N.J.

PAGAN, A.R., and NICHOLLS, D.F. (1976) Exact maximum Likelihood estimation of regression models with finite order moving average errors, *Rev. Econ. Studies*, XLIII, 383-387.

PANUSKA, V. (1968) A stochastic approximation method for identification of linear systems using adaptive filtering, *Proc. Joint Auto. Control Conf.*, 1014-1021.

PANUSKA, V. (1969) An adaptive recursive least squares identification algorithm, *Proc. 8th IEEE Symp. on Adaptive Processes, paper 6e.*

PENMAN, H.L. (1950) The water balance of the Stour catchment area, *Jnl. Inst. Water Eng.*, 4, 457-469.

PENROSE, R. (1955) A generalized inverse for matrices, *Proc. Phil. Soc.*, 51, 406-413 (see also ALBERT, A. (1972) Regression and the Moore-Penrose pseudoinverse, Acad. Press: New York).

PHADKE, M.S., and WU, S.M. (1974) Modelling of continuous stochastic processes from discrete observations with applications to sunspots data, *J. Am. Statist. Assoc.*, 69, 325.

PHILLIPS, A.W. (1958) The relationship between unemployment and rate of change of money wage rates in the United Kingdom, 1861-1957, *Economica*, 25, 283-299.

PHILLIPS, A.W. (1959) The estimation of parameters in systems of stochastic differential equations, *Biometrika*, 46, 67.

PIERCE, D.A. (1972) Least squares estimation in dynamic disturbance time-series models, *Biometrika*, 59, 73-78.

PITTOCK, A.B. (1975) Climatic change and patterns of variation in Australian rainfall, *Search*, 6, 498-504.

PLACKETT, R.L. (1950) Some theorems in least squares, *Biometrika*, 37, 149-157.

POLJAK, B.T., and TSYPKIN, Ja.Z. (1980) Robust identification, *Automatica*, 16, 53-63.

PRIESTLEY, M.B. (1980) State dependent models: a general approach to non-linear time-series analysis, *Time Series Anal.*, 1, 47-72.

QUANDT, R.E. (1960) Tests of the hypothesis that a linear regression system obeys two separate régimes, *Jnl. Am. Statist. Assoc.*, 55, 324-330.

RAUCH, H.E., TUNG, F., and STREIBEL, C.T. (1965) Maximum likelihood estimates of linear dynamic systems, *AIAA. Jnl.*, 3, 1445-1450.

ROBBINS, H., and MONRO, S. (1951) A stochastic approximation method, *Ann. Math. Statist.*, 22, 400-407.

ROSENBROCK, H.H., and STOREY, C. (1966) *Computational Techniques for Chemical Engineers*, Pergamon: Oxford.

ROWE, I.H. (1970) A bootstrap method for the statistical estimation of model parameters, *Int. Jnl. Control*, 12, 721-38.

SAGE, A.P. (1968) *Optimum Systems Control*, Prentice Hall: N.J.

SAGE, A.P., and HUSA, G.W. (1969) Algorithms for sequential adaptive estimation of prior statistics, *Proc. 8th IEEE Symp. on Adaptive Processes*, paper 6a.

SAKRISON, D. (1966) Stochastic approximation: a recursive method for solving regression problems, in A.V. Balakrishnan (ed.) *Advances in Communication Theory*, 2, Acad. Press: New York.

SARIDIS, G.N. (1974) Comparison of 6 on-line identification algorithms, *Automatica*, 10, 69-79.

SASTRY, D., and GAUVRIT, M. (1978) Some simplified algorithms for Bayesian identification of aircraft parameters, *Int. Jnl. Syst. Sci.*, 9, 1215.

SHELLSWELL, S.H. (1972) A Computer Aided Procedure for Time-series Analysis and Identification of Noisy Processes (CAPTAIN - original user manual), *Control Division, Dept. of Eng., Univ. of Cambridge, Rep. No. CUED/B-Control/TR25 (1972).*

SIDAR, M. (1976) Recursive identification and tracking of parameters for linear and non-linear multivariable systems, *Int. Jnl. Control*, 24, 361-78.

SMETS, A.J. (1970) The instrumental variable method and related identification schemes, *Dept. Elect. Eng., Univ. of Tech, Eindhoven, Netherlands, Internal Report.*

SODERSTROM, T. (1973) An on line algorithm for approximate maximum likelihood identification of linear dynamic systems, *Lund Inst. of Tech., Div. Auto. Control., Rep. No. 7308.*

SODERSTROM, T., and STOICA, P. (1980) Optimal instrumental variable estimation, Part I: optimal instruments, submitted to *IEEE Trans. Auto. Control.*

SODERSTROM, T., LJUNG, L., and GUSTAVSSON, I. (1974) A comparative study of recursive identification methods, *Lund Inst. of Tech., Div. Auto. Control, Rep. No. 7427.*

SOLO, V. (1978) A unified approach to recursive parameter estimation, *Centre for Resource and Environmental Studies, ANU., Rep. No. AS/R20 (1978).*

SOLO, V. (1980) Some aspects of recursive parameter estimation, *Int. Jnl. Control*, 32, 395-410.

SPROTT, D.A. (1978) Gauss's contributions to statistics, *Royal Soc. of Canada Symp. on Gauss's Contributions to Science and Mathematics.*

STALEY, R.M., and YUE, P.C. (1970) On system parameter identifiability, *Inf. Sciences*, 2, 127-138.

STEPNER, D.E., and MEHRA, R.K. (1973) Maximum likelihood identification and optimal input design for identifying aircraft stability and control derivatives, *NASA Rep. No. NASA CR-2200.*

TAKAHASHI, Y., RABINS, M.J., and AUSLANDER, D.M. (1970) *Control and Dynamic Systems,* Addison Wesley: Reading, Mass.

TALMON, J.L. (1971) Approximated Gauss-Markov estimators and related schemes, *Dept. of Elect. Eng., Univ. of Tech., Eindhaven Netherlands, Int. Report.*

TALMON, J.L., and VAN DEN BOOM, A.J.W. (1973) On the estimation of the transfer function parameters of process and noise dynamics using a single stage estimator, in P. Eykhoff (ed.) *Identification and System Parameter Estimation,* North Holland/American Elsevier: Amsterdam/New York.

TAYLOR, L.W., and ILLIFF, K.W. (1972) Systems identification using a modified Newton-Raphson method, *NASA Tech. Note, NASA TND-6734.*

TODINI, E. (1978) Mutually interactive state/parameter (MISP) estimation in hydrological applications, in G.C. Vansteenkiste (ed.) *Modeling, Identification and Control in Environmental Systems,* North Holland: Amsterdam.

TROTTER, H.F. (1957) Gauss's work 1803-1826 on theory of least squares; an English translation, *Statist. Techniques Research Group, Dept. of Maths., Univ. of Princeton, N.J.*

TRUXAL, T.G. (1955) *Control System Synthesis,* McGraw Hill: New York.

TSYPKIN, Ya.Z. (1971) *Adaption an Learning in Automatic Systems,* Acad. Press: New York.

VENN, M.W., and DAY, B. (1977) Computer Aided Procedure for Time-Series Analysis and Identification of Noisy Processes (CAPTAIN) - User Manual, *Inst. of Hydrology (U.K.) Rep. No. 39, National Environment Research Council* (User Manual for I.H. version of CAPTAIN).

WELLSTEAD, P.E., EDMUNDS, J.M., PRAGER, D., and ZANKER, P. (1979) Self-tuning pole/zero assignment regulators, *Int. Jnl. Control,* 30, 1-26.

WEYMAN, D.R. (1975) *Runoff Processes and Streamflow Modelling,* Oxford Univ. Press: Oxford.

WHITEHEAD, P.G., YOUNG, P.C., and HORNBERGER, G. (1979) A systems model of streamflow and water quality in the Bedford-Ouse River, I: Stream flow modelling, *Water Res.,* 13, 1155-1169.

WHITEHEAD, P.G., YOUNG, P.C., and MICHELL, P. (1978) Some hydrological and water quality modelling studies in the A.C.T. region, *Proc. Hydrology Symp. Canberra, Australia,* Inst. of Eng. Australia: Canberra.

WHITTLE, P. (1953) Estimation and information in stationary time-series, *Arkiv. für Mathematik,* 2, 423.

WIENER, N. (1949) *The extrapolation, interpolation and smoothing of stationary time-series,* Wiley: New York.

WILDE, D.J. (1964) *Optimum Seeking Methods,* Prentice-Hall: N.J.

WONG, K.Y., and POLAK, E. (1967) Identification of linear discrete-time systems using instrumental variables, *IEEE Trans. Auto. Control,* AC-12, 707.

YAGLOM, A.M. (1955) The correlation theory of processes whose nth. difference constitute a stationary process, *Matem. Sb.,* 37, 141 (see also Box G.E.P. and Jenkins, G.M. (1970) Chapter 4.)

YOUNG, P.C. (1965a) The determination of the parameters of a dynamic process, *Radio Electron. Engineer (J.Brit. IERE),* 29, 345-362.

YOUNG, P.C. (1965b) Process parameter estimation and self adaptive control, Proc. IFAC Symp. Teddington; appears in P.H. Hammond (ed.) *Theory of Self Adaptive Control Systems,* Plenum Press: New York, 1966.

YOUNG, P.C. (1965c) On a weighted steepest descent method of process parameter estimation, *Control Division, Dept. of Eng., Univ. of Cambridge, Int. Rep. No. PCY/TN(Camb)/1, Dec. 1965* (available from author).

YOUNG, P.C. (1968a) Process parameter estimation, *Control and Automation Progress*, 12, 931-937.

YOUNG, P.C. (1968b) The use of linear regression and related procedures for the identification of dynamic processes, *Proc. 7th IEEE Symp. on Adaptive Processes, San Antonia, Texas*, 501-505.

YOUNG, P.C. (1968c) Identification problem associated with the equation error approach to process parameter estimation, *Proc. 2nd Asilomar Conf. on Circ. and Systems*, 416-422.

YOUNG, P.C. (1969a) Applying parameter estimation to dynamic systems, Parts I and II, *Control Eng.*, 16: No. 10, 119-125; No. 11, 118-124.

YOUNG, P.C. (1969b) An instrumental variable method for real-time identification of a noisy process, *Proc. IFAC Congress, Warsaw.*

YOUNG, P.C. (1969c) *Ph.D. Thesis, Dept. of Eng., Univ. of Cambridge, England.*

YOUNG, P.C. (1970) An instrumental variable method for real-time identification of a noisy process, *Automatica*, 6, 271-287.

YOUNG, P.C. (1972) Comments on 'On-line identification of linear dynamic systems with applications to Kalman filtering', *IEEE Trans. Auto. Control*, AC-17, 269-70.

YOUNG, P.C. (1974) Recursive approaches to time-series analysis, *Bull. Inst. Maths. Appl.*, 10, 209-224.

YOUNG, P.C. (1975) Discussion of 'Techniques for assessing the constancy of a regression relationship over time' *Jnl. Royal Stat. Soc., Series B*, 37, 149-192.

YOUNG, P.C. (1976a) Some observations on instrumental variable methods of time-series analysis, *Int. Jnl. Control*, 23, 593-612.

YOUNG, P.C. (1976b) Optimization in the presence of noise - a guided tour, in L.C.W. Dixon, (ed.), *Optimization in Action*, Acad. Press: London, 517-573.

YOUNG, P.C. (1978) A general theory of modeling for badly defined systems, in G.C. Vansteenkiste (ed.) *Modeling, Identification and Control in Environmental Systems*, North Holland/American Elsevier: Amsterdam/New York.

YOUNG, P.C. (1979a) Parameter estimation for continuous-time models - a survey, in R. Isermann (ed.) *Identification and System Parameter Estimation*, Pergamon Press: Oxford (also *Automatica*, 17, 23-39, 1981).

YOUNG, P.C. (1979b) Self adaptive Kalman filter, *Electronics Letters*, 15, 358.

YOUNG, P.C. (1979c) A second generation adaptive autostabilization system for airborne vehicles, in R. Isermann (ed.) *Identification and System Parameter Estimation*, Pergamon Press: Oxford, 1073-1086, (*Automatica*, 17, 459-469, 1981).

YOUNG, P.C. (1983) The validity and credibility of models for badly defined systems, to appear in M.B. Beck and G. van Straten (eds.) *Uncertainty and Forecasting of Water Quality* Springer-Verlag: Berlin (presented at Task Force Mtg. Int. Inst. Applied Syst. Anal. (IIASA), Vienna, 1980).

YOUNG, P.C. (1984) *Recursive Estimation and Time-Series Analysis* (in preparation).

YOUNG, P.C., and BECK, M.B. (1974) The modelling and control of water quality in a river system, *Automatica*, 10, 455-468.

YOUNG, P.C., and HASTINGS-JAMES, R. (1970) Identification and control of discrete linear systems subject to disturbances with rational spectral density, *Proc. 9th IEEE Symp. on Adaptive Processes*, IV.6.1-IV.6.8.

YOUNG, P.C., and JAKEMAN, A.J. (1979a) Refined instrumental variable methods of recursive time-series analysis, Part I: single input, single output systems, *Int. Jnl. Control*, 29, 1-30.

YOUNG, P.C., and JAKEMAN, A.J. (1979b) The development of CAPTAIN: a Computer Aided Program for Time-Series Analysis and Identification of Noisy Systems, in M.A. Cuenod (ed.) *Computer Aided Design of Control Systems*, Pergamon Press: Oxford.

YOUNG, P.C., and JAKEMAN, A.J. (1979c) An inverse problem: the estimation of input variables in stochastic dynamic systems, *Centre for Resource and Environmental Studies, ANU, Rep. No. AS/R28(1979)* (later version entitled "Recursive filtering and smoothing procedures for inversion of ill-posed causal problems" to appear in *Utilitas Mathematica*).

YOUNG, P.C., and JAKEMAN, A.J. (1980) Refined instrumental variable methods of recursive time-series analysis, Part III: Extensions, *Int. Jnl. Control,* 31, 741-764.

YOUNG, P.C., and SHELLSWELL, S.H. (1972) Review of "Time-Series Analysis, Forecasting and Control" by Box, G.E.P. and Jenkins, G.M., *IEEE Trans. Auto. Control,* AC-17, 281-282.

YOUNG, P.C. and SIRAKOFF, C. (1981) A recursive smoothing approach to trend removal and seasonal adjustment, *Centre for Resource and Environmental Studies, ANU, Rep. No. AS/R44 (1981).*

YOUNG, P.C., and WHITEHEAD, P.G. (1977) A recursive approach to time-series analysis for multivariable systems, *Int. Jnl. Control,* 25, 457-482.

YOUNG, P.C., and YANCEY, C.B. (1971) A second generation adaptive pitch autostabilization system for a missile or aircraft, *Naval Weapons Center, China Lake, California, Tech. Note No. TN 404-109.*

YOUNG, P.C., HORNBERGER, G.M., and SPEAR, R.C. (1978) Modeling badly defined systems: some further thoughts, *Proc. SIMSIG Simulation Conf., Canberra, Australia.*

YOUNG, P.C., JAKEMAN, A.J., and McMURTRIE, R. (1980) An instrumental variable method for model order identification, *Automatica,* 16, 281-294.

YOUNG, P.C., SHELLSWELL, S.H., and NEETHLING, C.G. (1971) A recursive approach to time-series analysis, *Dept. of Eng., Univ. of Cambridge, England, Rep. No. CUED/B-Control/TR16.*

Omitted References:

GRANGER, C.W.J., and HATANAKA, M. (1964) *The Spectral Analysis of Economic Time Series,* Princeton Univ. Press: Princeton, N.J.

PRIESTLEY, M.B., and RAO, S.T. (1969) A test for nonstationarity of time-series, *Jnl. Royal Stat. Soc.,* Series B, 31, 140-149.

Author Index

Subject Index

This index is concerned predominantly with the main text. Mathematical and statistical topics covered in the Appendices are indexed in the Contents List, pages (iv) - (v).